Innovative Manufacturing Processes of Silicate Materials

Innovative Manufacturing Processes of Silicate Materials

Guest Editors

Maurice Gonon
Sandra Abdelouhab
Gisèle Laure Lecomte-Nana

MDPI

Basel • Beijing • Wuhan • Barcelona • Belgrade • Novi Sad • Cluj • Manchester

Guest Editors
Maurice Gonon
Material Institute
University of Mons
Mons
Belgium

Sandra Abdelouhab
Research and Development
Belgian Ceramic
Research Centre
Mons
Belgium

Gisèle Laure Lecomte-Nana
ENSIL-ENSCI
Université de Limoges
Limoges
France

Editorial Office
MDPI AG
Grosspeteranlage 5
4052 Basel, Switzerland

This is a reprint of the Special Issue, published open access by the journal *Ceramics* (ISSN 2571-6131), freely accessible at: https://www.mdpi.com/journal/ceramics/special_issues/R0SROX8C47.

For citation purposes, cite each article independently as indicated on the article page online and as indicated below:

Lastname, A.A.; Lastname, B.B. Article Title. *Journal Name* **Year**, *Volume Number*, Page Range.

ISBN 978-3-7258-4548-4 (Hbk)
ISBN 978-3-7258-4547-7 (PDF)
https://doi.org/10.3390/books978-3-7258-4547-7

Contents

About the Editors

Maurice Gonon

Maurice Gonon is an Associate Professor at the University of Mons, Belgium. He earned his Ph.D. in Materials Science and Engineering from the Institut National des Sciences Appliquées de Lyon (INSA-Lyon, France) in 1993, where he conducted research on the development of ceramic matrix composites from a polysilazane precursor under the supervision of Professor Gilbert Fantozzi. He subsequently held a first postdoctoral research position at the University of Limerick, Ireland (1994–1995), where he worked with Professor Stuart Hampshire. He then held two additional postdoctoral positions at the Faculté Polytechnique de Mons (FPMs), Belgium (1996–1998), and INSA-Lyon, France (1999–2001). In 2001, he joined the University of Mons, where he currently leads the Ceramics and Glass-Ceramics Research Unit within the Materials Science Department. His research interests focus on silicate ceramics and glass-ceramics, with a particular emphasis on (i) innovative processing techniques, (ii) the influence of processing conditions on microstructure and properties, and (iii) the development of functional ceramic-based devices.

Sandra Abdelouhab

Sandra Abdelouhab is a Programme Manager at the Belgian Ceramic Research Centre (BCRC) in Mons, Belgium, where she has led research activities on refractory and clay-based materials since 2022. She earned her Ph.D. in Chemistry of Solid Materials in 2005 from the Laboratoire de Chimie des Solides Minéraux (LCSM) in Nancy, France, where she investigated the basicity of oxide melts using electrochemical methods. Following her doctorate, she held research positions in the industrial sector, first at ArcelorMittal's research centre in Maizières-lès-Metz, France (2005–2006), and then at Glaverbel (now AGC Glass Europe) in Jumet, Belgium (2006–2007). She subsequently joined BCRC, where she has contributed since 2008 to numerous research and development projects. Her current work focuses on (i) the formulation and processing of circular refractory castables, (ii) the characterisation of these materials from the fresh state through to high-temperature performance, and (iii) the development of innovative and sustainable solutions for high-temperature industrial applications.

Gisèle Laure Lecomte-Nana

Gisèle Laure Lecomte-Nana is an Associate Professor at ENSIL-ENSCI, University of Limoges, France. She received an Engineer's degree in industrial ceramics together with a Master's degree in 2001 from the Engineering college ENSCI and the University of Limoges in France. This was followed by a PhD degree in Materials Science and Surface Treatments from the University of Limoges in 2004. Her PhD research topic was related to the structural changes, thermal behavior, and sintering mechanisms of kaolin–muscovite systems. From 2004 to 2008, she carried out post-doctoral work in the Heterogeneous Materials Group (GEMH) laboratory in Limoges and with ceramics enterprises (Saint-Gobain—CREE Cavaillon; Terreal—Castelnaudary) on the formulation and structural characterization of geo-materials and highly textured technical ceramics. In 2008, she became an Assistant Professor in the Engineering College of industrial ceramics (ENSCI). She is currently working at ENSIL-ENSCI, University of Limoges, where she is in charge of international relations (Ceramic Department) and the international Semester devoted to ceramics and related processes. Her research interests focus on silicate ceramics and geomaterials, with a particular emphasis on (i) innovative processing techniques, (ii) the formulation of sustainable hierarchical ceramics, and (iii) the thermal behavior and consolidating mechanisms of ceramics and geomaterials.

ceramics

MDPI

Editorial

Innovative Manufacturing Processes of Silicate Materials

Maurice Gonon [1,*], Sandra Abdelouhab [2] and Gisèle Laure Lecomte-Nana [3]

1 Materials Institute, University of Mons, Rue de l'Epargne 56, 7000 Mons, Belgium
2 BCRC, Avenue du Gouverneur Cornez, n°4, 7000 Mons, Belgium; s.abdelouhab@bcrc.be
3 IRCER, UMR CNRS 7315, ENSIL-ENSCI, Université de Limoges, 87065 Limoges, France; gisele.lecomte@unilim.fr
* Correspondence: maurice.gonon@umons.ac.be

1. Introduction

Silicate ceramic materials are likely the oldest manufactured materials in human history. For centuries, they have played a crucial role in domestic life, construction, and artisanal crafts. In the modern era, their relevance remains strong, largely due to their outstanding refractory and electrical insulating properties [1–3]. Consequently, silicate ceramics continue to represent the most widely produced class of ceramics by volume today.

Since the mid-twentieth century, however, the rise of technical ceramics has introduced growing competition for silicate ceramics in advanced technological applications. Nonetheless, silicate ceramics retain significant advantages in numerous domains, operating at ambient to moderate temperatures (up to 1200 °C). These materials are sourced from abundant natural resources, require relatively low to moderate sintering temperatures, and offer greater potential for recyclability.

Recent advancements in fabrication techniques, coupled with the emergence of innovative materials such as geopolymers [4], have sparked renewed interest in silicate ceramics. They are increasingly being explored in strategic sectors such as energy [5], biomedicine [6], and electronics [7]. This resurgence is particularly relevant in the context of current global priorities centered on sustainability and circular economy principles [8].

Unlocking the full potential of silicate ceramics across diverse industries requires a deep understanding and precise control of their properties and microstructures—an objective that hinges on the development and application of innovative manufacturing processes. In this light, the Special Issue of the journal *Ceramics*, entitled "Innovative Manufacturing Processes of Silicate Materials", aims to highlight recent progress in this evolving field. This issue comprises ten contributions that offer new insights into key research areas, including additive manufacturing, geopolymer processing, refractory technologies, and functional applications.

2. Overview of the Contributions

Since the 1990s, additive manufacturing has emerged as a revolutionary approach to producing objects. Initially confined to research laboratories, it has gradually transitioned to industrial-scale production and is now being applied across all classes of materials.

The additive manufacturing of silicate ceramics can be implemented through various techniques, depending on the desired characteristics of the final part and the type of feedstock used [9–11]. Among these, robocasting stands out as one of the most mature and widely studied approaches [12]. However, several challenges remain, notably the formation of defects caused by non-uniform shrinkage during the drying stage [13–15]. In their contribution, N. Lauro et al. [16] present innovative methods for measuring shrinkage using non-destructive optical vision techniques combined with computer-controlled data acquisition. Applied to the robocasting of a porcelain-based paste, these methods revealed anisotropic shrinkage, which was attributed to the preferred grain orientation induced by the extrusion process. This work is particularly valuable as it offers promising tools for the real-time monitoring and control of the drying process, potentially improving the dimensional accuracy and reliability of robocasted components.

Robocasting also shows great promise as a shaping technique for geopolymers. The geopolymerization process enables the formation of inorganic silicate materials that exhibit excellent mechanical strength, chemical resistance, and long-term durability [4,17,18]. As a result, there is a growing body of research dedicated to developing geopolymers for applications traditionally dominated by conventional ceramics [19,20]. Particularly, porous geopolymer materials are gaining attention for their suitability in diverse applications, including radioactive waste encapsulation, the adsorption of hazardous molecules and synthetic dyes, and heterogeneous catalysis [21–26]. In this context, C. Zoude et al. [27] explored the influence of curing parameters, specifically humidity and temperature, on the mechanical properties of metakaolin-based geopolymer filaments fabricated via 3D printing. Their study also compared the performance of samples produced through robocasting with those obtained via conventional molding. The results provide valuable insights into the optimization of the processing conditions for geopolymer-based additive manufacturing.

Beyond the curing conditions, the print quality of robocasted parts—regardless of the material used—depends critically on the rheological behavior of the extruded paste. In this regard, A. Gasmi et al. [28] conducted a comprehensive study on the fabrication of geopolymer composites, with particular emphasis on rheological characterization. Their research focused on how the amount and morphology of mineral fillers affect the paste's flow properties, aiming to ensure structural integrity, dimensional precision, and minimal sagging during printing.

Geopolymer composites represent a promising pathway for producing ceramic-like materials without the need for high-temperature sintering. In these systems, the geopolymer matrix provides a "cold" consolidation mechanism, binding the filler particles through chemical reactions that can also enhance the thermal and mechanical stability of the final material [29,30]. Expanding on this approach, F. Casarrubios et al. [31] investigated the development of cordierite–geopolymer composites for potential applications in filtration and catalysis at temperatures of up to 1000 °C. Their study placed particular emphasis on the recycling of cordierite powders from industrial waste and examined the effects of both the K/Al ratio and cordierite content on key performance parameters, including dimensional and porosity stability during heating, Young's modulus, and the coefficient of thermal expansion. These insights highlight the potential of geopolymer-based systems for sustainable, high-performance applications in harsh environments.

In addition to extrusion-based techniques such as robocasting, vat photopolymerization methods like stereolithography (SLA) and digital light processing (DLP) have gained increasing attention for the additive manufacturing of ceramics, particularly when complex geometries and high resolution are required. In this context, Yurii Milovanov et al. [32] employed DLP to fabricate mullite-based porous substrates intended for use in humidity sensors. Their work demonstrates that mullite is a promising functional material for the development of resistive sensors capable of operating under extreme environmental conditions.

More broadly, porous ceramics are of considerable interest across a wide array of fields, including filtration, catalysis, biomedical applications, and environmental technologies, due to their thermal stability, mechanical robustness, ease of cleaning, and extended service life [33]. As a result, the processing of porous ceramics has been extensively explored over the past few decades. Numerous fabrication strategies have been developed, such as partial sintering, freeze-casting, sacrificial template replication, and direct foaming. Notably, producing porous ceramics from natural silicate materials presents several advantages, particularly for water filtration applications, where cost, sustainability, and material availability are critical factors. In their review article, I-Q Maury Njoya et al. [34] provide a comprehensive overview of the techniques and raw material systems used for fabricating porous clay-based ceramics, focusing on key performance parameters such as porosity, mechanical strength, permeability, and filtration efficiency.

In the field of refractories, the increasing deployment of biomass combustion boilers has created a demand for materials with enhanced resistance to alkali corrosion. In particular, the aggressive conditions encountered during the combustion of certain biomass species—where temperatures can reach up to 1200 °C—render traditional castables containing aluminosilicate aggregates unsuitable in the long term. In their study, J. Malaiškienė et al. [35] demonstrate that the key properties of conventional and medium-cement fireclay refractory castables can be significantly improved through impregnation with liquid sodium silicate glass. This treatment results in microstructural densification and reduced porosity, while also promoting crack healing and facilitating the formation of anorthite. However, it is noted that the thermal shock resistance of the treated materials decreases due to an increase in Young's modulus, highlighting a trade-off between chemical durability and mechanical resilience.

Finally, silicate-based materials exhibit strong potential for functional applications. For example, in the biomedical field, Y. Sugiura et al. [36] investigated the synthesis and characterization of silica-substituted octacalcium phosphate (OCP, $Ca_8(PO_4)_4(HPO_4)_2{\cdot}5H_2O$) blocks incorporating dicarboxylic acids, specifically thiomalic acid (SH-malate). Their study demonstrates the successful fabrication of monophasic OCP blocks simultaneously substituted with silica and dicarboxylic acid molecules. Notably, the resulting composites exhibit enhanced mechanical strength compared to their non-substituted counterparts. These findings suggest promising avenues for the development of advanced bone replacement materials, offering improved mechanical performance and potential for more efficient bone regeneration and healing.

Beyond biomedical applications, silicate-based materials also hold significant promise in catalysis due to their tunable surface chemistry and porosity. In this context, A. Guerrero-Torres et al. [37] present a novel approach for dispersing metal particles on sepiolite supports via a microwave-assisted acid treatment. This process partially leaches the octahedral layer of the sepiolite structure, leading to an increased specific surface area and pore volume, while simultaneously altering the surface chemistry through the removal of Mg^{2+} species. The resulting modified sepiolite exhibits characteristics favorable for catalytic applications, particularly as a support for the dispersion of active phases. The use of microwave irradiation not only accelerates the treatment process, but also enables the use of significantly more diluted acid solutions, enhancing both efficiency and sustainability. The study further evaluates the impact of these structural modifications on catalytic performance, specifically in the gas-phase hydrogenation of furfural.

Within the diverse landscape of silicate materials, glasses and glass–ceramics represent a particularly attractive class for optical and electro-optical applications, due to their unique ability to incorporate and control functional nanostructures. In this context, G. Shakhgildyan et al. [38] investigate the precipitation of gold nanoparticles (NPs) in a $ZnO–Al_2O_3–SiO_2$ glass matrix as a means of generating tunable localized surface plasmon resonance (LSPR). Their work reveals that the spectral position and bandwidth of the LSPR can be precisely modulated through thermal treatment parameters. This controlled nanoparticle formation, leading to ultra-broad LSPR bands, opens up new avenues for enhancing the luminescence of rare-earth-doped systems and contributes to the development of advanced photonic components and devices.

3. Conclusions

The contributions compiled in this Special Issue illustrate the remarkable versatility and continued evolution of silicate-based materials across a broad spectrum of applications. From structural ceramics and refractories engineered for extreme environments, to additive manufacturing innovations, environmental remediation strategies, and advanced photonic systems, the articles reflect the dynamic interplay between composition, processing, and functionality. These works underscore the relevance of both traditional and emerging approaches—ranging from geopolymer technology to nanoengineering, expanding the performance boundaries of silicate materials. Collectively, they offer new insights and technological perspectives that will undoubtedly guide future research and industrial development in the field of ceramic science.

Conflicts of Interest: The authors declare no conflict of interest.

References

1. Kingery, W.D.; Bowen, H.K.; Uhlman, D.R. *Introduction to Ceramics*, 2nd ed.; Wiley Interscience: New York, NY, USA, 1976.
2. Schacht, C.A. *Refractories Handbook*; Marcel Dekker, Inc.: New York, NY, USA, 2004. [CrossRef]
3. Moulson, A.J.; Herbert, J.M. *Electroceramics: Materials, Properties and Applications*; John Wiley & Sons Ltd.: New York, NY, USA, 2003. [CrossRef]
4. Davidovits, J. Geopolymers: Ceramic-Like Inorganic Polymers. *J. Ceram. Sci. Technol.* **2017**, *8*, 335–350. [CrossRef]
5. Bera, B.; Das, N. Synthesis of High Surface Area Mesoporous Silica SBA-15 for Hydrogen Storage Application. *Int. J. Appl. Ceram. Technol.* **2019**, *16*, 294–303. [CrossRef]
6. Arango-Ospina, M.; Nawaz, Q.; Boccaccini, A.R. Silicate-Based Nanoceramics in Regenerative Medicine. In *Nanostructured Biomaterials for Regenerative Medicine*; Guarino, V., Iafisco, M., Spriano, S., Eds.; Woodhead Publishing: Duxford, UK, 2020; pp. 255–273, ISBN 9780081025949. [CrossRef]
7. Kamutzki, F.; Schneider, S.; Barowski, J.; Gurlo, A.; Hanaor, D.A.H. Silicate Dielectric Ceramics for Millimetre Wave Applications. *J. Eur. Ceram. Soc.* **2021**, *41*, 3879–3894. [CrossRef]
8. Vidak Vasić, M.; Muñoz Velasco, P.; Bueno-Rodríguez, S.; Netinger Grubeša, I.; Dondi, M.; Pérez Villarejo, L.; Eliche-Quesada, D.; Zanelli, C. State and Perspectives of Sustainable Production of Traditional Silicate Ceramics. *Open Ceram.* **2024**, *17*, 100537. [CrossRef]
9. Deckers, J.; Vleugels, J.; Kruth, J. Additive Manufacturing of Ceramics: A Review. *J. Ceram. Sci. Technol.* **2014**, *5*, 245–260.
10. Chen, Z.; Li, Z.; Li, J.; Liu, C.; Lao, C.; Fu, Y.; Liu, C.; Li, Y.; Wang, P.; He, Y. 3D Printing of Ceramics: A Review. *J. Eur. Ceram. Soc.* **2019**, *39*, 661–687. [CrossRef]
11. Zocca, A.; Franchin, G.; Colombo, P.; Günster, J. Additive Manufacturing. In *Encyclopedia of Materials: Technical Ceramics and Glasses*; Elsevier: Amsterdam, The Netherlands, 2021; Volume 1, pp. 203–221.
12. Cesarano, J.; Denham, H.; Stuecker, J.; Baer, T.; Griffith, M. *Freeforming of Ceramics and Composites from Colloidal Slurries*; U.S. Department of Energy, Office of Scientific and Technical Information: Oak Ridge, TN, USA, 1999.
13. Cesarano, J.; Baer, T.A.; Calvert, P. Recent Developments in Freeform Fabrication of Dense Ceramics from Slurry Deposition. In Proceedings of the 1997 International Solid Freeform Fabrication Symposium, Austin, TX, USA, 11–13 August 1997.
14. Lamnini, S.; Elsayed, H.; Lakhdar, Y.; Baino, F.; Smeacetto, F.; Bernardo, E. Robocasting of Advanced Ceramics: Ink Optimization and Protocol to Predict the Printing Parameters—A Review. *Heliyon* **2022**, *8*, e10651. [CrossRef]
15. Feilden, E.; Glymond, D.; Saiz, E.; Vandeperre, L. High Temperature Strength of an Ultra High Temperature Ceramic Produced by Additive Manufacturing. *Ceram. Int.* **2019**, *45*, 18210–18214. [CrossRef]
16. Lauro, N.; Alzina, A.; Nait-Ali, B.; Smith, D.S. Non-Uniform Drying Shrinkage in Robocasted Green Body Ceramic Products. *Ceramics* **2024**, *7*, 1122–1136. [CrossRef]
17. Duxson, P.; Fernández-Jiménez, A.; Provis, J.L.; Lukey, G.C.; Palomo, A.; van Deventer, J.S.J. Geopolymer Technology: The Current State of the Art. *J. Mater. Sci.* **2007**, *42*, 2917–2933. [CrossRef]
18. Provis, J.L.; Van Deventer, J.S.J. Introduction to Geopolymers. In *Geopolymers*; Provis, J.L., Van Deventer, J.S.J., Eds.; Woodhead Publishing: Cambridge, UK, 2009; pp. 1–11.

19. Djangang, C.N.; Tealdi, C.; Cattaneo, A.S.; Mustarelli, P.; Kamseu, E.; Leonelli, C. Cold-Setting Refractory Composites from Cordierite and Mullite–Cordierite Design with Geopolymer Paste as Binder: Thermal Behavior and Phase Evolution. *Mater. Chem. Phys.* **2015**, *154*, 66–77. [CrossRef]
20. Mohd Mortar, N.A.; Abdullah, M.M.; Abdul Razak, R.; Abd Rahim, S.Z.; Aziz, I.H.; Nabiałek, M.; Jaya, R.P.; Semenescu, A.; Mohamed, R.; Ghazali, M.F. Geopolymer Ceramic Application: A Review on Mix Design, Properties and Reinforcement Enhancement. *Materials* **2022**, *15*, 7567. [CrossRef] [PubMed]
21. Archez, J.; Texier-Mandoki, N.; Bourbon, X.; Caron, J.F.; Rossignol, S. Adaptation of the Geopolymer Composite Formulation Binder to the Shaping Process. *Mater. Today Commun.* **2020**, *25*, 101501. [CrossRef]
22. Ma, S.; Fu, S.; Yang, H.; He, P.; Sun, Z.; Duan, X.; Jia, D.; Colombo, P.; Zhou, Y. Exploiting Bifunctional 3D-Printed Geopolymers for Efficient Cesium Removal and Immobilization: An Approach for Hazardous Waste Management. *J. Clean. Prod.* **2024**, *437*, 140599. [CrossRef]
23. Jin, H.; Zhang, Y.; Zhang, X.; Chang, M.; Li, C.; Lu, X.; Wang, Q. 3D Printed Geopolymer Adsorption Sieve for Removal of Methylene Blue and Adsorption Mechanism. *Colloids Surf. A Physicochem. Eng. Asp.* **2022**, *648*, 129235. [CrossRef]
24. Gonçalves, N.P.F.; Olhero, S.M.; Labrincha, J.A.; Novais, R.M. 3D-Printed Red Mud/Metakaolin-Based Geopolymers as Water Pollutant Sorbents of Methylene Blue. *J. Clean. Prod.* **2023**, *383*, 135315. [CrossRef]
25. Oliveira, K.G.; Botti, R.; Kavun, V.; Gafiullina, A.; Franchin, G.; Repo, E.; Colombo, P. Geopolymer Beads and 3D Printed Lattices Containing Activated Carbon and Hydrotalcite for Anionic Dye Removal. *Catal. Today* **2022**, *390–391*, 57–68. [CrossRef]
26. Ma, S.; Liu, X.; Fu, S.; Zhao, S.; He, P.; Duan, X.; Yang, Z.; Jia, D.; Colombo, P.; Zhou, Y. Direct Ink Writing of Porous SiC Ceramics with Geopolymer as Binder. *J. Eur. Ceram. Soc.* **2022**, *42*, 6815–6826. [CrossRef]
27. Zoude, C.; Prud'homme, E.; Johannes, K.; Gremillard, L. The Mechanical Properties of Geopolymers as a Function of Their Shaping and Curing Parameters. *Ceramics* **2024**, *7*, 873–892. [CrossRef]
28. Gasmi, A.; Pélegris, C.; Davidovits, R.; Guessasma, M.; Tortajada, H.; Jean, F. Advanced Refinement of Geopolymer Composites for Enhanced 3D Printing via In-Depth Rheological Insights. *Ceramics* **2024**, *7*, 1316–1339. [CrossRef]
29. Kamseu, E.; Rizzuti, A.; Leonelli, C.; Perera, D. Enhanced Thermal Stability in K_2O-Metakaolin-Based Geopolymer Concretes by Al_2O_3 and SiO_2 Fillers Addition. *J. Mater. Sci.* **2010**, *45*, 1715–1724. [CrossRef]
30. Hemra, K.; Aungkavattana, P. Effect of Cordierite Addition on Compressive Strength and Thermal Stability of Metakaolin Based Geopolymer. *Adv. Powder Technol.* **2016**, *27*, 1021–1026. [CrossRef]
31. Casarrubios, F.; Marlier, A.; Lang, C.; Abdelouhab, S.; Mastroianni, I.; Bister, G.; Gonon, M.-F. Characterization of the Evolution with Temperature of the Structure and Properties of Geopolymer-Cordierite Composites. *Ceramics* **2024**, *7*, 1513–1532. [CrossRef]
32. Milovanov, Y.; Bertero, A.; Coppola, B.; Palmero, P.; Tulliani, J.-M. Mullite 3D Printed Humidity Sensors. *Ceramics* **2024**, *7*, 807–820. [CrossRef]
33. Chen, Y.; Wang, N.; Ola, O.; Xia, Y.; Zhu, Y. Porous Ceramics: Light in Weight but Heavy in Energy and Environment Technologies. *Mater. Sci. Eng. R Rep.* **2021**, *143*, 100589. [CrossRef]
34. Maury Njoya, I.Q.; Lecomte-Nana, G.L.; Barry, K.; Njoya, D.; El Hafiane, Y.; Peyratout, C. An Overview on the Manufacture and Properties of Clay-Based Porous Ceramics for Water Filtration. *Ceramics* **2025**, *8*, 3. [CrossRef]
35. Malaiškienė, J.; Antonovič, V.; Boris, R.; Kudžma, A.; Stonys, R. Analysis of the Structure and Durability of Refractory Castables Impregnated with Sodium Silicate Glass. *Ceramics* **2023**, *6*, 2320–2332. [CrossRef]
36. Sugiura, Y.; Saito, Y.; Yamada, E.; Horie, M. Fabrication of Dicarboxylic-Acid- and Silica-Substituted Octacalcium Phosphate Blocks with Stronger Mechanical Strength. *Ceramics* **2024**, *7*, 796–806. [CrossRef]
37. Guerrero-Torres, A.; Maderuelo-Solera, R.; García-Sancho, C.; Quirante-Sánchez, J.J.; Moreno-Tost, R.; Maireles-Torres, P.J.; Cecilia, J.A. Modification of Sepiolite for Its Catalytic Upgrading in the Hydrogenation of Furfural. *Ceramics* **2025**, *8*, 21. [CrossRef]
38. Shakhgildyan, G.; Avakyan, L.; Atroshchenko, G.; Vetchinnikov, M.; Zolikova, A.; Ignat'eva, E.; Ziyatdinova, M.; Subcheva, E.; Bugaev, L.; Sigaev, V. Ultra-Broadband Plasmon Resonance in Gold Nanoparticles Precipitated in ZnO-Al_2O_3-SiO_2 Glass. *Ceramics* **2024**, *7*, 562–578. [CrossRef]

ceramics

MDPI

Article

Non-Uniform Drying Shrinkage in Robocasted Green Body Ceramic Products

Nicolas Lauro, Arnaud Alzina, Benoit Nait-Ali * and David S. Smith

University of Limoges, IRCER, UMR CNRS 7315, 12 rue Atlantis, F-87068 Limoges, France;
nicolas.lauro@unilim.fr (N.L.); arnaud.alzina@unilim.fr (A.A.); david.smith@unilim.fr (D.S.S.)
* Correspondence: benoit.nait-ali@unilim.fr

Abstract: The formation of defects, due to drying, in robocasted ceramic objects is an important issue arising from non-uniform shrinkage of the material during this step in the process. Common methods for shrinkage measurement are not well suited to the small size of robocasted cords or the complexity of robocasted objects. Innovative methods for shrinkage measurement were developed using non-destructive optical vision techniques with computer-controlled data acquisition, allowing measurement on millimetric cords and on specific zones of a product. The study of a single porcelain cord revealed an anisometric shrinkage related to the orientation of grains during extrusion. A differential shrinkage at the macroscopic scale was also measured on a robocasted object, indicating a moisture content gradient in the material. The methods presented in this paper are of particular relevance to real-time control of the drying process for robocasted objects.

Keywords: drying; shrinkage; robocasting; image analysis

1. Introduction

Additive manufacturing of ceramics covers a large number of techniques which have been well described in the literature [1–3]. The choice of a processing technique depends on parameters such as the size of the object, the dimensional resolution and the cost as well as the ability to produce ceramic parts with the required microstructure and properties in relation to the intended application. The variety of additive manufacturing techniques also corresponds to very different feedstock forms such as powders, slurries, pastes or filaments of ceramic polymer composites. The preparation of raw materials to meet the process requirements is another factor to be considered in the choice of a 3D fabrication technique.

Among these techniques, robocasting, which is a Direct Ink Writing (DIW) technique, was developed in the period 1995–2000 to print complex ceramic parts [4]. The raw material is a paste which is extruded through a small diameter nozzle of typically a few hundred micrometres. The position of the nozzle is controlled by a computer based on a GCODE file generated from a CAD file to produce the ceramic object layer by layer. Robocasting has several advantages, mainly its versatility, ease of use, speed of manufacturing and low equipment and operating costs. Today, a wide variety of ceramic materials can be printed by robocasting for different applications in electronics, energy and medicine [2,5]. In recent years, there has been strong interest in the fabrication of complex ceramic parts with high-performance materials for specialised applications: HAP for bio-medical applications (bone scaffolds) [6,7], SiC for highly thermal-resistant parts [8], Al_2O_3 for its mechanical strength [9,10] and AlN for heat dissipation in electronic devices [11]. Although a lot of studies focus on technical ceramics due to their remarkable properties, robocasting is also of interest for silicate ceramic materials. Recent work has investigated the formulation of porcelain pastes for 3D printing applications in the tableware industry [12,13]. Robocasting has now become one of the major ceramic printing technologies on the market. Formerly used for rapid prototyping, recent technical breakthroughs, both in terms of paste formulation and process optimisation, have allowed the development of robocasting for the industrial-scale fabrication of finished products.

Even though robocasting offers many advantages, some challenges remain when printing ceramic parts. The major one is the formulation of a ceramic paste to ensure that the rheological properties required for printing remain constant during the entire fabrication process. Another important challenge pointed out by some authors including Cesarano is the drying step [14–16]. Robocasting requires ceramic pastes with sufficient plasticity to be extruded through the printer nozzle. The initial amount of water in the pastes must then be sufficiently high to separate grains by a thin film of water so that they can slide in relation to each other. Once the shaping is achieved, complete drying is necessary to ensure that water is removed before firing. During drying, grains approach each other until they are in contact, yielding a shrinkage at the macroscopic scale. This is known to be the most delicate stage of drying because if the shrinkage is not uniform, it can be responsible for drying defects such as warping or cracks. After the shrinkage stage is completed or when it becomes negligible, drying still operates to remove water in the porous network [17–19]. Furthermore, an important difference between robocasting and conventional shaping methods is that drying starts during the casting process, meaning that the bottom part of the object starts to dry before the top of the object is extruded. For this reason, Cesarano et al. recommend optimising the drying rate, which needs to be sufficiently high to favour consolidation of the extruded cord and avoid the collapse of the object during printing, but sufficiently low to avoid macroscopic defects due to non-uniform shrinkage [14]. Different strategies have been proposed for a better control of the drying operation in robocasting. With this objective, the simplest way is to adjust and ensure reproducible drying conditions by controlling the air characteristics, i.e., the temperature and the amount of water vapour. Another strategy which has been proposed consists of printing in a liquid oil to prevent evaporation during the layer deposition and to use infrared radiation to dry the deposited layer before extruding a new one. This is called the Ceramic On-Demand Extrusion process [20]. Baltazar et al. compared these two different strategies to control the drying rate. They printed alumina parts in a liquid paraffin bath (followed by infrared drying of each layer) and in air with a relative humidity of 85%. Both methods gave alumina parts with high relative density and good mechanical properties [21].

In addition to the drying conditions, the behaviour of the ceramic paste is of strong importance and should be well characterised to achieve robust and reproducible processing. This requires methods for measuring drying shrinkage. A common practice for measuring shrinkage is to use a position sensor to track the evolution of the height and/or length of the product in 1D. However, recent work by Oummadi et al. presented a method based on tracking marks with an optical camera that allowed shrinkage to be measured precisely in different directions at the same time [22]. The shrinkage evaluated using this image analysis method was compared with values measured using a Linear Variable Differential Transformer (LVDT) sensor for alumina and kaolin pastes. Closely similar curves were obtained when plotting the shrinkage as a function of the water content with the two methods. Nevertheless, existing methods using sensors are not well suited for measurements on cords printed by robocasting due to their small size, low mechanical strength and convex surface. The tracking of marks method would also be difficult to reproduce as it requires placing sub-millimetric marks without altering the surface of the cord. Using marks for measuring shrinkage on robocasted products could cause irreversible damage and yields spatially limited shrinkage measurements.

In view of the challenge of a controlled drying operation in the robocasting process and the importance of the shrinkage that occurs during water removal, in this work we have developed methods for monitoring dimensional variations at different scales, based on image analysis without using marks. At the scale of an extruded cord, the approach provides basic information on the behaviour of the paste during drying which is useful for process parameter adjustment and for the formulation of ceramic pastes. At the scale of an extruded object, differential shrinkage rates are demonstrated. Such information could be used to optimise drying conditions. The article presents the methods and illustrates them with some results of particular interest.

2. Materials and Methods

2.1. Materials and Devices

Two cameras were used for image acquisition during drying. The first one is an IDS UI-1490SE (Obersulm, Germany) with a resolution of 3840 × 2748 pixels mounted with a Kowa LM50JC10M lens (Nagoya, Japan) and the second is an IDS UI-1480SE with a resolution of 2560 × 1920 pixels mounted with an OPTEM Fusion 12.5:1 lens. The scripts used for controlling the cameras and analysing the images were developed using Python and the OpenCV library [23].

The experiments presented in the following sections were made using a commercial porcelain paste formulated for robocasting (Imerys EZPrint3D-E, Paris, France). Thermogravimetric analysis, conducted on dried powder from the paste using a TG-DSC Labsys Evo Setaram system, reveals a 7% mass loss between 450 °C and 700 °C. This can be attributed to the dehydroxylation of kaolinite. Compared with pure kaolinite, where the mass loss due to dehydroxylation is 14%, the weight loss indicates that the powder is made of 50 wt% of kaolinite. Samples (powder and extruded cords) were observed in a Scanning Electron Microscope (SEM) FEI Quanta 450 FEG (Hillsboro, OR, USA). The micrograph of the powder in Figure 1 shows the strongly anisometric kaolinite platelets.

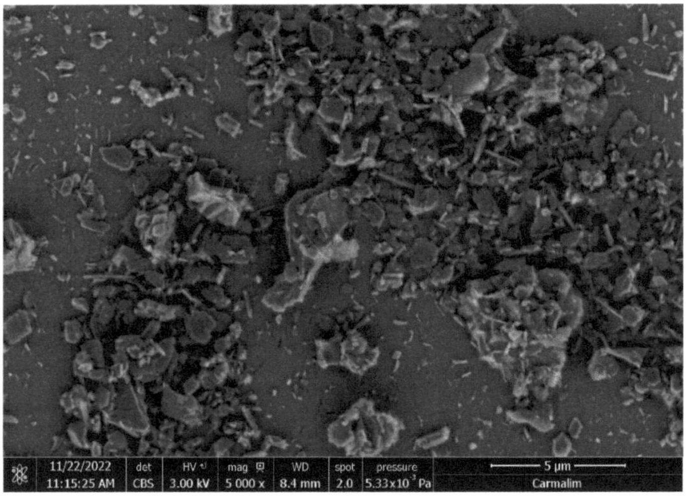

Figure 1. SEM micrograph of the powder taken from dried porcelain paste EZPrint3D-E.

The products studied in this paper were printed using a ceramic 3D printer WASP 2040 Clay (Massa Lombarda, Italy), mounted with a LDM Wasp Extruder 3.0 and a nozzle with an internal diameter of 1 mm.

2.2. Measuring the Drying Shrinkage of an Individual Robocasted Ceramic Cord

A method was first developed for measuring the drying shrinkage of single ceramic cords obtained through robocasting. Due to their small radius, ranging from 0.4 mm to 2 mm, conventional shrinkage measuring methods were not optimal for obtaining satisfying results on these cords. The new method is non-destructive and relies on taking digital images and analysing them using automated image analysis scripts to obtain shrinkage-over-time plots for both the longitudinal and radial directions.

The experimental apparatus used for measuring the drying shrinkage of the cord is illustrated in Figure 2. The two cameras take pictures of the cord at regular time intervals for the duration of the experiment. Examples of the images taken by camera 1 (IDS UI-1490SE) and camera 2 (IDS UI-1480SE) are shown, respectively, in Figures 3a and 3b. The cord was deposited on a PVC plate coated with a thin layer of graphite by spraying. The graphite coating has two main advantages in this experiment: its mat black finish yields strong contrast between the cord and the plate, which helps when performing image analysis; graphite is also a solid lubricant which reduces the interaction between the cord and the plate [24], thus allowing the cord to shrink without sticking to the plate. A light source is used to illuminate the cord and obtain a better contrast with the plate.

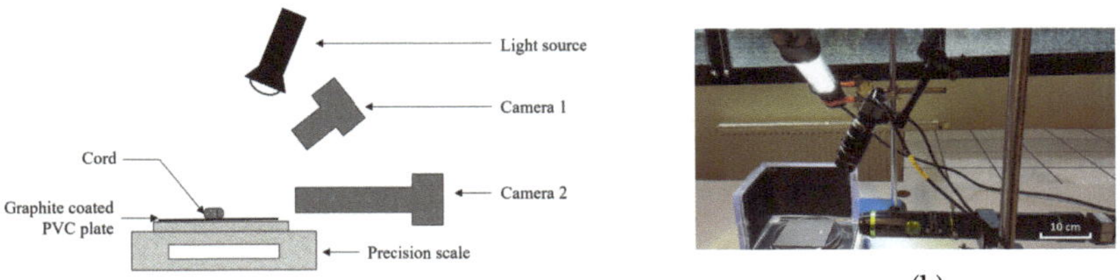

(a) (b)

Figure 2. Experimental apparatus used for measuring the drying shrinkage of an individual robocasted ceramic cord: (**a**) schematic representation and (**b**) photograph of the experimental setup.

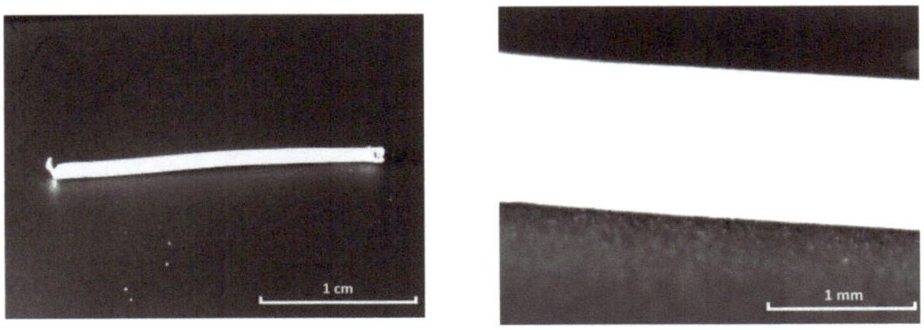

(a) (b)

Figure 3. Examples of images obtained using (**a**) camera 1 and (**b**) camera 2.

For this experiment, the cord is extruded freely and separated from the nozzle using a spatula. It is then deposited on a graphite coated plate with particular attention not to add any strain. This method ensures that little to no extra deformation is incurred by the cord that could potentially reduce the quality of the results. The cord is extruded at a rate of 1 cm/s for the preliminary results, and at 1.5 cm/s and 2 cm/s to investigate the effect of the extrusion rate on drying shrinkage.

After drying is complete, two sets of images are obtained: one for each camera. A Python script is then used to measure the dimensions of the cord on each image. The algorithms used to analyse each set of images are different, but they both start by processing the images in the same manner. Each image is first truncated, converted to grayscale and then blurred. A threshold is then applied before performing a contour analysis using the algorithm proposed by Suzuki et al. [25]. These steps are illustrated in Figure 4. The length of the cord is obtained by placing the minimum bounding rectangle on the contour (red), as shown in Figure 5, and then measuring its length.

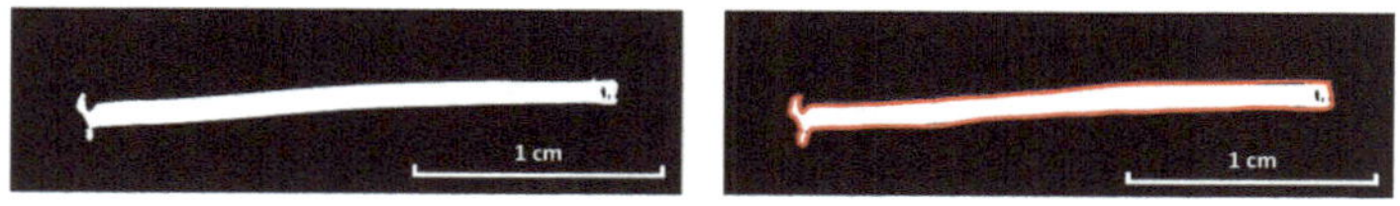

(**a**) (**b**)

Figure 4. Examples of images obtained after (**a**) thresholding and (**b**) after contour analysis (the contour of the cord is highlighted in red).

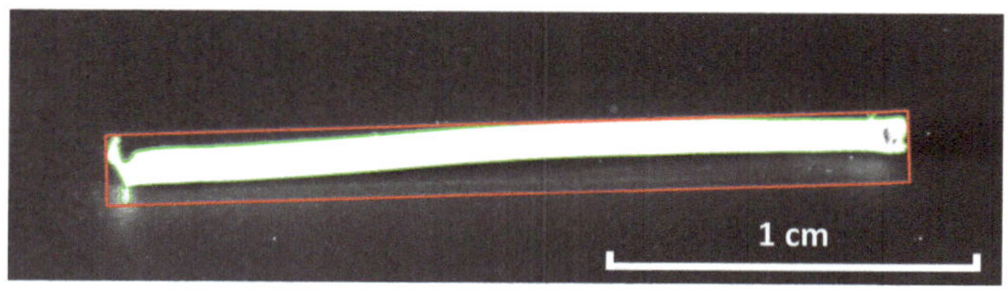

Figure 5. Example of a minimum bounding rectangle (in red) placed over the image of a robocasted cord, used to evaluate the length.

To obtain an accurate value for the diameter of the cord, five measurements are made, distributed over the visible part of the cord. To achieve this, the script places five equally spaced virtual markers on the top part of the contour, searches for the closest point on the bottom part of the contour for each of them, and computes the distance between the top and bottom points as is illustrated in Figure 6. The value for the diameter is obtained using the mean of the five measurements. This spatially averaged value significantly reduces the measurement noise.

Once the values for both length and diameter are obtained for the whole duration of drying, the longitudinal and radial shrinkages are calculated using the following relation:

$$S_{lin} = \frac{L_i - L}{L_i} \quad (1)$$

where L is the dimension at time t and L_i is the initial dimension.

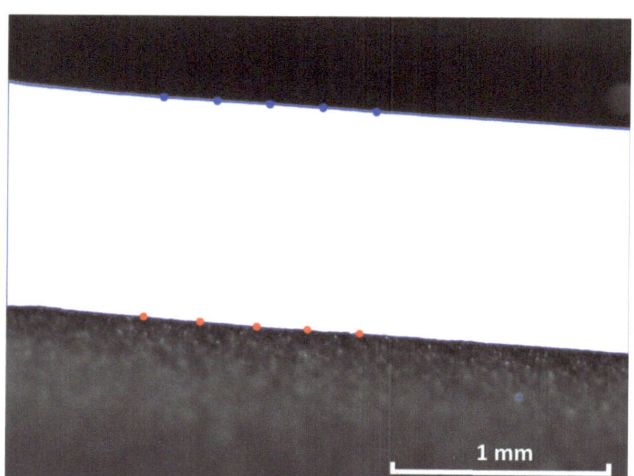

Figure 6. Example of virtual markers placed over the image of a robocasted cord to evaluate the diameter.

2.3. Measuring the Drying Shrinkage of a Robocasted Ceramic Product

The method presented in this section was developed to measure the drying shrinkage of complete robocasted ceramic products throughout the drying process in a non-destructive manner. Similarly to the previously described method, this one also relies on image analysis using a camera and Python scripts. The measurement of shrinkage relies on an automatic detection of cords by taking advantage of the unique surface topology of robocasted products, allowing the size of the cords to be tracked over time.

The experimental apparatus used to measure the shrinkage of the products is illustrated in Figure 7. The camera (IDS UI-1490SE) captures images of the product at regular time intervals throughout its drying period and a light source is used to illuminate the product and highlight the interfaces between the individual cords. An example of an image taken during drying is shown in Figure 8. Each image is analysed using an algorithm written in Python to compute the size of each cord.

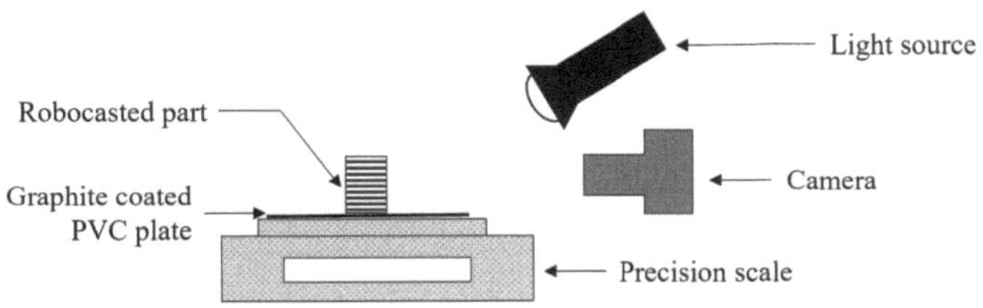

Figure 7. Experimental apparatus used to capture images of the drying product.

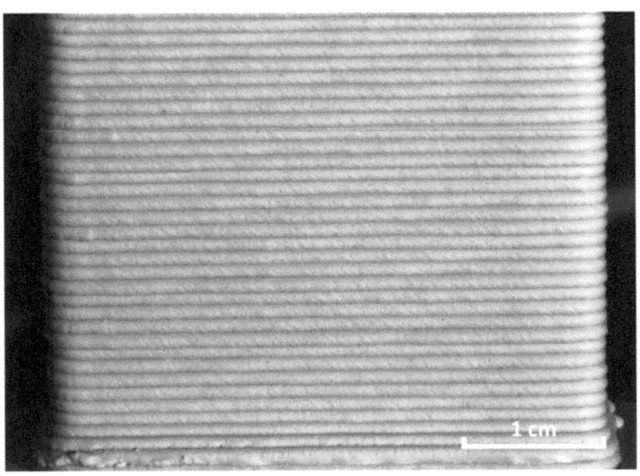

Figure 8. Example of an image of the drying product captured using the apparatus described in Figure 7.

Before analysis, each image is processed to highlight the interfaces between cords. The image is first rotated so that the cords are horizontal. The zone of interest for the study is then selected by the user and will be applied to all subsequent images. The image is then converted to grayscale and blurred in order to reduce the noise and facilitate detection of the interfaces. The interfaces are identified using the canny edge detection technique [26]. This algorithm detects brutal variations in the pixel intensity gradient to extract the edges in an image. The canny edge detection technique is used in a wide variety of computer vision systems and provides a reliable tool for investigating, e.g., micro-cracks or particle size distribution [27,28]. An example of an image after processing is shown in Figure 9.

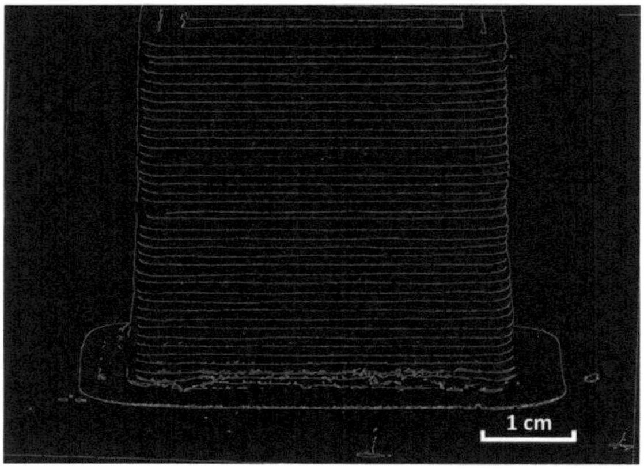

Figure 9. Example of an image of the drying product after processing.

The interfaces are then detected using an algorithm that computes a histogram of the white pixel density over the height of the image. Each peak in the histogram corresponds to an interface in the image and the height of each cord corresponds to the distance between the successive interfaces. The analysis of each image taken during the drying of a product is used for tracking the drying shrinkage of either the product as a whole or only for a zone of interest on its surface.

For this experiment, the products are printed with a 1 mm nozzle and a layer height of 0.8 mm. The extrusion rate is set at 1 cm/s.

3. Results and Discussion

The methods presented in the previous section were used to study the drying shrinkage of several robocasted green-body porcelain parts.

3.1. Drying Shrinkage of a Robocasted Ceramic Cord

The cord was dried at room temperature and ambient relative humidity. Its dimensions were measured throughout drying using the previously described method. The drying shrinkages in both the longitudinal and radial directions are shown in Figure 10. A difference in final shrinkage can be observed in both directions with 3.8% in longitudinal shrinkage and 5.7% in radial shrinkage. These results are consistent with analysis by hand of the starting and finishing images of the drying surface using an image editor to evaluate the total amount of shrinkage. A hypothesis to explain this difference in behaviour depending on the measuring direction could be the orientation of the kaolinite platelets during extrusion due to the shear stress created by the nozzle. The platelets are preferentially oriented to lie parallel to the longitudinal direction [29]. Consequently, the number of liquid/solid interfaces per unit length is higher in the radial direction. Thus, as water evaporates from the material and the platelets approach each other, the shrinkage is higher in the radial direction. Additional experiments were conducted to test the effect of adhesion with the support. A very similar behaviour was observed for a cord deposited on a support coated with oil instead of graphite. However, when the cord is extruded directly on the support without coating, the longitudinal shrinkage is strongly inhibited (less than 1%) while the radial shrinkage remains identical to approx. 6%. This leads to the conclusion that the adhesion interaction between the cord and the support has a strong influence on the longitudinal shrinkage but a negligible effect in the radial direction.

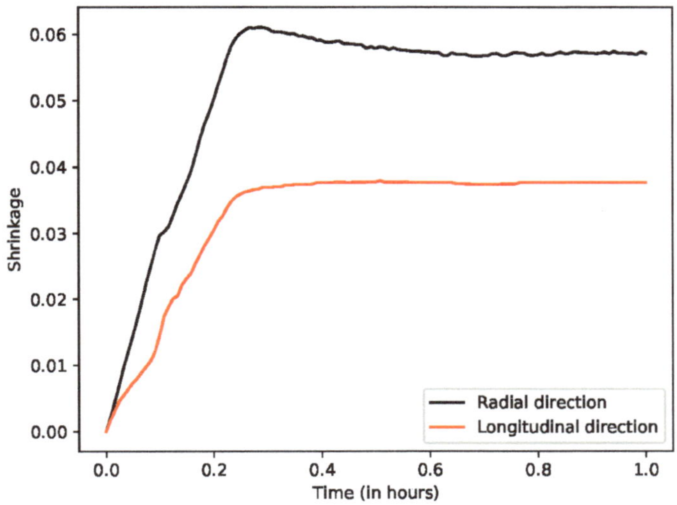

Figure 10. Drying shrinkage of a porcelain cord in both the longitudinal and radial direction as a function of time.

SEM images of a slice across the cord and of the external surface of the cord are shown in Figure 11. The slice corresponds to a fracture of a cord, made by hand, where the observation surface is perpendicular to the direction of extrusion while the external surface of the cord is parallel to the extrusion direction. The slice, displayed in Figure 11a, reveals a general orientation of the platelets in the extrusion direction as mostly edges are visible. This orientation is confirmed by the micrograph of the surface of the cord in Figure 11b, where only the basal surfaces of the platelets can be observed. Similar results on platelet orientation have been reported by Feilden et al. for cords with 70 wt% of alumina platelets, with the advantage of improving the mechanical behaviour of ceramic products [30]. The effect of kaolinite grain orientation on drying shrinkage anisotropy was also reported by Oummadi et al. when comparing the shrinkage in three orthogonal directions measured on pressed kaolin paste samples prepared from a powder containing 78 wt% of kaolinite [22]. The total amount of shrinkage in the pressing direction (11% related to the dry length) is reported to be 1.8 times greater than in directions perpendicular to the pressing axis (6%). Compared to this ratio of 1.8, the value presented in Figure 10 leads to an anisotropy ratio of 1.5. This lower value could be explained by the lower amount of kaolinite (50 wt% in the present study) and differences in the aspect ratio of the individual kaolinite platelets.

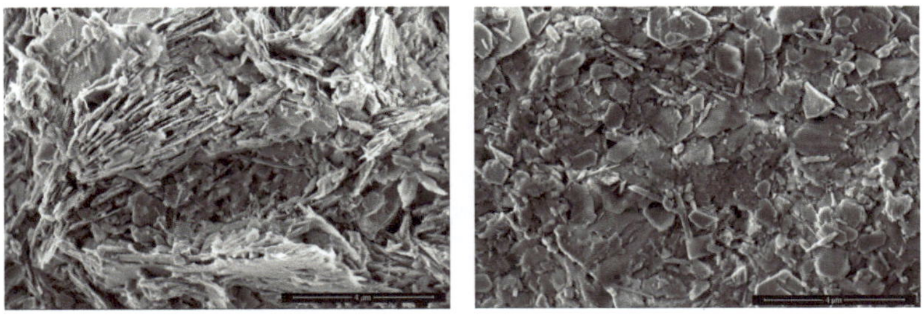

(**a**) (**b**)

Figure 11. SEM micrographs of a dried porcelain cord. (**a**) A cut perpendicular to the extrusion axis and (**b**) along its surface.

It can also be noted in Figure 10 that a swelling occurs at the end of the drying shrinkage, after 0.3 h, in the radial direction. One hypothesis is that this behaviour is due to the orientation of the platelets in the direction of extrusion. As water evaporates, liquid bridges appear in between the platelets, as illustrated in Figure 12. The pressure difference at the meniscus interfaces and the forces due to the water surface tension (at the solid/liquid/air line) produce the capillary forces that pull the grains together [31]. During the shrinkage stage, these forces are continuously balanced by repulsive forces that originate from surface charges on the solid grains [32]. After the major shrinkage is achieved, water in the bridges evaporates and the capillary forces decrease, causing a local reorganisation of the grains, due to repulsive contribution, leading to the swelling observed in the experiment [33]. Figure 12 illustrates this hypothesis. This phenomenon was also observed by other authors and reported by Onoda et al. when analysing the volume–weight relations in ceramic particulate systems encountered in processing [34,35].

Further experiments were conducted by varying the extrusion rate. Due to the extrusion technology used by the machine (a screw extruder fed by an external tank under 6 bar of pressure) a stable extrusion rate is difficult to obtain outside of a working range of 1.0 to 2.0 cm/s. Three extrusion rates were investigated and each measurement was repeated three times to account for the difficulty in obtaining stable extrusion rates. The radial shrinkage as a function of the extrusion rate is reported in Figure 13. Longitudinal shrinkage is not shown here as there was too much variation due to interactions with the support.

Radial shrinkage seems to be constant up to an extrusion rate of 1.5 cm/s and then diminishes at 2.0 cm/s. Taking into account that all the cords were dried on the same support at similar ambient drying conditions, variations in drying shrinkage can be attributed to the effect of the extrusion rate. Assuming that volumic shrinkage is identical for all samples, these results indicate a decrease in drying shrinkage anisotropy in the cords extruded at a rate greater than 1.5 cm/s. In their work, Feilden et al. explain that the orientation of anisometric grains during extrusion is dependent on the duration of exposure to the velocity gradient in the extrusion die [30]. As extrusion rate increases, the duration for which any given grain is inside of the extrusion die decreases, thus reducing the probability of it being oriented in the extrusion direction. The decrease in shrinkage anisotropy may also be caused by a disappearance of the velocity gradient. When the extrusion rate exceeds a threshold value, the laminar flow with a velocity gradient can transform into a plug flow where the velocity is almost uniform.

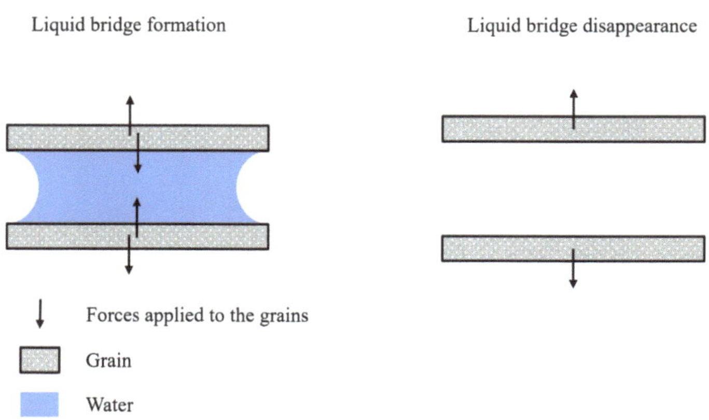

Figure 12. Schematic representation of a liquid bridge and the resulting forces.

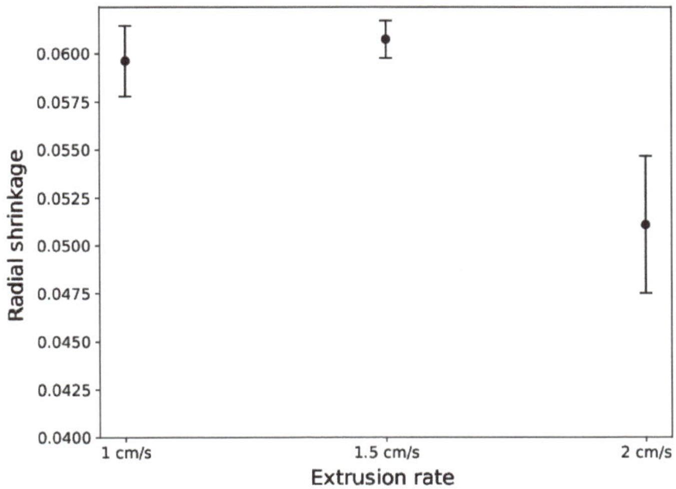

Figure 13. Radial shrinkage of an extruded cord as a function of the extrusion rate.

3.2. Drying of a Complete Robocasted Product

A tower-shaped product of $40 \times 20 \times 40$ mm^3 with 5 mm thick walls was printed and then dried in the experimental apparatus shown in Figure 7. The product was subjected to ambient drying conditions with an air temperature of 18 °C and a relative humidity of 55%. Its mass was tracked using a precision balance and images were taken every minute until the mass was stabilised. The images were then analysed using the methodology described in Section 2.3. Drying shrinkage was measured in two zones of interest: the first one close to the bottom of the product and the second one close to the top, displayed in Figure 14a. One hypothesis is that the shape of the product, displaying a cavity, would yield a difference in drying rate along its height.

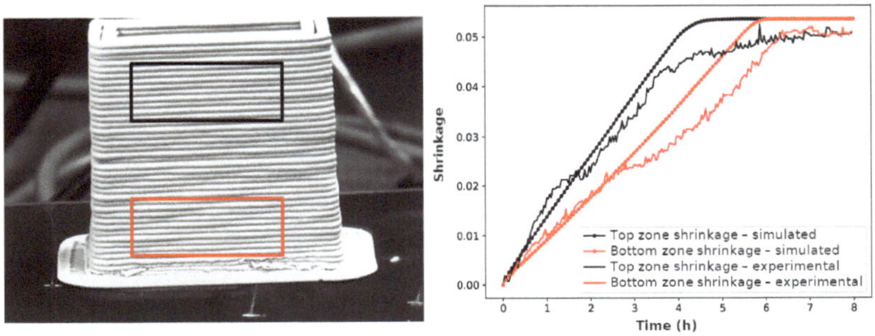

(**a**) (**b**)

Figure 14. (**a**) Image of the studied ceramic product with highlighted zones of interest (in red for the bottom and in black for the top of the object) and (**b**) drying shrinkage over time for both zones of interest.

Drying shrinkage over time is shown in Figure 14b for both zones of interest. The total drying shrinkage of the product is around 5%, and was achieved after 6 h 30 min of drying time. However, a clear difference in drying rate can be observed between the top and bottom zones. As drying shrinkage is intimately linked to the moisture content of the material, this disparity could reflect the existence of a moisture content gradient in the vertical direction. It can be noted that for the first 15 min both zones of interest display similar drying rates before diverging. This differential shrinkage between the bottom and the top of the object might be explained by the difference in the drying conditions over the height of the internal surface. It can be assumed that the relative humidity inside the cavity increases from the top to the bottom due to evaporation and poor air circulation.

In a previous work, a computer model of the drying behaviour of ceramic green bodies combining heat and mass transfers, boundary conditions for water evaporation and moisture-dependent properties was developed [36]. This model was expanded to include drying shrinkage using the Lagrangian method, which allows the geometry of the drying body to change according to its water content [37]. The experiment presented in this section was reproduced using this computer model. For these calculations, the external surfaces were subjected to convective drying conditions with the experimental ambient air characteristics while the internal surfaces were subjected to increasing relative humidity starting from 55% at the top to 90% at the bottom of the cavity in an effort to simulate the stagnant air inside the cavity. The drying shrinkage was tracked over time in the same two zones as in the experiment and the results of the simulation are displayed in Figure 14b. The predicted shrinkage curves are in good agreement with the experimental data, which is consistent with the assumption that differences in drying conditions can lead to a differential shrinkage along the height of the object.

The measurement noise observed in Figure 14b is explained by the small size of the cords in relation to the resolution of the camera. For a given image, each cord has a height of around 100 pixels meaning that small variations in size on a single cord may not be picked up by the algorithm. However, the general drying shrinkage behaviour as well as the key values are clearly visible in the graph.

It would now be interesting to place the methods developed in a more general context of existing tools for monitoring drying shrinkage. Several methods and devices have been described in the literature to evaluate drying shrinkage. The simplest method consists of measuring the distance between marks before and after drying using a caliper. The marks can be made with a gauge or a knife on a sufficiently long sample (~100 mm) to achieve a good accuracy, as described in the standard test method recommended for measuring drying shrinkage of ceramic whiteware clays [38]. The barelatograph is another device that can be used to follow dimensional variations in one dimension, by means of a small metal bar in contact with the sample. It is used for plotting Bigot curves representing the water content as a function of shrinkage. This method has been used to investigate drying behaviour in traditional but also technical ceramics [34,39]. Other devices which can be implemented for computer monitoring have also been used to track the evolution of the dimension of ceramic green bodies in 1D, for example LVDT or laser sensors [22,40]. In recent years, image analysis techniques have been used to investigate shrinkage during drying. Tracking marks and Digital Image Correlation (DIC) have successfully been implemented to measure drying shrinkage in concrete materials [22,41,42]. The DIC technique yields evaluation of shrinkage fields with a high accuracy, but it requires a speckled pattern to be coated on the sample surface. For green bodies made of several materials with different colours, the natural contrast of the surface can provide the required speckle; otherwise, a powder or a paint has to be sprinkled over the surface of the sample, which makes this technique more complicated because the added material can also interact with the drying process. Therefore, the proposed method seems to be a good compromise to follow shrinkage in different zones of a printed object without any interaction with the product and with a sufficient accuracy to highlight differences in shrinkage rate.

4. Perspectives on Real-Time Measurements of Drying Shrinkage for Robocasted Products

The previous methods were developed to measure drying shrinkage once the drying process is over and all the images have been taken. However, one of the main advantages of this kind of method is its ability to be deployed to measure shrinkage in real time. A precise control of the drying shrinkage could then be achieved by taking preventive actions during drying in order to avoid the occurrence of defects such as cracking or warping. A Python script was developed to measure the shrinkage of a product during drying and provide a real-time feedback to the user in the form of an updating shrinkage-over-time plot. This method is largely based on the experimental apparatus presented in the previous section for image acquisition and analysis with a modified workflow. The operating principle of this method is summarized in Figure 15.

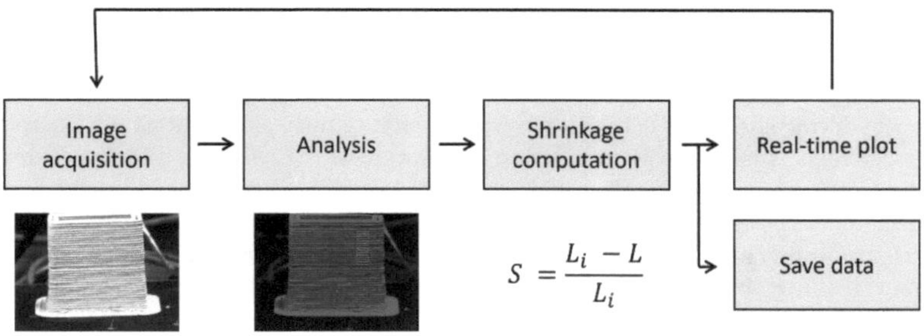

Figure 15. Operating principle of the method used to measure the drying shrinkage in real-time.

The real-time measurement method was used to track the drying shrinkage of a complete product during drying in a climatic chamber. A tower-shaped product of $40 \times 20 \times 40$ mm^3 with 5 mm thick walls was placed in the chamber and the camera was placed in front of the window. The climatic chamber was modified so that mass measurements of the sample, placed inside, could be made.

The product was dried at a temperature of 30 °C and a relative humidity of 40%. The shrinkage-versus-time plot, displayed in Figure 16, was updated in real time. This graph gives valuable information to a user about the drying state of the product as well as the exact time at which drying shrinkage is complete. Once shrinkage is completed, the risk of defects in the product, such as cracking or warping due to uncontrolled drying, is severely reduced.

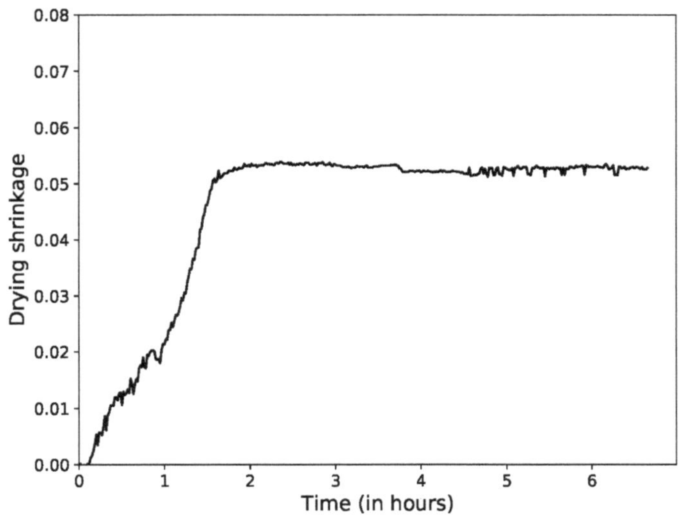

Figure 16. Drying shrinkage versus time of a tower-shaped product, measured and displayed in real time.

Another method for detecting the end of drying shrinkage in real time was proposed in a recent paper by Nait-Ali et al., making use of a convolutional neural network. The algorithm was trained using images taken from experiments on drying robocasted objects to identify the presence of water on the drying surface. This absence of water on the surface corresponds to the end of the drying shrinkage stage. Such precise information can then be used to modify the drying process [43].

5. Conclusions

New methods relying on digital images for measuring drying shrinkage in robocasted ceramic parts have been developed. Each of the two methods presented allow information to be gathered at different scales, from a single cord to a complete part, in a non-destructive manner. A single robocasted porcelain cord of 1mm radius was studied, revealing a clear anisotropy in drying shrinkage between the radial and longitudinal directions at 5.7% and 3.8%, respectively. This was explained by an orientation of the kaolinite platelets in the paste during its extrusion. A difference in drying shrinkage was also observed for different zones of a complete product exhibiting an open cavity. This was explained by the geometry of the part inhibiting airflow and causing unequal drying conditions along the height of the product. Having observed this phenomenon in a simple product, we conclude that drying shrinkage must be taken into account in the design phase of robocasted parts.

The digital methods for measuring shrinkage have the important advantage that they do not harm the products or disturb the drying process. In situ measurement of shrinkage could offer robust information for real-time fine control of the drying process for robocasted products, reducing the appearance of defects and the cost of fabrication. Specific adjustments to drying conditions, in terms of temperature, relative humidity or air flow, could be made by leveraging the methods presented in this paper to identify differences in shrinkage and correct them in real time.

Author Contributions: Conceptualisation, N.L., A.A., B.N.-A. and D.S.S.; methodology, N.L.; software, N.L. and A.A.; investigation, N.L.; resources, B.N.-A. and D.S.S.; data curation, N.L.; writing—original draft preparation, N.L.; writing—review and editing, B.N.-A., D.S.S. and A.A.; visualisation, N.L.; supervision, D.S.S., B.N.-A. and A.A.; project administration, D.S.S.; funding acquisition, D.S.S., A.A. and B.N.-A. All authors have read and agreed to the published version of the manuscript.

Funding: This research was funded by the Nouvelle-Aquitaine region, grant number 2019-1R10309.

Institutional Review Board Statement: Not applicable.

Informed Consent Statement: Not applicable.

Data Availability Statement: All data will be made available upon request.

Acknowledgments: We would like to thank Imerys Tableware France for the donation of EZPrint3D-E porcelain paste for robocasting.

Conflicts of Interest: The authors declare no conflicts of interest. The funders had no role in the design of the study; in the collection, analyses or interpretation of data; in the writing of the manuscript; or in the decision to publish the results.

References

1. Deckers, J.; Vleugels, J.; Kruthl, J. Additive Manufacturing of Ceramics: A Review. *J. Ceram. Sci. Technol.* **2014**, *5*, 245–260. [CrossRef]
2. Chen, Z.; Li, Z.; Li, J.; Liu, C.; Lao, C.; Fu, Y.; Liu, C.; Li, Y.; Wang, P.; He, Y. 3D printing of ceramics: A review. *J. Eur. Ceram. Soc.* **2019**, *39*, 661–687. [CrossRef]
3. Zocca, A.; Franchin, G.; Colombo, P.; Günster, J. Additive Manufacturing. *Encycl. Mater. Tech. Ceram. Glas.* **2021**, *1*, 203–221. [CrossRef]
4. Cesarano, J.; Denham, H.; Stuecker, J.; Baer, T.; Griffith, M. *Freeforming of Ceramics and Composites from Colloidal Slurries*; U.S. Department of Energy, Office of Scientific and Technical Information: Oak Ridge, TN, USA, 1999. [CrossRef]
5. Paterlini, A.; Le Grill, S.; Brouillet, F.; Combes, C.; Grossin, D.; Bertrand, G. Robocasting of self-setting bioceramics: From paste formulation to 3D part characteristics. *Open Ceram.* **2021**, *5*, 100070. [CrossRef]
6. Mirkhalaf, M.; Men, Y.; Wang, R.; No, Y.; Zreiqat, H. Personalized 3D printed bone scaffolds: A review. *Acta Biomater.* **2022**, *156*, 110–124. [CrossRef]
7. Monfared, M.H.; Nemati, A.; Loghman, F.; Ghasemian, M.; Farzin, A.; Beheshtizadeh, N.; Azami, M. A deep insight into the preparation of ceramic bone scaffolds utilizing robocasting technique. *Ceram. Int.* **2022**, *48*, 5939–5954. [CrossRef]

8. Boretti, A.; Castelletto, S. A perspective on 3D printing of silicon carbide. *J. Eur. Ceram. Soc.* **2024**, *44*, 1351–1360. [CrossRef]
9. Gourdonnaud, D.; Pateloup, V.; Junger, A.; Bourret, J.; Chartier, T.; Geffroy, P.M. Correlation between filament deposition path, microstructural and mechanical properties of dense alumina parts printed by robocasting. *J. Eur. Ceram. Soc.* **2024**, *44*, 1027–1035. [CrossRef]
10. Maillard, M.; Chevalier, J.; Gremillard, L.; Baeza, G.P.; Courtial, E.J.; Marion, S.; Garnier, V. Optimization of mechanical properties of robocast alumina parts through control of the paste rheology. *J. Eur. Ceram. Soc.* **2023**, *43*, 2805–2817. [CrossRef]
11. Belmonte, M.; Lopez-Navarrete, G.; Osendi, M.I.; Miranzo, P. Heat dissipation in 3D printed cellular aluminum nitride structures. *J. Eur. Ceram. Soc.* **2021**, *41*, 2407–2414. [CrossRef]
12. Pires, L.; Luís, J.; Fernandes, M.; Oliveira, M. Controlling properties of ceramic formulations for porcelain robocasting. *Ceram. Int.* **2023**, *49*, 4764–4774. [CrossRef]
13. Wu, Y.; Lan, J.; Wu, M.; Zhou, W.; Zhou, S.; Yang, H.; Zhang, M.; Li, Y. Rheology and Printability of a Porcelain Clay Paste for DIW 3D Printing of Ceramics with Complex Geometric Structures. *ACS Omega* **2024**, *9*, 26450–26457. [CrossRef]
14. Cesarano, J.; Baer, T.A.; Calvert, P. Recent developments in freeform fabrication of dense ceramics from slurry deposition. In Proceedings of the 1997 International Solid Freeform Fabrication Symposium, Austin, TX, USA, 11–13 August 1997.
15. Lamnini, S.; Elsayed, H.; Lakhdar, Y.; Baino, F.; Smeacetto, F.; Bernardo, E. Robocasting of advanced ceramics: Ink optimization and protocol to predict the printing parameters—A review. *Heliyon* **2022**, *8*, e10651. [CrossRef]
16. Feilden, E.; Glymond, D.; Saiz, E.; Vandeperre, L. High temperature strength of an ultra high temperature ceramic produced by additive manufacturing. *Ceram. Int.* **2019**, *45*, 18210–18214. [CrossRef]
17. Scherer, G.W. Theory of Drying. *J. Am. Ceram. Soc.* **1990**, *73*, 3–14. [CrossRef]
18. Brosnan, D.; Robinson, G. *Introduction to Drying of Ceramics: With Laboratory Exercises*; Wiley: Hoboken, NJ, USA, 2003.
19. Ford, R.W. *Ceramics Drying*; Elsevier: Amsterdam, The Netherlands, 1986.
20. Ghazanfari, A.; Li, W.; Leu, M.; Hilmas, G. A novel extrusion-based additive manufacturing process for ceramic parts. In Proceedings of the 27th Annual International Solid Freeform Fabrication Symposium, Austin, TX, USA, 8–10 August 2016.
21. Baltazar, J.; Torres, P.; de Oliveira, J.D.; da Cruz, J.P.; Gouveia, S.; Olhero, S. Influence of filament patterning in structural properties of dense alumina ceramics printed by robocasting. *J. Manuf. Process.* **2021**, *68*, 569–582. [CrossRef]
22. Oummadi, S.; Nait-Ali, B.; Alzina, A.; Paya, M.C.; Gaillard, J.M.; Smith, D.S. Optical method for evaluation of shrinkage in two dimensions during drying of ceramic green bodies. *Open Ceram.* **2020**, *2*, 100016. [CrossRef]
23. Itseez. Open Source Computer Vision Library. 2015. Available online: https://github.com/itseez/opencv (accessed on 1 January 2020).
24. Shaji, S.; Radhakrishnan, V. An investigation on surface grinding using graphite as lubricant. *Int. J. Mach. Tools Manuf.* **2002**, *42*, 733–740. [CrossRef]
25. Suzuki, S.; be, K. Topological structural analysis of digitized binary images by border following. *Comput. Vis. Graph. Image Process.* **1985**, *30*, 32–46. [CrossRef]
26. Canny, J. A Computational Approach to Edge Detection. In *Readings in Computer Vision*; Fischler, M.A., Firschein, O., Eds.; Morgan Kaufmann: San Francisco, CA, USA, 1987; pp. 184–203.
27. Gayathri Monicka, S.; Manimegalai, D.; Karthikeyan, M. Detection of microcracks in silicon solar cells using Otsu-Canny edge detection algorithm. *Renew. Energy Focus* **2022**, *43*, 183–190. [CrossRef]
28. Meng, Y.; Zhang, Z.; Yin, H.; Ma, T. Automatic detection of particle size distribution by image analysis based on local adaptive canny edge detection and modified circular Hough transform. *Micron* **2018**, *106*, 34–41. [CrossRef]
29. Dietemann, B.; Wahl, L.; Travitzky, N.; Kruggel-Emden, H.; Kraft, T.; Bierwisch, C. Reorientation of Suspended Ceramic Particles in Robocasted Green Filaments during Drying. *Materials* **2022**, *15*, 2100. [CrossRef]
30. Feilden, E.; Ferraro, C.; Zhang, Q.; García-Tuñón, E.; D'Elia, E.; Giuliani, F.; Vandeperre, L.; Saiz, E. 3D Printing Bioinspired Ceramic Composites. *Sci. Rep.* **2017**, *7*, 13759. [CrossRef]
31. Rossetti, D.; Pepin, X.; Simons, S.J. Rupture energy and wetting behavior of pendular liquid bridges in relation to the spherical agglomeration process. *J. Colloid Interface Sci.* **2003**, *261*, 161–169. [CrossRef]
32. Norton, F.H. Fundamental Study of Clay: VIII, a New Theory for the Plasticity of Clay-Water Masses. *J. Am. Ceram. Soc.* **1948**, *31*, 236–240. [CrossRef]
33. Mielniczuk, B.; Hueckel, T.; El Youssoufi, M.S. Laplace pressure evolution and four instabilities in evaporating two-grain liquid bridges. *Powder Technol.* **2015**, *283*, 137–151. [CrossRef]
34. Tarì, G.; Ferreira, J.M.F.; Lyckfeldt, O. Influence of the Stabilising Mechanism and Solid Loading on Slip Casting of Alumina. *J. Eur. Ceram. Soc.* **1998**, *18*, 479–486. [CrossRef]
35. Onoda, G.Y., Jr. Specific Volume Diagrams for Ceramic Processing. *J. Am. Ceram. Soc.* **1983**, *66*, 297–301. [CrossRef]
36. Lauro, N.; Oummadi, S.; Alzina, A.; Nait-Ali, B.; Smith, D.S. Computer model of drying behaviour of ceramic green bodies with particular reference to moisture content dependent properties. *J. Eur. Ceram. Soc.* **2021**, *41*, 7321–7329. [CrossRef]
37. Lauro, N. Caractérisation et Modélisation des Retraits de Séchage dans le Procédé de Microextrusion d'une Pâte Céramique. Ph.D. Thesis, IRCER—Institut de Recherche sur les CERamiques, Limoges, France, 2023.
38. *ASTM C326-09*; Standard Test Method for Drying and Firing Shrinkages of Ceramic Whiteware Clays ASTM C326-09. ASTM: West Conshohocken, PA, USA, 2014.
39. Pardo, F.; Jordan, M.M.; Montero, M.A. Ceramic behaviour of clays in Central Chile. *Appl. Clay Sci.* **2018**, *157*, 158–164. [CrossRef]

40. Kiennemann, J.; Chartier, T.; Pagnoux, C.; Baumard, J.; Huger, M.; Lamérant, J. Drying mechanisms and stress development in aqueous alumina tape casting. *J. Eur. Ceram. Soc.* **2005**, *25*, 1551–1564. [CrossRef]
41. Mauroux, T.; Benboudjema, F.; Turcry, P.; Aït-Mokhtar, A.; Deves, O. Study of cracking due to drying in coating mortars by digital image correlation. *Cem. Concr. Res.* **2012**, *42*, 1014–1023. [CrossRef]
42. Ylmaztürk, F.; Kulur, S.; Pekmezci, B.Y. Measurement of shrinkage in concrete samples by using digital photogrammetric methods. In Proceedings of the International Archives of the Photogrammetry, Remote Sensing and Spatial Information Sciences—ISPRS Archives, Istanbul, Turkey, 12–23 July 2004; Volume 35, pp. 258–261.
43. Nait-Ali, B.; Alzina, A.; Lauro, N.; Smith, D.S. Perspectives in drying of ceramics. *Open Ceram.* **2024**, *17*, 100554. [CrossRef]

ceramics

MDPI

Article

The Mechanical Properties of Geopolymers as a Function of Their Shaping and Curing Parameters

Camille Zoude [1,*], Elodie Prud'homme [1], Kévyn Johannes [2] and Laurent Gremillard [1]

[1] INSA Lyon, Université Claude Bernard Lyon 1, CNRS, MATEIS, UMR5510, 69621 Villeurbanne, France; elodie.prudhomme@insa-lyon.fr (E.P.); laurent.gremillard@insa-lyon.fr (L.G.)
[2] INSA Lyon, Université Claude Bernard Lyon 1, CNRS, CETHIL, UMR5008, 69621 Villeurbanne, France
* Correspondence: camille.zoude@insa-lyon.fr

Abstract: This study investigates the impact of curing conditions, porosity and shaping techniques on the mechanical properties of metakaolin-based geopolymers. Geopolymers offer versatility in shaping, including 3D printing, yet the influence of curing conditions after printing on mechanical properties remains unclear. This is assessed by measuring the bending properties of 3D-printed metakaolin-based geopolymer filaments cured under varied humidity and temperature conditions. The influences of porosity and of shaping technique are observed by comparing the compression properties of molded and 3D-printed samples of various porosity. Samples cured at low humidity exhibit unusually high mechanical properties, which decrease when moved from a dry to a humid environment. This behavior may be due to the presence of PEG within the composition and/or to residual stresses due to the too rapid evacuation of water. High humidity is therefore necessary to ensure optimal curing and stable properties. Increasing the curing temperature helps accelerate geopolymerization without significantly compromising mechanical properties. Direct ink writing offers design flexibility and suitable porosity, but the samples appear to exhibit different failure mechanisms than the molded samples. Additional studies are necessary to understand the interactions between PEG and the geopolymer as well as to better identify the fracture mechanisms within the different samples.

Keywords: direct ink writing; porosity; bending strength; compressive strength; molding

1. Introduction

Geopolymers are inorganic materials obtained generally by the alkaline activation of an aluminosilicate source at temperatures below 100 °C.

The geopolymer formation reaction, called geopolymerization, begins with the dissolution of the aluminosilicate source in the alkaline solution to form monomers. These monomers restructure to form oligomers which, beyond a critical concentration, polycondense, leading to the formation of a gel and the development of an amorphous, three-dimensional aluminosilicate network [1]. Metakaolin is a widely used aluminosilicate source for geopolymer manufacturing because it is a common industrial mineral that can be obtained in large quantities with consistent properties [2].

Generally, geopolymers are shaped by molding. During curing, molds are kept hermetically sealed to avoid the rapid evaporation of water which could lead to incomplete geopolymerization and poor mechanical characteristics. Perera et al. [3] observed that metakaolin-based geopolymers cured between 30% and 70% relative humidity in unsealed containers at 40 °C exhibited cracks, unlike those left sealed and cured between 22 and 60 °C. Samuel and Kriven [4] found in a recent study that curing the metakaolin-based geopolymer in low relative humidity from the time of molding resulted in a gradient in the extent of the reaction from the exposed surface to the bulk: the surface of geopolymers presented a mixture of geopolymer, alkaline silicate gel and undissolved metakaolin.

Several authors have demonstrated the significant influence that curing temperature can have on the mechanical properties of metakaolin-based geopolymers placed in sealed molds. Curing at elevated temperatures (up to 80 °C) accelerates the development of geopolymer strength [2,5,6].

Recently, 3D printing has emerged as a popular method for shaping geopolymers, offering significant flexibility in design. This shaping technique is applied mostly in the field of construction. Theoretically, it offers several advantages over traditional construction methods, such as shortened construction timelines, architectural design freedom, better safety at work and reduced labor, costs and waste [7]. However, this technique also enables the production of geopolymers with controlled porosity, suitable for diverse applications such as encapsulating radioactive waste, adsorbing hazardous molecules and synthetic dyes or heterogeneous catalysis [8–13]. These 3D-printed geopolymers are often printed under ambient temperature and humid conditions, and their size does not always allow for curing under controlled conditions. Furthermore, the resulting shapes often diverge substantially from traditional casts, particularly in the case of 3D printing through material extrusion, where filaments are assembled layer by layer, resulting in distinct mechanical properties. Nevertheless, there is no comprehensive study on the impact of environmental humidity and temperature on the mechanical properties of additively manufactured geopolymers.

This work is divided into two complementary studies. The first study investigates the effect of curing parameters (humidity and temperature) on the bending mechanical properties of metakaolin-based geopolymer filaments extruded using a 3D printer. These tests enable the rapid assessment of the mechanical properties of multiple filaments cured under varied conditions, rather than printing more intricate components.

The second part of this work compares the compression mechanical properties of geopolymer samples molded and 3D printed via extrusion and filament assembly (direct ink writing of lattices). The curing conditions for these samples remained constant to isolate the influence of the shaping technique and sample porosity on mechanical properties. The compositions of these samples were predetermined in a prior study focusing on characterizing the porosity of different metakaolin-based geopolymer compositions shaped by molding or 3D printing [14].

2. Materials and Methods

2.1. Raw Materials

The primary mix for the geopolymer includes metakaolin ARGICAL-M 1000, provided by Imerys, Paris, France (aluminosilicate source), a commercial sodium silicate solution (molar ratio: Si/Na = 1.1, density: 1.5 g.cm^{-3}, 60 wt.% water) and sodium hydroxide pellets (97% purity), both obtained from Fisher Scientific, Loughborough, UK, and used as alkaline activators. The geopolymers obtained are made of 43.7 wt% metakaolin, 51.8% sodium silicate and 4.5% sodium hydroxide.

Additives were added to this mixture to achieve specific characteristics [14]:

- Polyethylene glycol (PEG) with an average molecular weight of 1500 g·mol^{-1} (PROLABO, Paris, France) was used to induce a shear thinning behavior necessary for printing.
- To study the impact of porosity on the mechanical properties of geopolymers, metallic aluminum powder (Thermo Fisher, 99.5% purity, particle size less than 45 μm), referred to as Al, was introduced to generated dihydrogen and therefore porosity inside the paste, thanks to the following reaction:

$$2Al + 2NaOH + 2H_2O \rightarrow 2NaAlO_2 + 3H_2 \quad (1)$$

- Hexadecyltrimethylammonium bromide (CTAB, Sigma-Aldrich, St. Louis, MI, USA), a surfactant, was added to stabilize the gas bubbles.

2.2. Paste Preparation

The activation solution was first prepared by dissolving sodium hydroxide pellets and, depending on the composition, polyethylene glycol flakes, into sodium silicate solution. The mixture was stirred until all solids were fully dissolved and the solution returned to room temperature. Subsequently, it underwent homogenization using a planetary mixer (SpeedMixer, Synergy Devices Limited, High Wycombe, UK) for 30 s at 800 rpm.

Metakaolin, and CTAB when needed, was then gradually introduced and manually mixed for 30 s. The mixture was then placed for 1 min in the planetary mixer at 800 rpm.

For mixtures containing aluminum powder, this duration of mixing was reduced to 30 s because the mixture was subjected to a final mixing step, during which the aluminum powder was added. The paste containing the aluminum powder was mixed again in the planetary mixer for 30 s.

All the geopolymers in this study present the following molar ratios: SiO_2/Al_2O_3 = 3.73, Na_2O/Al_2O_3 = 0.95 and H_2O/Na_2O = 10.5.

2.3. Sample Preparation

The final mixture was poured into different containers depending on the tests to be performed:

- For compression tests, $16 \times 16 \times 20$ mm^3 rectangular molds;
- To print lattices and filaments containing PEG for compression and three point bending tests, 5 mL syringes (Nordson EFD, USA);
- For three point bending tests of filaments without PEG (which cannot be printed), Teflon mold pierced with 2.5 mm diameter holes.

If the mixture did not contain aluminum powder, the containers were directly sealed to prevent any evaporation of water. If the mixture contained aluminum powder, the samples were left in the open air for 20 min to allow the foam to form. Excess foam was removed using a glass slide, and the containers were sealed.

All containers remained for one hour (from the addition of metakaolin) at room temperature.

After that time, the paste contained in the syringes was shaped by direct ink writing using a Robocasting printer (3D-Inks, Tulsa, OK, USA) equipped with tips of 840 µm inner diameter (Nordson EFD, USA). This step took place at room temperature (between 20 and 24 °C) and ambient humidity.

Two types of parts were printed: 17 mm long filaments (on a single layer) and porous lattices of $13.7 \times 13.7 \times 11.8$ mm^3.

To print filaments, the needle was placed 1 mm above the printing substrate to avoid any compression of the filament on the support. Extrusion was completed in less than 10 min.

In the case of lattices, filaments were assembled layer by layer, with an interpenetration of 12% of the tip diameter between two layers to ensure the cohesion of the structure. A disorientation of 90° between successive layers and an 840 µm hollow space (1.68 mm center to center) between filaments of the same layer was imposed.

Lattices and filaments were deposited on a printing substrate made of a flat glass plate covered with a sheet of smooth polymer (copier polyester transparent film).

Once printing was completed, the samples (molded and printed) were subjected to the curing conditions described in the following part.

2.4. Curing of the Samples

In the case of lattices and molded geopolymer, the samples were subjected to the curing protocol described in Figure 1. The last step allows for the elimination of water potentially contained within the samples before the testing protocols.

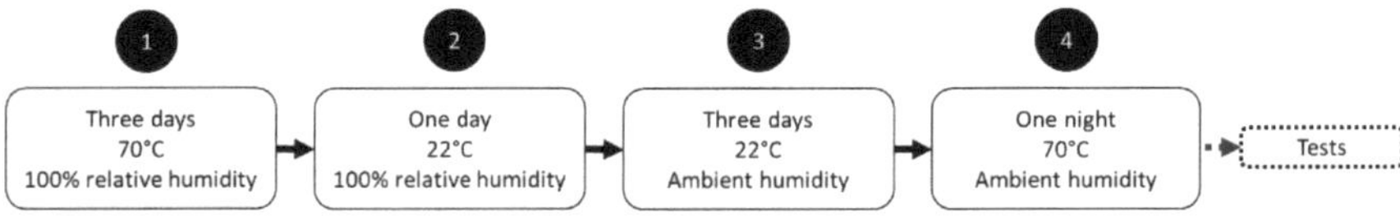

Figure 1. Curing protocol for molded geopolymers and lattice geopolymers.

In the case of 3D-printed filaments containing PEG, several curing conditions were tested. Filaments were subjected to six different curing conditions for periods up to 28 days. Three different temperatures were tested: 22 °C (room temperature), 50 °C and 70 °C. They were combined with two relative humidity (RH) conditions, 100% relative humidity and relative humidity less than 30%. At room temperature, curing took place in sealed boxes, one with the bottom covered with distilled water to reach 100% relative humidity and the other with silica gel at the bottom to achieve relative humidity below 30%. Curing at 50 °C and 70 °C took place in ovens. Samples were simply placed in the ovens in the case of curing at relative humidity below 30% (ambient humidity). For curing at 100% relative humidity, samples were placed in a carefully closed desiccator above a layer of distilled water.

Filaments without PEG, molded in Teflon molds, were subjected to only two different curing conditions: 70 °C at 100% relative humidity and 70 °C at less than 30% relative humidity. The mold was kept closed for two hours of curing to maintain the shape of the filaments. As a result, the low humidity curing condition was not completely observed during this time.

2.5. Samples Designation

The present article uses the same nomenclature as [14]. Geopolymers (G) are designated as follows:

$$\text{GPC } x\text{Al M/R/F}$$

where

- P signals the presence of 4.5 wt.% polyethylene glycol (with respect to the mass of metakaolin + sodium silicate + sodium hydroxide).
- C indicates the presence of 0.5 wt.% CTAB (with respect to the mass of metakaolin + sodium silicate + sodium hydroxide).
- xAl is the quantity of aluminum powder (0, 0.1, 0.2, 0.5 or 1 wt.% of metakaolin mass).
- M/R/F is the geopolymer shaping technique: molded (M), robocast as lattice (R) or robocast as filament (F).
- An absent letter (P or C) means that the composition does not contain the corresponding additive.

As an example, GPC 0.5Al R designates a geopolymer containing 4.5 wt.% PEG, 0.5 wt.% CTAB and 0.5 wt.% aluminum powder and shaped by robocasting (a lattice).

2.6. Samples Characterization

Porosity characterization is described in detail in a previous article [14]. This characterization was performed using three techniques: mercury intrusion porosimetry, helium pycnometry and X-ray tomography.

The mechanical bending properties of the filaments were tested via a three-point bending test on an Electroforce 3200 device (Bose, Eden Prairie, MN, USA) equipped with a 22 N force sensor at 0.3 mm.min^{-1} crosshead speed. At least 10 filaments were tested for each curing condition or composition.

Three-dimensional-printed filaments present a flat surface due to contact with the printing support (see Figure 2a). To minimize the impact of this flat surface on the test results, it was systematically placed on the compression side (in direct contact with the crosshead). Additionally, as the cross-sectional dimensions can vary significantly from sample to sample, depending on the spread of the filament, the actual cross-sectional dimensions of each sample were measured using an optical microscope (Hirox Europe, Limonest, France) to calculate quadratic moment to each filament. The flexural strength to each filament was then calculated thanks to the following equation:

$$\sigma_f = \frac{F \times L}{4} \times \frac{x}{I_y} \tag{2}$$

where

- F: load applied to the center of the filament (N);
- L: distance between the two lower supports (set at 0.015 m);
- x: distance to neutral fiber in meters;
- I_y: quadratic moment in m^4.

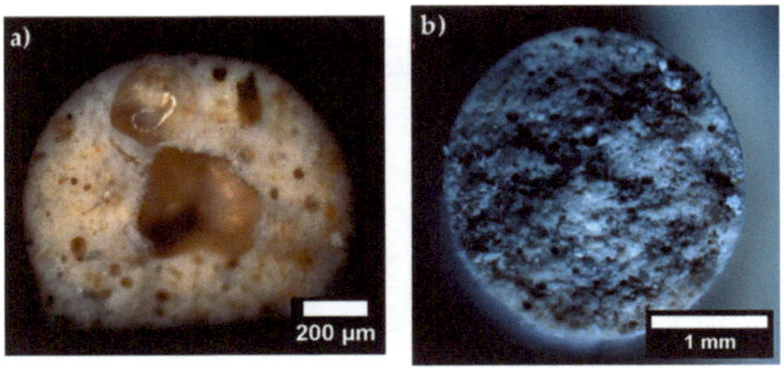

Figure 2. Optical microscope photographs of the fracture surfaces of (**a**) GPC 1Al F and (**b**) G 0Al F.

Filaments, which were molded (not containing PEG), are perfectly cylindrical (Figure 2b). Their section was compared to a disk, and their diameter was measured using a caliper. The equations can be therefore simplified as follows:

$$\sigma_f = \frac{8 \times F \times L}{\pi \times d^3} \tag{3}$$

where

- d: filament diameter (m).

Compressive mechanical properties of the lattices and molded geopolymers were tested using an Instron electro-mechanic universal testing machine (Instron 8500 +, Elancourt, France) equipped with a 25 kN (for sample G and GP) or 5 kN load cell (for other samples) crosshead speed of 0.3 mm $\cdot$ min^{-1}. Load/displacement curves were recorded and converted in stress/strain curves using the apparent section of each sample.

To highlight certain bonds and chemical groups, FTIR spectra were obtained using a Thermo Fisher Scientific (Waltham, MA, USA) IS50 infrared spectrometer via the attenuated total reflection (ATR) mode. The IR spectra were gathered between 500 and 4000 cm^{-1} with a resolution of 0.482 cm^{-1} on 32 scans. The atmospheric CO_2 contribution was replaced by a straight line between 2400 and 1900 cm^{-1}, and the spectra were corrected using a baseline. They were normalized on the peak of the Si-O-Na bond.

The filament fracture surfaces were observed using a SEM Vega 3 scanning electron microscope (Tescan, Brno, Czech Republic) equipped with a tungsten filament. Observations were carried out in secondary electron (SE) mode at 10 kV, with a working distance of 10 mm and a beam intensity of 8.

X-ray microtomography was performed on extruded filaments with a diameter of 840 µm using an RX Solutions (Chavanod, France) EasyTom Nano tomograph equipped with a Hamamatsu tube, at a resolution of 0.7 µm. Images were extracted using Fiji software.

3. Results

Bending tests, infrared spectroscopy and SEM observations were performed on dense filaments to evaluate the impact of curing conditions on mechanical properties and the microstructure. Compression tests were implemented on scaffolds and molded samples with one single curing condition to evaluate the impact of the shaping technique and porosity on compression properties.

3.1. Three Point Bending Tests on Dense Filaments

GP 0Al F were printed by direct ink writing, cured under different conditions and then subjected to the test. Six storage conditions were tested: three different temperatures (room temperature, 50 °C, 70 °C) combined with two humidity conditions (relative humidity less than 30% and relative humidity 100%). The evolution of mechanical properties was recorded at 24 h, 3 days, 14 days and 28 days. Figure 3 presents the evolution of the flexural strength according to the different conditions.

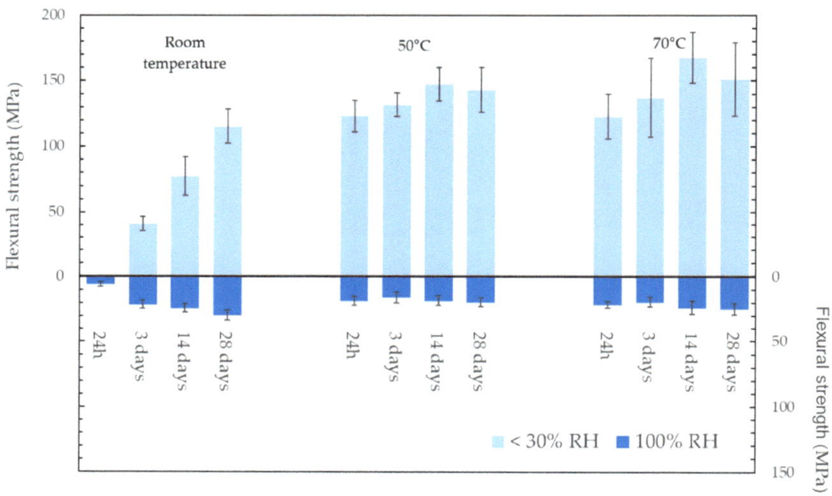

Figure 3. The flexural strength evolution of GP 0Al F as a function of time, relative humidity and the temperature of the environment.

Every filament exhibits brittle fracture (Figure 4), except filaments subjected to a relative humidity lower than 30% at room temperature and tested after 24 h and 3 days of curing. Indeed, under these conditions, after 24 h of curing, the filaments do not break but instead deform plastically until they are no longer in contact with the test supports (Figure 5). No cracks are observed on the surface. The filaments break after 3 days of storage but still show strong plastic deformation. However, this deformation is no longer visible after 14 days of curing (Figure 4a).

A significant difference is observed between the flexural strength of filaments stored at low relative humidity and those stored at high relative humidity, regardless of the temperature (Figure 3). Filaments cured at low humidity have an average flexural strength 5.5 times higher than filaments stored at high humidity after 28 days.

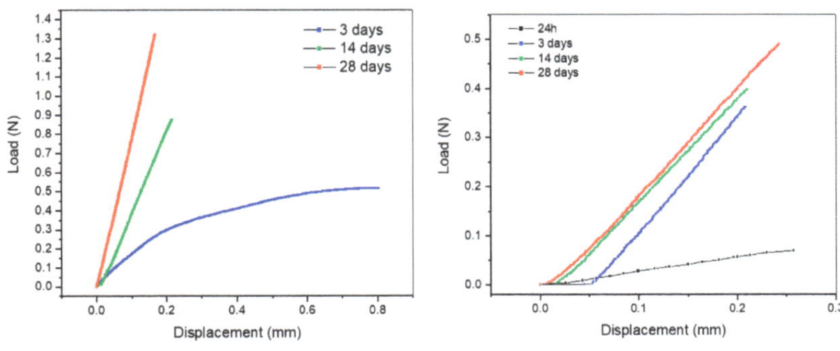

Figure 4. Typical load/displacement curves of GP 0Al F cured at room temperature and (**a**) relative humidity below 30% and (**b**) 100% relative humidity.

Figure 5. Digital (**a**) and electronic (**b**,**c**) micrographs of GP 0Al F following three-point bending test after 24 h of curing at room temperature below 30% relative humidity.

Samples stored at high humidity and at a temperature of 50 °C or 70 °C reach a flexural strength very close to the flexural strength obtained after 28 days within 24 h (about 20 MPa). It takes between 24 h and 3 days of curing for samples stored at room temperature to approach the 28-day flexural strength. It appears that the higher the temperature, the higher the average flexural strength. The same trend is noted in the literature on compression tests [15]. However, the results presented here are subject to high uncertainty, and these tests should be conducted on a larger sample population to confirm the trend.

Samples subjected to low humidity during curing exhibit a slower attainment of their maximum mechanical properties, especially at room temperature, where the flexural strength consistently increases throughout the 28-day testing period. The increase in temperature accelerates the increase in the strength since after 24 h at 50 °C or 70 °C, the flexural strength is already very close to the flexural strength reached at room temperature after 28 days (36 GPa). They continue to increase, reaching in both cases approximately 42 GPa after 28 days.

To assess the stability of these mechanical properties against humidity variations, GP 0Al filaments conditioned at a specific humidity and temperature for 28 days undergo two additional days of storage in inverted humidity conditions at room temperature. They are then subjected to the three-point bending test. Figure 6 presents the results before and after the change of condition.

The flexural strength of samples stored for 28 days at high humidity is not modified after 48 h at low humidity.

However, after 48 h at high humidity, the very high flexural strength of samples cured for 28 days at low humidity decreases. Filaments initially stored at room temperature display a flexural strength even lower than those stored at high humidity for 28 days. The final flexural strength of samples initially stored at 50 °C and 70 °C is slightly higher on average than that of samples stored at high humidity for 28 days at the same temperatures.

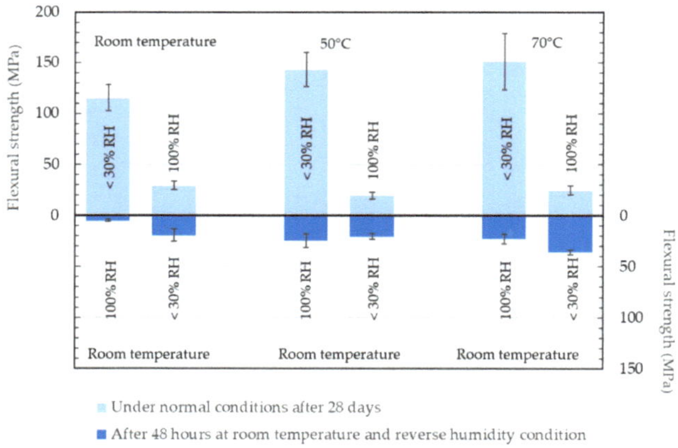

Figure 6. Flexural strength evolution of GP 0Al F before and after modification of curing conditions.

To determine if the addition of PEG impacts the mechanical properties of the filaments, a three-point bending test was performed on PEG-free filaments (G 0Al F). As the paste is not suitable for filament extrusion by direct ink writing, the filaments were molded into Teflon molds. Given the delicate nature of implementing this method and the very limited number of filaments obtained, only two conservation conditions were tested: 70 °C at 100% relative humidity and 70 °C at less than 30% relative humidity. Measurements were carried out at 14 days and are summarized in Table 1.

Table 1. Flexural strength of G 0Al F under different humidity conditions after 14 days at 70 °C.

	Flexural Strength (MPa)	
	$<\sigma_f>$	SD
<30% RH	29.7	6.7
100% RH	20.1	8.7

The significant difference in flexural strength observed as a function of humidity is not evident. The flexural strength of PEG-free filaments at low humidity is slightly higher than that at high humidity, but considering measurement uncertainties, they are quite similar. These values are also very close to those of the GP 0Al filaments stored at 100% relative humidity at the same temperature and after 14 days.

Thus, the high mechanical properties observed at low humidity seem to be influenced by the presence of polyethylene glycol. A comprehensive discussion of the role of PEG, considering additional results, will be discussed in Part 4.

However, it is already apparent that, in the case of geopolymer compositions containing polyethylene glycol, it is more prudent to print and store the samples under high relative humidity to ensure the stability of mechanical properties.

3.2. *Infrared Spectroscopy of Dense Filaments*

After being stored under the different conditions described above, GP 0Al F were ground and analyzed using infrared spectroscopy (ATR mode).

3.2.1. Filaments Stored 28 Days at Low Humidity

PEG is constantly detected in filaments that have been stored 28 days at low humidity regardless of the conservation temperature (Figure 7).

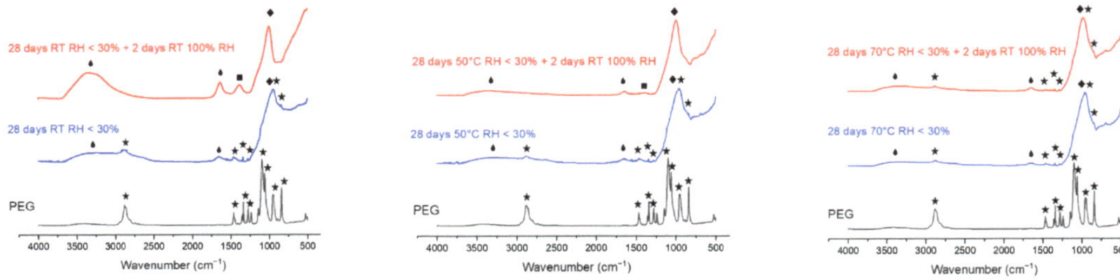

(**a**) (**b**) (**c**)

Figure 7. Normalized infrared spectra of crushed filaments GP 0Al F and PEG 1500. GP 0Al filaments were cured 28 days at relative humidity below 30% and (**a**) room temperature or (**b**) 50 °C or (**c**) 70 °C before being placed 2 days at room temperature and high relative humidity. 💧: Vibrational bands of water binding [16,17]; ★ vibrational bands of PEG binding [18]; ■ vibrational bands of carbonate binding group [17]; ◆ vibrations of the characteristic binding of the geopolymer network [19].

After 28 days, when the humidity conditions are reversed for two days, PEG is no longer detected, except in the case where the filaments have been stored 28 days at 70 °C. A peak associated with the presence of carbonate is detected during this humidity condition inversion for filaments stored at room temperature and 50 °C. It is particularly pronounced for filaments stored at room temperature.

3.2.2. Filaments Stored 28 Days at High Humidity

PEG is not detected in filaments that have been stored 28 days at high humidity (Figure 8). All filaments present a significant peak associated with carbonates. When reversing the humidity conditions, PEG was still not detected in filaments that have been stored at room temperature and 50 °C. On the other hand, it is detected in filaments that have been stored at 70 °C. The carbonate peak is consistently present, whether before or after reversing humidity conditions.

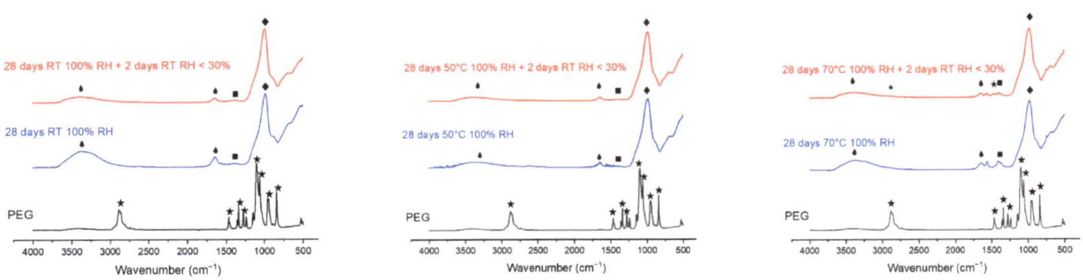

(**a**) (**b**) (**c**)

Figure 8. Normalized infrared spectra of crushed filaments GP 0Al F and PEG 1500. GP 0Al filaments were cured 28 days at high relative humidity and (**a**) room temperature or (**b**) 50 °C or (**c**) 70 °C before being placed 2 days at room temperature and relative humidity below 30%. 💧: Vibrational bands of water binding [16,17]; ★ vibrational bands of PEG binding [18]; ■ vibrational bands of carbonate binding group [17]; ◆ vibrations of the characteristic binding of the geopolymer network [19].

3.3. Fracture Surface of Dense Filaments

The morphology of the fracture surfaces of filaments cured at room temperature varies depending on the environmental humidity (Figure 9). Filaments stored in high humidity conditions exhibit a more spread-out structure with a substantial flat area, while those cured under low humidity are rounder with a smaller flat region, indicating a better preservation of their original shape. As the filaments were extruded on the same day under consistent conditions, this phenomenon is likely to occur during the curing process.

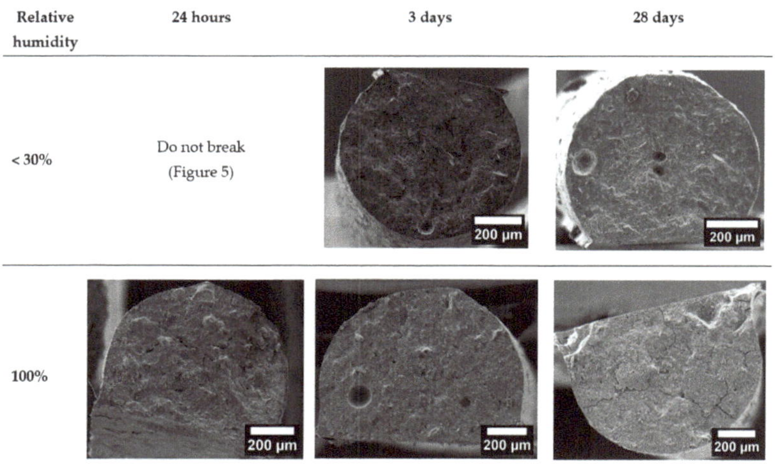

Figure 9. Fracture surface of GP 0Al cured at room temperature for 24 h, 3 days and 28 days under high and low relative humidity.

At high relative humidity, increasing the curing temperature appears to limit the spreading of the filaments (Figure 10).

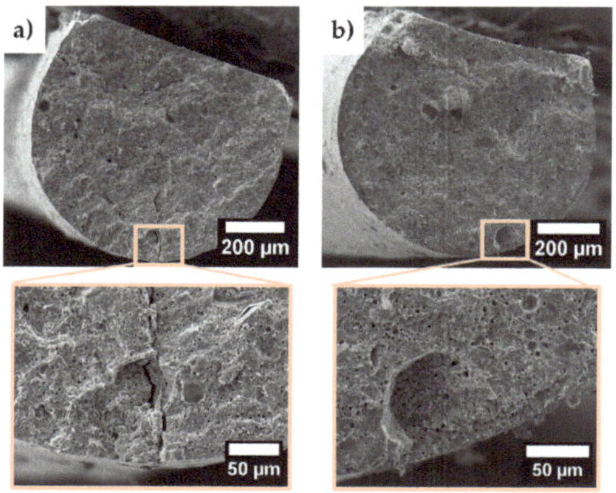

Figure 10. SEM micrographs of fracture surfaces of GP 0Al F after 24 h under 100% RH at (**a**) 50 °C and (**b**) 70 °C.

In addition, filaments stored at room temperature have different microstructures depending on the humidity of the environment.

The fracture surface of filaments stored at high humidity for 24 h exhibits large cracks and a microstructure with visible pores within a homogeneous mass (Figure 11a). After three days of curing, these pores become numerous and take on a well-defined spherical shape, similar to the microstructure observed after 28 days of curing (Figure 11b,c). Each time, more or less significant cracks are observed on the fracture surface. Increasing the curing temperature (50 °C and 70 °C) while maintaining high relative humidity results in a microstructure like that obtained after three days at room temperature and high humidity in only 24 h (Figure 10).

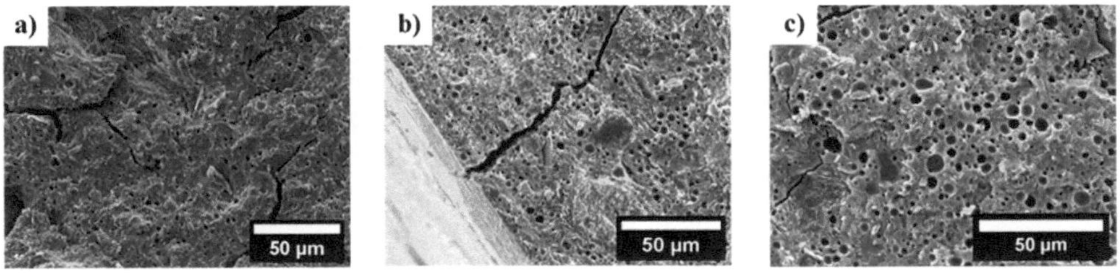

Figure 11. SEM micrographs of fracture surface of GP 0Al filaments cured at room temperature under 100% RH after (**a**) 24 h, (**b**) 3 days and (**c**) 28 days.

As mentioned previously, filaments cured at low humidity for 24 h are strongly deformed by the three-point bending test. Their surface shows no cracks (Figure 12). After three days of curing, the fracture surfaces exhibit irregularities, visually indicating the plasticity observed in the fracture curves (Figure 12a). Unlike samples stored at high humidity, only fine cracks are visible. Pores are also visible, but the geopolymer matrix seems less homogeneous. After 28 days of curing, the fracture surfaces are more clearly defined but seem to have a double microstructure. The interior of the filaments presents a microstructure similar to that observed after 24 h of curing in the case of filaments cured under high humidity (Figure 12b). However, the exterior of the filament has a very dense microstructure, with no visible porous network (Figure 12c).

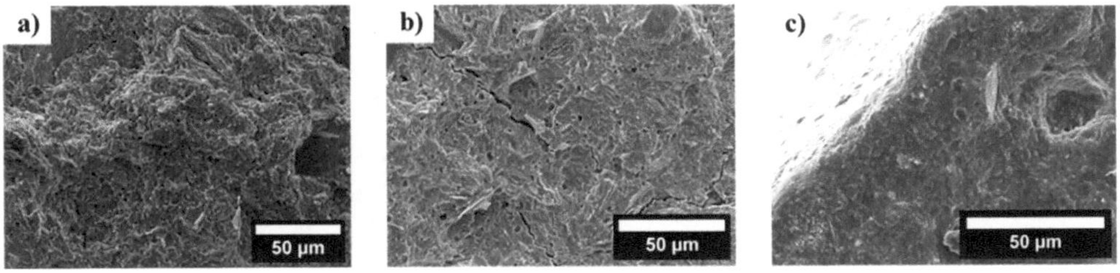

Figure 12. SEM micrographs of fracture surface of GP 0Al F cured at room temperature under low relative humidity after (**a**) 3 days, (**b**) 28 days (filament core) and (**c**) 28 days (filament periphery).

3.4. Compressive Stress of Molded Geopolymers and Lattices

3.4.1. Molded Geopolymers

Table 2 gathers the main properties of the geopolymer compositions studied in the previous work. Figure 13 shows one of their typical stress/strain curves.

It is evident that the compressive strength decreases with increasing porosity. All curves exhibit brittle behavior, with a linear elastic domain at low stresses followed by a more or less significant plateau typical of fragile foams [20]. The extent of the linear elastic phase, representing undamaged material, diminishes as porosity increases. The plateau following the linear phase is attributed to the collapse of the solid phase inside the material by the crushing of the walls between the pores. In GPC compositions (Figure 13b), higher quantities of aluminum powder led to the initiation of material damage at progressively lower stresses, but there is a notable increase in the allowed deformation before reaching fracture.

Table 2. Envelope density, total porosity and maximum compressive stress of molded geopolymers.

Designation	Envelope Density (g.cm^{-3}) ± SD	Total Porosity (vol.%)	Compressive Strength (MPa) ± SD
G 0Al M	1.43 ± 0.01	34.2 ± 3.1	50.8 ± 5.2
GP 0Al M	1.40 ± 0.01	36.7 ± 1.6	41.6 ± 9.9
GPC 0.1Al M	0.90 ± 0.06	58.3 ± 3.4	5.0 ± 0.4
GPC 0.2Al M	0.74 ± 0.03	66.0 ± 2.3	3.4 ± 0.8
GPC 0.5Al M	0.65 ± 0.07	70.1 ± 3.1	2.4 ± 0.8
GPC 1Al M	0.61 ± 0.02	71.6 ± 1.7	2.1 ± 0.4
GP 0.5Al M	0.43 ± 0.03	79.7 ± 1.7	1.3 ± 0.3
GC 0.5Al M	0.28 ± 0.02	87.0 ± 1.2	0.21 ± 0.05

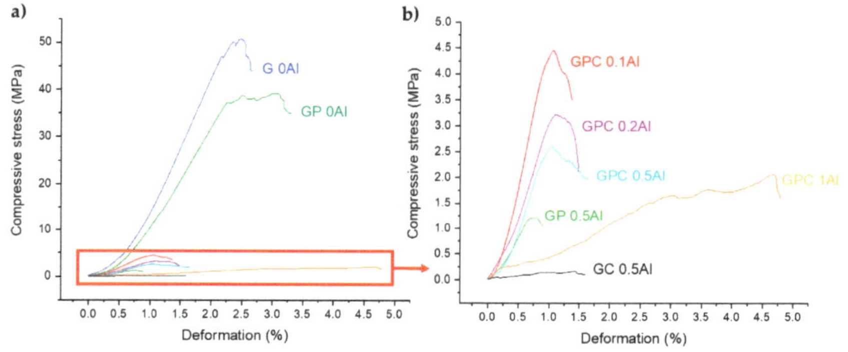

Figure 13. Stress/strain curves of all molded geopolymers (**a**) and composition manufactured with aluminum powder ((**b**): detail of (**a**)).

Figure 14 shows the evolution of the compressive strength as a function of the envelope density of the samples. A power law with a good correlation coefficient was chosen to represent this evolution. However, it tends to underestimate compressive strengths at lower envelope densities. Despite this limitation, this relationship provides an estimate of the expected compressive strength for molded geopolymers with similar composition.

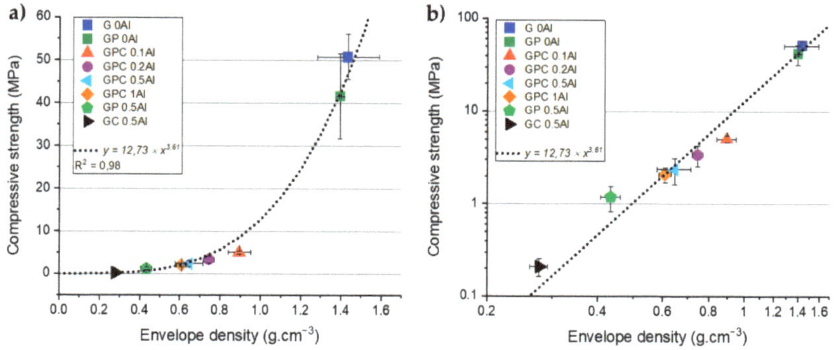

Figure 14. The evolution of the compressive strength as a function of the envelope density of the molded geopolymer samples in (**a**) linear and (**b**) logarithmic scale. Error bars represent standard deviations.

3.4.2. Three-dimensional-Printed Geopolymer Lattices

The main properties of the geopolymer lattices obtained by direct ink writing are summarized in Table 3. G 0Al R have greater porosity—and therefore lower compressive strength—compared to G 0Al M, thanks to the spacing between the filaments. The lattices made with 0.1, 0.2, 0.5 and 1% aluminum powder have similar total porosities (for more information, see [14]) and therefore similar compressive strengths.

Table 3. Envelope density, total porosity and maximum compressive stress of lattices.

Designation	Envelope Density ($g.cm^{-3}$) ± SD	Total Porosity (vol.%)	Compressive Strength (MPa) ± SD
GP 0Al R	1.08 ± 0.04	51.1 ± 2.7	18.3 ± 5.3
GPC 0.1Al R	0.83 ± 0.04	61.6 ± 2.4	2.7 ± 0.5
GPC 0.2Al R	0.70 ± 0.03	67.9 ± 2.3	3.3 ± 0.3
GPC 0.5Al R	0.75 ± 0.05	65.6 ± 3.3	3.2 ± 0.6
GPC 1Al R	0.74 ± 0.04	65.5 ± 2.8	3.0 ± 0.6

The stress/strain curves do not always present a linear elastic domain, likely due to local defects and the difficulty of obtaining perfectly parallel surfaces on these samples (Figure 15). The irregular decreases in stress could also be due to the progressive collapse of the layers by the fracture of the filaments, possibly at the junctions of the filaments, or by the propagation of cracks already present within the material.

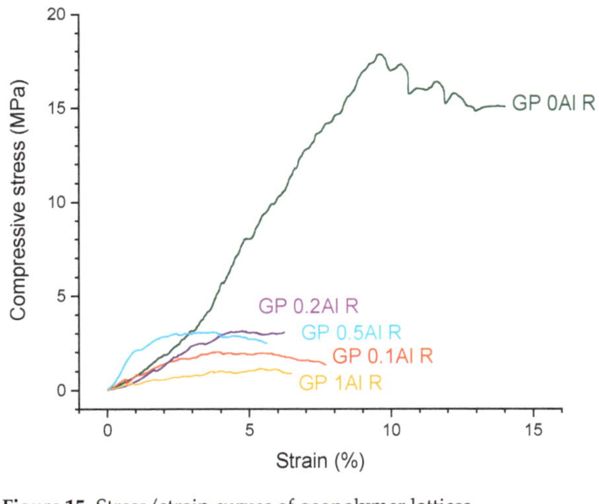

Figure 15. Stress/strain curves of geopolymer lattices.

The addition of the points corresponding to lattices on the evolution of the compressive strength as a function of the envelope density leads to a lower correlation coefficient (drop from 0.98 to 0.70, Figure 16). However, the points respect the general trend.

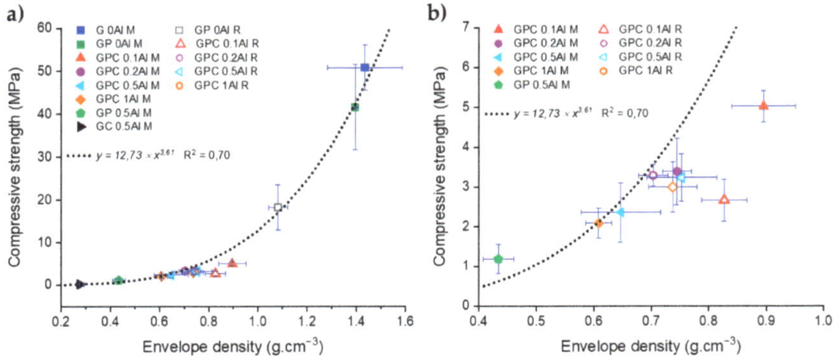

Figure 16. (**a**) The evolution of the compressive strength as a function of the envelope density of different geopolymer compositions shaped by molding and direct ink writing; (**b**) zoom of 16 (**a**). Error bars represent standard deviations.

4. Discussion

4.1. Influence of PEG and Curing Parameters on Filament Bending Stress

4.1.1. Impact of Storage at High Relative Humidity on the Properties of Geopolymers

At high relative humidity, Baird et al. [21] demonstrated that PEG 1450 absorbs atmospheric humidity through the phenomenon of deliquescence, transitioning from a solid to a solution by sorbing significant amounts of atmospheric moisture. Although this phenomenon depends on the molar mass of PEG and temperature, the study shows that the deliquescence of PEG 1450 at 25 °C begins at 66% relative humidity and at 49% relative humidity at 40 °C. Although our work involves a slightly different PEG (PEG 1500) and different temperatures (room temperature, 50 °C and 70 °C), this study's findings suggest that under our conditions, at 100% relative humidity, the polyethylene glycol in geopolymers is likely to undergo deliquescence. Part of PEG is probably in the liquid state, especially since the temperatures 50 °C and 70 °C are higher than the melting temperature of PEG 1500 (45 °C). This high relative humidity would allow for the conservation of a large quantity of water within the capillary network.

This presence of water explains the systematic detection of sodium carbonates in infrared spectroscopy after 28 days of storage at high humidity (Figure 8). Carbonate formation indeed requires water to dissolve the CO_2 present in the air [22]. This observation suggests that sodium is found in excess in the geopolymer network and is free to react to form carbonates. Thus, PEG is immersed in an extremely basic environment which could, over time, degrade it and explain why it is not detected in infrared spectroscopy after 28 days. But this could also be explained by the fact that the bonds constituting PEG are probably looser since it is in the liquid state, and its presence could be masked by the spectrum of water.

The liquid state of PEG at high humidity explains the greater spreading of the filaments cured in these conditions (Figure 9). The filament spreading diminishes progressively with rising temperatures (Figure 10), likely due to the accelerated geopolymerization process. Indeed, it takes between 24 h and 3 days for filaments stored at room temperature to reach mechanical properties close to the final properties (Figure 3). For storage of filaments at 50 and 70 °C, this time is reduced to less than 24 h. Increasing the curing temperature accelerates the geopolymerization at high humidity.

The presence of PEG does not affect the mechanical properties in the bending of geopolymer filaments cured at high humidity (Figure 3 and Table 1). Filaments with PEG present mechanical properties equivalent to compositions without PEG and close to the values reported in the literature, with a flexural strength between 10 and 20 MPa [2,23].

But the introduction of PEG has an impact on the mechanical properties in the compression of molded samples cured under high relative humidity. The compressive strength decreases on average from 50.1 to 41.6 MPa after the addition of PEG (Table 2). This difference is more likely due to a difference in density as the molded samples containing PEG are more porous than the molded samples without PEG. Given the higher viscosity of pastes containing PEG [14], they are more prone to trapping air bubbles, potentially leading to sample failure. Despite this difference in mechanical properties in compression, the orders of magnitude obtained for these mechanical properties are consistent with those found in the literature (compressive strength around 50 MPa) [24].

4.1.2. Impact of Low Relative Humidity Storage on Geopolymer Properties

After curing at low relative humidity, the filaments exhibit unusually high mechanical properties, regardless of temperature and curing duration (Figure 3). However, filaments without PEG maintain normal mechanical properties regardless of humidity conditions (Table 1).

The first hypothesis to explain this phenomenon would be that PEG, which does not undergo deliquescence at low humidity, might act as a mechanical binder by solidifying and forming a continuous 3D network within the geopolymer. Its solid form is always detected in infrared spectroscopy after grinding filaments cured under these conditions (Figure 7). Despite curing temperatures of 50 °C and 70 °C exceeding PEG's melting point, there is a possibility that PEG rapidly solidifies within the structure upon returning to room temperature for the tests.

An alternative hypothesis emerges based on recent findings by Samuel and Kriven [4]. They conducted microhardness measurements on geopolymer bars cured at 20 °C and 40% RH, revealing notably high surface microhardness over an approximate thickness of 2 mm. The authors attribute this phenomenon to potential residual stresses near the surface induced by capillary stresses from rapid water evacuation and/or density variations within the material due to these stresses. Given the small size of the filaments in this study (840 μm in diameter), it is plausible that these residual stresses contribute to the observed significant mechanical properties, especially considering the filaments exhibit a density gradient after 28 days of curing (Figure 12b,c). The rapid evacuation of water and the stresses generated may be sufficient to prevent filament spreading. The fact that the PEG-free sample did not exhibit high properties at low humidity (Table 1) could be since the mold was opened two hours after molding the filaments, to allow them to retain their shape. These two hours in endogenous conditions were perhaps enough to modify the constraints within the filament. In this scenario, PEG does not act as a mechanical binder, although present in the structure (because detected in infrared).

Finally, it is possible that these unusually high mechanical properties at low humidity can be explained by a combination of these two hypotheses, with PEG acting as a mechanical binder, residual stresses and density variation. Additional tests must be carried out to achieve a more comprehensive understanding of the phenomena.

These hypotheses are consistent with the fact that filaments cured at low humidity and room temperature for up to 3 days deform plastically, while filaments cured at high humidity do not (Figures 4 and 5). Samuel and Kriven [4] observe a similar behavior on geopolymer bars without PEG. This behavior becomes more pronounced with the decrease in the amount of water introduced into geopolymer pastes. The curing process at low humidity may prevent the retention of a sufficient water quantity within the capillary network. The hygroscopic nature of polyethylene glycol could contribute to this effect, resulting in insufficient water availability for proper species dissolution and network formation. Additionally, unlike filaments cured at high humidity, there is no discernible carbonation peak (Figure 7), indicating insufficient water for carbonate formation (where sodium is necessarily present). The geopolymerization phenomenon, already slower at room temperature than at 50 and 70 °C, would therefore be further slowed down. The material would be composed of partially dried alkaline silicate, partially solid polyethylene glycol and geopolymer/metakaolin particles. These components would form a deformable gel, which can be further reinforced by the presence of PEG, and which would constitute the mechanical skeleton of the geopolymer during the first 24 h, while the geopolymerization takes place slowly until influencing the fracture behavior. However, this geopolymerization at room temperature and low humidity is clearly incomplete. When the storage conditions are reversed, the flexural strength of the filaments collapses below the anticipated level, equivalent to the flexural strength of filaments stored at high humidity, which formed their own geopolymer network without being influenced by rapid water drainage (Figure 6). In addition, a very significant carbonation peak is detected in infrared spectroscopy (Figure 7a). This is probably due to the rehydration of the sodium-containing species which did not have time to react to form the geopolymer network before curing. Consequently, sodium returns to solution due to high ambient humidity and reacts to form carbonates. Moreover, the microstructure of the filament is very dense after 3 days of storage, and only the core of the filament shows the signs of a porous network characteristic of a geopolymer after 28 days of curing (Figures 9 and 12).

Samuel and Kriven [4] also observed incomplete geopolymerization on the surface of their samples without PEG stored at 20 °C and 40% RH.

Filaments cured at low humidity at 50 °C and 70 °C do not exhibit plasticity during the first curing days (Figure 4a). But these samples also exhibit unusually high mechanical properties. When humidity conditions are reversed, these properties collapse to levels comparable to filaments cured at high humidity (Figure 6). Following this reversal, the infrared band of PEG is nearly absent, and the carbonation signal is notably reduced compared to filaments cured at low humidity and room temperature. These observations suggest that the geopolymer network was well formed and that the transition to high humidity caused either the deliquescence of the PEG, rendering it ineffective as a binder (resulting in a mechanical structure purely made of geopolymer), or the relaxation of the constraints within the structure or both.

Several hypotheses can be considered and combined to explain this more complete formation of the geopolymer network at 50 and 70 °C at low humidity. First, the temperature increase accelerates geopolymerization reactions sufficiently to occur, while there is still enough water in the medium. Then, the melting temperature of PEG might be exceeded in the case of curing at 50 °C and 70 °C. The PEG being in a liquid form, its hygroscopic nature might be limited, and more water could be found within the structure in the early stages. Despite efforts to minimize the time between exiting the oven and conducting the bending test, the rapid cooling of the small filaments (840 μm in diameter) may result in the quick return of PEG to a solid state, resuming its role as a binder during the test.

These observations indicate that it is not the presence of PEG that limits geopolymerization and confers plasticity properties at room temperature but rather the lack of water within the geopolymer during curing at low humidity. However, it is difficult to assert that PEG is responsible for the unusually high mechanical properties. These could also be explained by the presence of residual stresses following the rapid evacuation of water.

Storing samples at room temperature and low humidity after 28 days of storage at high humidity does not result in a significant increase in mechanical properties (Figure 6). The properties remain stable. The PEG band is almost not detected in infrared spectroscopy (Figure 8). In the case of a hypothesis where PEG acts as a mechanical binder, it is possible that it has been degraded by the basic environment at high humidity and can no longer play a binding role during the transition to low humidity. Alternatively, PEG may have been evacuated from the structure during the gradual formation of the geopolymer gel, or two days at low humidity might be insufficient for its solidification, particularly within the filament's core. Whatever remains inside the structure might not be present in a sufficient quantity to be detectable. Since the geopolymer network has already formed, and the PEG in liquid form at high humidity has been excluded from this network, it can no longer act as a binder at low humidity. In the scenario where high mechanical properties at low humidity result from residual stresses, it is not possible to achieve higher mechanical properties by lowering the humidity. The geopolymer network is already formed without being influenced by the rapid drainage of water. These residual constraints have no reason to reappear. The non-detection of PEG may be due to degradation or its presence in too small a quantity.

4.2. Influence of Porosity and Shaping on Mechanical Properties

The results presented in Section 3.4 highlight the correlation between the compressive mechanical properties and sample porosity. Molded geopolymers exhibit a power law relationship between the two parameters, enabling the estimation of compressive strength based on envelope densities (Figure 14). However, this law proves less effective for the most porous lattices. This less good agreement for the samples obtained by direct ink writing could be explained by the inherent difficulty in achieving a flat surface for compression testing, leading to increased uncertainties in measurements. But it could also be due to the different fracture mechanisms of the molded foams, linked to the progressive fracture of the filaments and the presence of the millimeter scale of porosity. The study of the fracture surfaces and the tomography of the filaments containing PEG reveal the existence of cracks prior to any testing (Figure 17, cracks are highlighted by red circles), potentially influencing structural failure. The compression curves of architectural structures display numerous irregularities, likely attributed to the propagation of these pre-existing cracks and the gradual fracture of filaments.

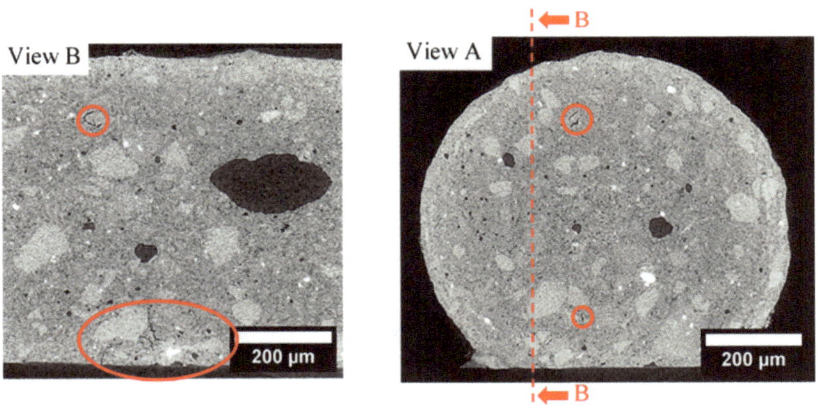

Figure 17. The microtomography of a filament of composition GP 0Al F having been stored for 3 days at 70 °C and 100% relative humidity. View B is the right view of the section placed on view A. The cracks are highlighted by red circles.

An exact comparison of these results with the literature proves challenging due to the dependence of these mechanical properties on various parameters, including raw materials, additives, curing conditions, and the type and morphology of blowing agents.

However, the average compressive strength of G 0Al M (50.8 MPa) closely aligns with that reported by Rowles and O'Connor (51.3 MPa) for geopolymers of similar composition, although the article does not specify the density of the samples [24].

Regarding molded foams, those produced in previous studies by incorporating aluminum powder typically exhibit envelope densities ranging from 0.4 to 0.8 g.cm^{-3} for a compressive strength between 0.4 and 3.7 MPa [25,26]. The foams generated in this work fall within this range and are therefore satisfactory.

For lattices, GP 0Al R has a compressive strength a little higher than that obtained by Franchin et al. [27] for geopolymer lattices of similar composition and porosity (11.5 MPa for 49.8% total porosity versus 18.3 MPA for 51.1% total porosity in this study). Gonçalves et al. [11] produced structures with a compressive strength of 17.3 $\pm$ 1.1 MPa for a total porosity of 65.9%. It is important to highlight that in those studies, porosity was induced by adjusting the spacing between filaments, whereas in our research, a significant part of the porosity is located inside the filament itself. A study closer to ours was carried out by

Oliveira et al. [12] who added porous fillers (activated carbon and hydrotalcite) into the geopolymer paste and printed structures with a compressive strength of 5.2 ± 0.8 MPa for a total porosity of 67.2 ± 0.2%.

5. Conclusions

This article studies the mechanical properties of metakaolin-based geopolymers shaped by molding and 3D printing.

It appears that the mechanical properties of extruded geopolymers in bending vary significantly depending on the humidity and temperature of the environment. Filaments containing PEG cured under low humidity conditions demonstrate unusually high mechanical properties (>115 MPa after 28 days) compared to filaments cured at high humidity (between 20 and 30 MPa after 28 days). These elevated mechanical properties are not stable and decrease when samples transition from dry to humid environments (<30 MPa).

Polyethylene glycol, added to enable the 3D printing of geopolymers, may be responsible for the abnormally high mechanical properties observed under low humidity. However, it is difficult to determine whether these observations are really due to a mechanical binder role played by PEG and/or to the presence of residual stresses. Additional studies are needed to clarify this point and to better understand the interactions of PEG with the geopolymer and the water available in the environment.

Under room temperature and low humidity curing conditions, the insufficient presence of water impedes complete geopolymerization, leading to mechanical properties that are both unstable and significantly low when exposed to high humidity (5 MPa). Conversely, storing samples at room temperature and high humidity results in consistently stable mechanical properties. Elevating the storage temperature accelerates geopolymerization, appearing to be rapid enough to offset water evacuation at low humidity and achieve complete geopolymerization. However, despite this, the samples cured at low humidity still display mechanical instability in response to humidity fluctuations.

Opting for printing and curing under high humidity conditions appears to be the most judicious decision for achieving samples with both complete geopolymerization and consistently stable mechanical properties. The strategic combination of high humidity with storage temperatures of 50 °C or 70 °C emerges as an effective approach to accelerate geopolymerization without compromising the mechanical integrity of the samples.

This work also focused on the influence of the shaping technique and porosity on the mechanical properties of geopolymers.

For molded geopolymers, a significant correlation emerged between compressive strength and envelope density. However, this correlation proves less pertinent for lattices, particularly those containing aluminum powder. The divergence suggests potential differences in fracture mechanisms between the two sample types. omplementary in situ monitoring of the compression test by X-ray tomography would be interesting to follow the initiation and propagation of cracks.

Author Contributions: Conceptualization, C.Z., E.P. and L.G.; methodology, C.Z., E.P. and L.G.; validation, C.Z., E.P. and L.G.; formal analysis, C.Z., E.P. and L.G.; investigation, C.Z.; resources, C.Z., E.P. and L.G.; data curation, C.Z.; writing—original draft preparation, C.Z.; writing—review and editing, C.Z., E.P., K.J. and L.G.; visualization, C.Z.; supervision, E.P., K.J. and L.G.; project administration, E.P.; funding acquisition, E.P., L.G. and K.J. All authors have read and agreed to the published version of the manuscript.

Funding: This work was supported by the LABEX iMUST of the University of Lyon (ANR-10-LABX-0064), created within the "Plan France 2030" set up by the French government and managed by the French National Research Agency (ANR). INSA Lyon through "Enjeu Energie" and IngéLyse (FR CNRS 3411) are acknowledged for their financial support.

Institutional Review Board Statement: Not applicable.

Informed Consent Statement: Not applicable.

Data Availability Statement: Data are contained within the article. Raw data can be made available upon request to the authors.

Acknowledgments: Special thanks to Sylvain MEILLE (MATEIS) for helpful discussion.

Conflicts of Interest: The authors declare no conflicts of interest. The funders had no role in the design of the study; in the collection, analyses, or interpretation of data; in the writing of the manuscript; or in the decision to publish the results.

References

1. Duxson, P.; Fernández-Jiménez, A.; Provis, J.L.; Lukey, G.C.; Palomo, A.; van Deventer, J.S.J. Geopolymer technology: The current state of the art. *J. Mater. Sci.* **2007**, *42*, 2917–2933. [CrossRef]
2. Rovnaník, P. Effect of curing temperature on the development of hard structure of metakaolin-based geopolymer. *Constr. Build. Mater.* **2010**, *24*, 1176–1183. [CrossRef]
3. Perera, D.S.; Uchida, O.; Vance, E.R.; Finnie, K.S. Influence of curing schedule on the integrity of geopolymers. *J. Mater. Sci.* **2007**, *42*, 3099–3106. [CrossRef]
4. Samuel, D.; Kriven, W.M. Depth dependence of hardness and reaction in metakaolin-based geopolymers cured at low humidity. *J. Am. Ceram. Soc.* **2024**, *107*, 543–560. [CrossRef]
5. Kirschner, A.V.; Harmuth, H. Investigation of geopolymers binders with respect to their application for building materials. *Ceramics-Silikáty* **2004**, *48*, 117–120.
6. Novais, R.M.; Ascensão, G.; Ferreira, N.; Seabra, M.P.; Labrincha, J.A. Influence of water and aluminium powder content on the properties of waste-containing geopolymer foams. *Ceram. Int.* **2018**, *44*, 6242–6249. [CrossRef]
7. Al-Noaimat, Y.A.; Ghaffar, S.H.; Chougan, M.; Al-Kheetan, M.J. A review of 3D printing low-carbon concrete with one-part geopolymer: Engineering, environmental and economic feasibility. *Case Stud. Constr. Mater.* **2023**, *18*, e01818. [CrossRef]
8. Archez, J.; Texier-Mandoki, N.; Bourbon, X.; Caron, J.F.; Rossignol, S. Adaptation of the geopolymer composite formulation binder to the shaping process. *Mater. Today Commun.* **2020**, *25*, 101501. [CrossRef]
9. Ma, S.; Fu, S.; Yang, H.; He, P.; Sun, Z.; Duan, X.; Jia, D.; Colombo, P.; Zhou, Y. Exploiting bifunctional 3D-Printed geopolymers for efficient cesium removal and immobilization: An approach for hazardous waste management. *J. Clean. Prod.* **2024**, *437*, 140599. [CrossRef]
10. Jin, H.; Zhang, Y.; Zhang, X.; Chang, M.; Li, C.; Lu, X.; Wang, Q. 3D printed geopolymer adsorption sieve for removal of methylene blue and adsorption mechanism. *Colloids Surfaces A Physicochem. Eng. Asp.* **2022**, *648*, 129235. [CrossRef]
11. Gonçalves, N.P.F.; Olhero, S.M.; Labrincha, J.A.; Novais, R.M. 3D-printed red mud/metakaolin-based geopolymers as water pollutant sorbents of methylene blue. *J. Clean. Prod.* **2023**, *383*, 135315. [CrossRef]
12. Oliveira, K.G.; Botti, R.; Kavun, V.; Gafiullina, A.; Franchin, G.; Repo, E.; Colombo, P. Geopolymer beads and 3D printed lattices containing activated carbon and hydrotalcite for anionic dye removal. *Catal. Today* **2022**, *390–391*, 57–68. [CrossRef]
13. Ma, S.; Liu, X.; Fu, S.; Zhao, S.; He, P.; Duan, X.; Yang, Z.; Jia, D.; Colombo, P.; Zhou, Y. Direct ink writing of porous SiC ceramics with geopolymer as binder. *J. Eur. Ceram. Soc.* **2022**, *42*, 6815–6826. [CrossRef]
14. Zoude, C.; Gremillard, L.; Prud'Homme, E. Combination of chemical foaming and direct ink writing for lightweight geopolymers. *Open Ceram.* **2023**, *16*, 100478. [CrossRef]
15. Yuan, J.; He, P.; Jia, D.; Yang, C.; Zhang, Y.; Yan, S.; Yang, Z.; Duan, X.; Wang, S.; Zhou, Y. Effect of curing temperature and SiO2/K2O molar ratio on the performance of metakaolin-based geopolymers. *Ceram. Int.* **2016**, *42*, 16184–16190. [CrossRef]
16. Prud'Homme, E.; Michaud, P.; Joussein, E.; Clacens, J.M.; Rossignol, S. Role of alkaline cations and water content on geomaterial foams: Monitoring during formation. *J. Non. Cryst. Solids* **2011**, *357*, 1270–1278. [CrossRef]
17. Pouhet, R. Formulation and Durability of Metakaolin-Based Geopolymers. Ph.D. Thesis, Université de Toulouse, Toulouse, France, 2015.
18. Khairuddin; Pramono, E.; Utomo, S.B.; Wulandari, V.; Zahrotul, A.W.; Clegg, F. FTIR studies on the effect of concentration of polyethylene glycol on polimerization of Shellac. *J. Phys. Conf. Ser.* **2016**, *776*, 012053. [CrossRef]
19. Prud'homme, E. Rôles du Cation Alcalin Et Des Renforts Minéraux Et Végétaux Sur Les Mécanismes de Formation de Géopolymères Poreux OU Denses. Ph.D. Thesis, Université de Limoges, Limoges, France, 2011.
20. Gibson, L.J.; Ashby, M.F. *Cellular Solids: Structure and Properties*, 2nd ed.; Cambridge University Press: Cambridge, UK, 1999.
21. Baird, J.A.; Olayo-Valles, R.; Rinaldi, C.; Taylor, L.S. Effect of molecular weight, temperature, and additives on the moisture sorption properties of polyethylene glycol. *J. Pharm. Sci.* **2010**, *99*, 154–168. [CrossRef] [PubMed]
22. Cyr, M.; Pouhet, R. Carbonation in the pore solution of metakaolin-based geopolymer. *Cem. Concr. Res.* **2016**, *88*, 227–235. [CrossRef]
23. Davidovits, J. *Geopolymer Chemistry and Applications*, 4th ed.; Institut Géopolymère: Saint-Quentin, France, 2015.
24. Rowles, M.; O'Connor, B. Chemical optimisation of the compressive strength of aluminosilicate geopolymers synthesised by sodium silicate activation of metakaolinite. *J. Mater. Chem.* **2003**, *13*, 1161–1165. [CrossRef]
25. Bai, C.; Colombo, P. Processing, properties and applications of highly porous geopolymers: A review. *Ceram. Int.* **2018**, *44*, 16103–16118. [CrossRef]

26. Kioupis, D.; Zisimopoulou, A.; Tsivilis, S.; Kakali, G. Development of porous geopolymers foamed by aluminum and zinc powders. *Ceram. Int.* **2021**, *47*, 26280–26292. [CrossRef]
27. Franchin, G.; Scanferla, P.; Zeffiro, L.; Elsayed, H.; Baliello, A.; Giacomello, G.; Pasetto, M.; Colombo, P. Direct ink writing of geopolymeric inks. *J. Eur. Ceram. Soc.* **2017**, *37*, 2481–2489. [CrossRef]

ceramics

MDPI

Article

Advanced Refinement of Geopolymer Composites for Enhanced 3D Printing via In-Depth Rheological Insights

Abrar Gasmi [1,*], Christine Pélegris [1], Ralph Davidovits [1,2], Mohamed Guessasma [1], Hugues Tortajada [3] and Florian Jean [4]

1 UPJV-LTI-UR 3899, 48 rue d'Ostende, 02315 Saint-Quentin, France; christine.pelegris@u-picardie.fr (C.P.)
2 Institut Géopolymère, 02100 Saint-Quentin, France
3 UPJV-LTI-UR 3899, Avenue des Facultés-Le Bailly, 80025 Amiens, France
4 UPHF, CERAMATHS, Boulevard Charles de Gaulle, 59600 Maubeuge, France
* Correspondence: abrar.gasmi@u-picardie.fr

Abstract: The advancement of 3D printing technology has been remarkable, yet the quality of printed prototypes heavily relies on the rheological behavior of the materials used. This study focuses on optimizing geopolymer-based composite formulas to achieve high-quality 3D printing, with particular attention given to rheological analysis. Three metakaolins, Argical M1200s, Metamax, and Tempozz M88, were used as alumino-silicate precursors for the preparation of the geopolymer binders. Rheological studies were conducted on viscosity, shear stress, and responses to oscillations in amplitude and frequency. The Tempozz M88-based binder was identified as the most effective for the extrusion due to its optimal rheological properties. Subsequently, the study investigated the influence of the amount, up to 55%, and morphology of the fillers, comprising feldspar and wollastonite, on the rheology of the pastes. Also, the addition of Xanthan gum, a gelling agent in the geopolymer paste, was analyzed, revealing improved extrusion quality and more stable bead structures. Finally, a comprehensive comparison was carried out between two formulations chosen according to rheological observations, utilizing image sequences captured during 3D printing. This comparison highlighted the formulation that ensures structural stability, design accuracy, and minimized sagging. This study underscores the significance of geopolymer formula optimization, leveraging rheology as a pivotal tool to enhance 3D printing quality, thereby facilitating more precise and reliable applications of additive manufacturing.

Keywords: geopolymer-based composites; 3D printing; rheological characterization; rheological modifier additive

1. Introduction

In an age marked by escalating environmental concerns and the imperative for sustainable practices, the importance of green materials has become increasingly evident [1,2]. According to OECD data, the materials sector contributed 23% of global CO_2 emissions in 2023, with 19% stemming from cement, steel, and the production of chemicals [3]. Green materials, such as geopolymers, play a vital role in mitigating the environmental impact and enhancing quality of life [4,5]. Geopolymers, coined by Joseph Davidovits in the 1970s, offer a promising alternative to Portland cement, boasting eco-friendliness, versatility, and cost-effectiveness [6]. Their properties, including straightforward preparation and impressive physico-chemical attributes, have attracted significant attention from researchers [7,8].

However, the adoption of green materials alone cannot ensure ecological preservation; manufacturing processes also play a critical role. Additive manufacturing, or 3D printing, offers advantages such as reduced manpower and space requirements, leading to improved workforce efficiency and productivity [9]. Moreover, it enables customization opportunities, resulting in cost and time savings [10,11]. The synergy between green materials and additive manufacturing, exemplified by "extrusion-based geopolymer composites", represents a powerful force for environmental preservation and underscores technology's role in fostering sustainability for a greener future.

The success of 3D printing, particularly for complex geometries, is heavily dependent on the rheology of the paste used. Rheological properties such as viscosity, yield stress, and viscoelasticity are critical in ensuring that the material flows smoothly during extrusion, maintains its shape post-deposition, adheres to previous layers, and solidifies into a durable structure. A high recovery rate after shear is essential to minimize filament sagging during extrusion, preserving the structural integrity of the printed part. Optimizing viscosity and thixotropy is especially challenging: the paste must be fluid enough to extrude easily, yet regain its strength quickly after deposition to support subsequent layers. Thixotropy helps the material flow under stress and recover its viscosity once the stress is removed, ensuring precision in layer placement. At the same time, yield stress must be finely controlled to prevent the material from deforming under the weight of additional layers while still allowing flow during the extrusion process. These factors are crucial to overcoming common issues like poor interlayer adhesion or material collapse in intricate designs. Thus, achieving the right balance of rheological parameters is vital not only for successful 3D printing of complex shapes but also for ensuring the printed structure's mechanical stability once cured [12,13].

Zoude et al. [14] investigate how the mechanical properties of geopolymers vary based on their shaping and curing parameters. The findings reveal that the shaping method and curing conditions significantly impact the strength and durability of geopolymers. Understanding this connection is important for ensuring that printed geopolymer structures can achieve the necessary mechanical integrity for practical applications, especially in construction.

Franchin et al. [15] contributed to the understanding of rheological properties specific to 3D printing by focusing on geopolymer ink formulations. Their work aimed to balance printability with the functional performance of the material. However, the ink formulations in this study consisted solely of the geopolymer binder, without the inclusion of reinforcing materials. Despite producing highly porous ceramic lattices, up to 70% porosity, this work highlighted the need for further improvements in formulation to achieve more robust and functional prints.

In contrast, Ma et al. [16] emphasized the ability to fine-tune the mechanical properties of printed geopolymer structures through direct ink writing, achieving high spatial resolution. Their research demonstrated that the printing resolution and material composition could be adjusted to produce components with specific mechanical characteristics, such as stiffness or flexibility, making the technology suitable for creating highly customized parts. Pramanik et al. [17] examined the effect of extrusion parameters on the structural performance of geopolymer-based prints, reinforcing the notion that extrusion speed, nozzle geometry, and other processing factors must be finely tuned to achieve desired outcomes in printed parts. Other studies, such as those by Ricciotti et al. [18] and Wang et al. [19], have highlighted the evolution of extrusion-based 3D printing technologies using geopolymers. These works explored critical aspects such as the production process, printability, mix design, and early-age properties, alongside the sustainability benefits of using geopolymers. Geopolymers, known for their reduced carbon footprint compared to traditional cement, are positioned as an environmentally friendly material suitable for both construction and industrial applications.

Panda et al. [20] made significant strides in developing a one-part geopolymer mix designed specifically for 3D printing. This research provided insights into how material formulation can be optimized to allow for better shape retention during the printing process, ensuring that layers do not collapse or deform as subsequent layers are added. Their findings emphasize the importance of rheological control in ensuring the stability of 3D-printed layers.

Further studies on rheological behavior [21,22] have investigated critical factors such as shear-thinning behavior and shape retention during printing. These properties are crucial for maintaining the dimensional accuracy of printed prototypes, particularly in intricate designs. Ma et al. [23] underscored the role of rheology modifiers in enhancing both the printability and mechanical properties of the final product. Additionally, the role of fillers and additives has been well-documented in the literature. For instance, fibers and other reinforcements [24] improve the rheological performance of geopolymer pastes, making them more suitable for extrusion. Gelling agents, such as Xanthan gum, have been studied for their ability to improve the structural integrity of printed components in various applications. Chen et al. [25] and Cofelice et al. [26] have also shown its ability to enhance the structural integrity of printed food components and other applications.

Finally, recent printability assessments [27,28] highlight various factors that influence printing efficiency and quality, including material composition and process parameters. Structural performance analyses [29,30] provide further insights into the mechanical properties and durability of 3D-printed geopolymers, which are essential for applications requiring strength and stability. Despite these advancements, a significant gap remains in quantitatively linking rheological properties to printability [31]. This presents an opportunity for further research to optimize extrusion-based geopolymer formulations for enhanced performance and broader application versatility.

The primary objective of this research is to enhance 3D printing technology by focusing on producing small, high-precision, and complex parts. As part of the GEOFAB project, these prototypes will be used in industrial applications, specifically for creating molding inserts used in thermoforming curved components in the aeronautical and automotive industries. Although the literature primarily focuses on large-scale construction applications and small-scale porous geopolymers or foams, this research targets heavily filled ceramic-like geopolymer composites to mitigate shrinkage, enhance mechanical properties, and investigate the effects of size distribution and morphology of the mineral filler elements.

The originality of this study lies in the development and optimization of geopolymer composites that are highly reinforced with mineral fillers. We aim to identify the optimal composition for high-quality extruded cord production and understand the influence of filler morphology and distribution. We began with a comprehensive assessment of different binders, including metakaolin variants (Tempozz M88, Metamax, and Argical M1200s) combined with potassium silicate, formulated through an alkaline process. Our selection was based on an in-depth analysis of rheological attributes.

After identifying the most suitable binder, we explored combinations filled up to 55% by weight with feldspar and wollastonite, chosen for their superior rheological performance. We also examined the effects of adding xanthan gum as a rheological modifier and gelling agent, focusing on its impact on rheological characteristics. Following the formulation of the optimal mix, we conducted comprehensive 3D printing trials to validate the functionality and performance of the geopolymer composite. These trials were essential to ensure that the material not only met the theoretical expectations derived from rheological testing but also performed reliably in practical applications.

2. Geopolymer-Based Composite Formulations

2.1. Material Characterization

The geopolymer binders were prepared from Woellner potassium silicate solution (Geosil 14517, Ludwigshafen, Germany) with a molar ratio of SiO_2/K_2O of 1.7. To complement the silico-aluminate matrix synergistically with potassium silicate, we opted for metakaolin as the fundamental building block source. Three distinct types of metakaolins were examined: Tempozz M88 (Temcon, Germany/China), Argical M1200s (Imerys, Paris, France), and Metamax (KAMIN, Macon, GA, USA), referred to as M88, M1200s, MX, respectively, each presenting specific characteristics. This initial selection process was guided by an assessment of their rheological properties, aiming to identify the most suitable binder for our 3D printing objectives.

Following the identification of the optimal binder, the next step involved optimizing the filler ratio within the binder matrix. The incorporation of fillers serves to fine-tune the material's characteristics, resulting in a final product with enhanced mechanical strength and reduced shrinkage. The chosen fillers include Feldspar IMX424 (Imerys, France) and Wollastonite 1250 mesh (Xatico, Luxembourg). Additionally, an extra step involved the introduction of a gelling agent, Xanthan gum, to the formulation.

In the scope of our investigation, supplementary experiments were conducted to further characterize our raw material powders. Particular emphasis was placed on examining the morphology of both our metakaolins and fillers, as depicted in Figures 1 and 2, respectively, utilizing Hitachi SU5000 (Hitachi High-Tech, Tokyo, Japan) and images acquired through Scanning Electron Microscopy (SEM). These SEM images offered invaluable visual insights, facilitating a detailed examination of the material morphology and enhancing our comprehension of microscopic properties. Following this, we proceeded to analyze the chemical composition of these constituents, as outlined in Table 1, employing Energy-Dispersive Spectroscopy (Thermo Scientific™ UltraDry EDS Detector, Waltham, MA, USA) to identify and quantify the elements present in the samples.

(a) (b) (c)

Figure 1. SEM micrographs of metakaolins: (**a**) M1200s, (**b**) M88, (**c**) MX.

(a) (b) (c)

Figure 2. SEM micrographs of fillers: (**a**) Feldspar IMX424, (**b**) Wollastonite 1250 mesh, and additive: (**c**) Xanthan Gum.

Table 1. Chemical composition of constituent materials by weight (%).

Sample	O	Na	Al	Si	P	S	K	Ca	Ti	Fe	Zr	Mg	Cl
Tempozz M88	51.52	0.14	22.38	22.88	0.30	0.15	0.01	0.12	0.86	0.40	1.24	0.00	0.00
Metamax	52.40	0.27	19.36	20.79	0.65	0.42	0.00	0.00	0.95	0.19	4.66	0.28	0.04
Argical M1200s	51.11	0.08	18.52	22.07	0.60	0.15	0.70	0.12	0.52	1.24	4.03	0.84	0.01
Feldspar IMX424	52.20	0.25	4.41	38.50	0.00	0.00	4.65	0.00	0.00	0.00	0.00	0.00	0.00
Wollastonite 1250 mesh	48.99	0.00	0.28	20.16	0.00	0.00	0.00	30.27	0.00	0.00	0.00	0.31	0.00

The particle size distribution of metakaolins and fillers was thoroughly investigated through laser analysis using the Malvern Mastersizer 3000 (Malvern Panalytical, Malvern, UK), and the findings of this study are summarized and visually presented in Figure 3 alongside the median particle size data presented in Table 2.

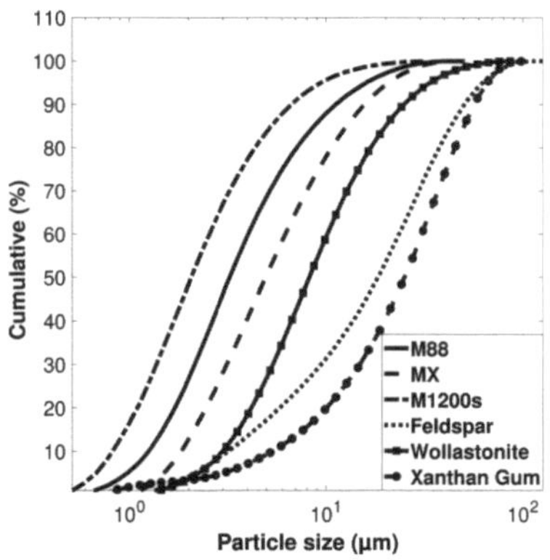

Figure 3. Particle size distribution of raw materials.

Finally, in-depth analyses were conducted to determine the specific surface area of metakaolins using Micromeritics TriStar II (Micromeritics Instrument, Norcross, GA, USA), revealing that M1200s exhibits the highest specific surface area, measured at 29.673 m^2/g. Next, M88 follows with a surface area of 25.104 m^2/g, and finally, Metamax displays a surface area of 22.864 m^2/g. The specific surface areas of the metakaolins, along with those of the fillers, are summarized in Table 2 for a comprehensive comparison. These results have yielded crucial insights into the potential reactivity of the materials and their physical properties, thereby enhancing the database and characterizing our raw materials within the context of our study.

Table 2. Characterization of constituent materials.

Sample	Argical M1200s	Metamax	Tempozz M88	Feldspar IMX4224	Wollastonite 1250 Mesh
Median particle size (μm)	2	5	3	20	10
Specific surface Area (m^2/g)	29.673	22.864	25.104	3.851	51.523

2.2. Mix Design Method

The process of mixing geopolymer components involves several key steps to ensure thorough blending and uniform distribution. In the initial stage, M1200s, M88, or MX-metakaolin, along with potassium silicate, are combined in a molar ratio of K:Al = 1 [32]. This blending process is carried out using a Hei-Torque 200 mixer (Heidolph Instruments, Schwabach, Germany) at 350 rpm for 10 min to ensure thorough mixing of the base materials. These mixtures will be denoted as B-M1200s, B-M88, and B-MX, respectively.

Subsequently, a meticulous manual blending process is undertaken to ensure the uniform distribution of various fillers. Within our geopolymer-based composite paste formulations, which comprise either 40% or 55% of fillers added to 75 g of metakaolin-based binder, we explored four distinct options. Initially, 'GP-F' integrates 100% feldspar as the filler to understand its impact on the binder, while 'GP-FW' combines 80% feldspar with 20% wollastonite to evaluate the effect of incorporating a needle-shaped filler. Moreover, 'GP-FWMax', aimed at understanding the implications of adding more fillers, constitutes a blend of 80% feldspar and 20% wollastonite, transitioning from 40% to approximately 55% filler in the total weight of the geopolymer paste. Lastly, in the 'GP-FWXg' formula, we aimed to assess the influence of adding an additive to the formulation, where 79.5% of feldspar and 20% of wollastonite are combined with 0.5% of Xanthan gum additive to obtain 40% of fillers in the total weight of the geopolymer paste, as specified in Table 3. Following the filler mixing step, the mixture undergoes further blending for one minute at 200 rpm, with the duration of this final step varying depending on the additives used, typically ranging from 10 to 15 min. This systematic approach ensures the meticulous preparation of geopolymer paste formulations tailored specifically for 3D printing, thereby guaranteeing reliability and consistency throughout the printing process.

Table 3. Proportions of fillers (%).

Formula	Feldspar/Fillers (%)	Wollastonite/Fillers (%)	Xanthan Gum/Fillers (%)	Total Weight Fillers (%)
GP-F	100	0	0	40
GP-FW	80	20	0	40
GP-FWMax	80	20	0	55
GP-FWXg	79.5	20	0.5	40

2.3. Rheological Protocol

The comprehension of rheological traits, particularly in materials such as geopolymer-based composites, involves introducing intentional disturbances to establish relationships between stimuli and responses [33]. In the realm of rheology, geopolymers often exhibit non-Newtonian behavior, viscoelasticity, and thixotropy [34,35].

The rheological characterization of geopolymer-based composite was carried out using the rotational Kinexus Lab+ rheometer (NETZSCH, Germany), employing vane geometry with a 25 mm diameter and 61 mm blade length. Measurements were consistently performed at 20 °C, with a 5 min material rest period when the upward plate reached the appropriate position. A pre-shear step is utilized in rheological experiments to standardize the rheological history of samples, ensuring a uniform initial state for all tests.

2.3.1. Shearing Tests

Continuous shear rheology involves applying a constant, steady shear rate to a material over a sustained period. This testing method is valuable for understanding the long-term behavior of materials, encompassing aspects such as flow properties, viscosity and shear-thinning behavior [36]. The first test focuses on the viscosity–shear rate relationship, highlighting variations in the viscosity of geopolymer composite material during 3D printing. To characterize our formulations, a shear rate ramp protocol ranging from 0.01 to 100 s^{-1} over a 10 min period was employed. The second test, known as the three interval-time-thixotropy (3ITT) test, involves a series of stages: an initial low shearing interval at 0.001 s^{-1} to assess the preserved geopolymer-based composite structure's viscosity, followed by a phase of structural breakdown at elevated shear rates of 100 s^{-1}. Subsequently, the material returns to a low shear interval of 0.001 s^{-1} to observe the process of structural reformation.

2.3.2. Oscillatory Tests

Oscillatory shear rheology involves the periodic variation in shear stress or the rate of strain in shear flow kinematics over time [36]. The first test is the amplitude sweep, where a material undergoes increasing strain or deformation amplitudes at a fixed frequency [37]. To determine the transition of the material's state from linear to non-linear viscoelastic regions, a stress sweep ranging from 0.001% to 100% was conducted, with a fixed frequency of 1 Hz for all samples [31,35,38]. The second test is the frequency sweep, a rheological testing technique characterizing the viscoelastic behavior of materials across a range of frequencies. The frequency range covers at least three decades, specifically from 0.1 Hz to 10 Hz, and the stress sweep was set at 0.01% to capture the material's full behavioral spectrum [31].

2.4. Three-Dimensional Printing Protocol

In order to delve deeper into the correlation and validation of rheology test results, we proceeded with printing two distinct formulations to compare and validate our conclusions drawn from the previous section.

Our custom-built 3-axis gantry printer, developed in-house, is designed for building 3D-printed geopolymer-based composites. As shown in Figure 4, this printer can construct specimens up to 360 mm (Length) × 240 mm (Width) × 40 mm (Height). It operates using a piston extrusion mechanism powered by compressed air, reaching pressures up to 10 bars. This system is crucial for handling the viscous and shear-thinning properties of geopolymer pastes, which require controlled extrusion to maintain structural integrity [39]. Air pump pressure was carefully controlled between 1.5 and 4.5 bars, depending on the rheology of the mixture, to ensure consistent extrusion.

Figure 4. The 3D printing machine.

The extrusion system features a screw auger, allowing for precise control of both pressure and material flow rate. This level of control ensures consistent layer deposition and reduces common defects like air entrapment and inconsistent flow during printing. The auger's ability to handle varying material viscosities makes it particularly effective for geopolymer pastes, which often require fine-tuned adjustments during the extrusion process to balance flow and stability [40]. Compared to traditional piston systems, the screw auger imposes fewer mechanical constraints and reduces maintenance needs, enhancing long-term reliability and lowering operational costs. Additionally, its adaptability to complex geopolymer formulations, such as those with fillers or fibers, ensures even distribution of materials, which is critical for achieving uniform structural properties throughout the printed part. This flexibility is essential in ensuring the quality and performance of geopolymer composites in 3D printing applications [40].

This study employed nozzle diameters, spanning from 0.6 mm to 9.5 mm, to thoroughly assess the printability of geopolymer mixtures on a polypropylene surface. A range of shapes, including straight-wall, cylindrical and complex designs, were printed to evaluate extrusion behavior and structural integrity. Figure 5 presents the geometries after slicing, prepared using PrusaSlicer. These sliced models showcase the layer-by-layer breakdown of the structures, ready for 3D printing. The first test, involving a four-layer wall printed with a 9.5 mm nozzle (Figure 5a), focused on layer bonding and its effect on the stability of the first layer. Image analysis and MATLAB-based automation were used to standardize print quality assessment, monitoring geometric accuracy and the impact of layer deposition on factors such as deformation and sagging. The second test, using a 2 mm nozzle (Figure 5b), aimed to assess precision and stability when printing finer geometries, while also observing how curved trajectories influenced the elasticity of the formulations. Finally, the third test involved printing a complex geometry (Figure 5c) to study the impact of material flow on structural stability, particularly in intricate shapes. Additionally, we aimed to examine how printing at various angles impacted the deposition and quality of the printed strand, focusing on how these orientations influenced the overall performance and consistency of the formulation. These tests provided a comprehensive understanding of geopolymer performance, optimizing both material composition and extrusion parameters to enhance 3D printing quality by addressing key factors like deformation, sagging, and structural stability [39,40].

(a) (b) (c)

Figure 5. (**a**) Four-layer wall, (**b**) cylinder, (**c**) complex shape.

3. Results and Discussion

The initial crucial factor involves thoughtful material selection, encompassing binders (M88, M1200s, MX), fillers (Feldspar IMX424, Wollastonite 1250 mesh), and additive (Xanthan Gum), each contributing distinct rheological properties that must synergize in the formulation.

3.1. Rheometry Development for Extrusion-Based 3D Printing of Geopolymer Composites

3.1.1. Comparative Analysis of Binders: Rheological Behavior Evaluation

In this comparative analysis, we investigated three binders: B-M1200s, B-M88, and B-MX, each representing a combination of metakaolin (M1200s, M88, and MX) with potassium silicate. Each material presents distinct rheological characteristics, shedding light on their suitability for various applications, particularly in 3D printing.

Viscosity Shear Tests

B-M1200s stands out due to its remarkably high viscosity, which measures at 969.1 Pa.s, surpassing that of other examined binders (Figure 6). This distinct viscosity profile is attributed to B-M1200s's shorter reaction time, facilitating the formation of intricate polymeric structures. The particle size distribution of M1200s metakaolin is notably different, with a median size (d50) of approximately 2 µm, the smallest among the compared metakaolins, as depicted in Table 2. A particle distribution implies more uniform particle sizing, fostering increased proximity between particles, as shown in Figure 1a. This proximity enhances interparticle attraction forces, including van der Waals and electrostatic forces, contributing significantly to the binder's overall viscosity [41]. Moreover, the viscosity of the binder is influenced by the size and shape of the dissolved oligomers. Longer and more entangled oligomers offer greater resistance to fluid shear [42]. As M1200s metakaolin exhibits the highest specific surface area of 29.673 m^2/g, as shown in Table 2, compared to other metakaolins, the interactions between molecules within the binder would be accentuated, enhancing fluid cohesion and consequently increasing viscosity [43].

Subsequently, B-M88 exhibits lower initial viscosity compared to B-M1200s, starting at approximately 80% lower, measuring at 180 Pa.s (Figure 6). It is well-established that an increased proportion of solids in metakaolin-based aluminosilicate slurries typically results in higher initial yield stress and apparent viscosity [44,45]. The viscosity contrast between B-M88 and B-M1200s can be attributed to several factors. Firstly, M88 metakaolin has a slightly higher d50 of 3 µm, as shown in Figure 3, and a lower specific surface area measured at 25.104 m^2/g compared to M1200s, as shown in Table 2. These differences reduce the impact of interparticle forces, leading to lower viscosity. Additionally, variations in the chemical composition of the two materials can alter particle interactions, further influencing viscosity. However, what truly distinguishes B-M88 is its rapid transition to fluidity, reaching a viscosity of 0.3 Pa.s at a shear rate of 100 s^{-1}. This notable shift can be attributed to the spherical particle shape, smooth surface, and ease of particle movement in the binder colloidal suspension, as illustrated in Figure 1b. Consequently, it facilitates swift alignment due to the impact of hydrodynamic forces and the untangling and alignment of polymer chains during flow [44]. As a result, the rate of viscosity reduction accelerates, enhancing the suitability of this binder for 3D printing applications.

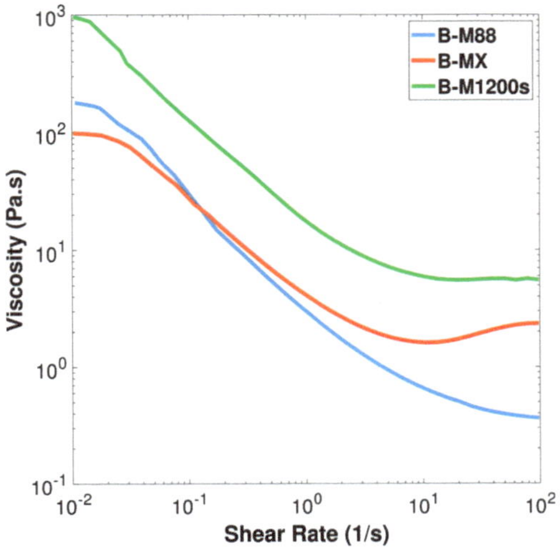

Figure 6. Shear viscosity tests of the binders.

B-MX displays an initial shear-thinning behavior, starting at 100 Pa.s and reaching 3.8 Pa.s at a shear rate of 1 s^{-1}. This lower viscosity is primarily attributed to its broader particle size distribution, approximately 5 µm, compared to M1200s and M88 (Figure 3). The diverse particle sizes in MX metakaolin diminish the impact of interparticle forces, leading to reduced viscosity. Additionally, MX has the smallest specific surface area among the three metakaolins, measured at 22.864 m^2/g, as shown in Table 2, indicating less surface available for interparticle interactions. This combination of a broader particle size distribution and a smaller specific surface area accounts for B-MX having the lowest viscosity among the examined binders. An intriguing turning point is observed when the viscosity begins to increase slightly, transitioning into a shear-thickening behavior, ultimately reaching 2.35 Pa.s at 100 s^{-1}. This transition could be attributed to the rise in viscosity, likely driven by turbulent flow, as elevated shear rates can induce such effects [46].

The three binders exhibited shear-thinning behavior, attributed to the transformation of non-crosslinked polymer molecule shapes under shear stress [47]. Shear thinning is crucial for geopolymer-based composites in 3D printing, facilitating extrusion by reducing viscosity under applied shear stress and ensuring smoother flow [36,48].

Oscillation shear tests

Figure 7 illustrates rheological changes in geopolymer binders under shear strain, presenting storage modulus G' and loss modulus G''. In rheology, the intersection point between G' and G'' represents the yield stress [23,49]. This yield stress indicates the point at which a material transitions from solid-like behavior to fluid-like behavior under stress. In the context of 3D printing, understanding the yield stress is crucial as it determines the minimum force required to initiate material flow through the printer nozzle, ensuring proper deposition of layers and overall print quality.

It is observed that all the binders exhibit a solid viscoelastic behavior, as evidenced by the fact that the storage component G' is greater than the loss component G''. Moreover, they seem to have surpassed the point of intersection of the curves, indicating that they are beyond the transition region which refers to the crossover point where the material behavior shifts from being more elastic (able to hold shape) to more viscous (tending to flow) [50].

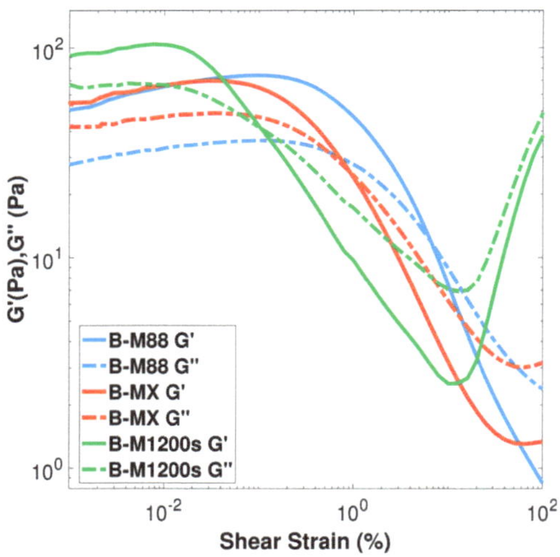

Figure 7. Oscillation amplitude sweeps of binders.

The order of G' values among the three binders, consistent with viscosity evaluation, highlights B-M1200s as distinctive with a G' of 94.1 Pa. This elevated storage modulus indicates heightened resistance to elastic deformation, underscoring M1200s' pivotal role in maintaining the material's solid viscoelastic structure. The yield stress, indicative of the point where the internal structure's network starts to break down, offers enhanced control at the onset of extrusion, preventing unintended material flow prior to intentional initiation. For B-M88, the determined yield stress was 8.1%, which was notably higher compared to B-MX at 1% and B-M1200s at 0.1%. This indicates that B-M88 necessitates a greater force during extrusion to initiate material flow. A high yield stress is not only associated with the material's ability to maintain its shape under external constraints but is crucial for ensuring precision and fidelity during the extrusion process, especially when printing successive layers. This property becomes essential in guaranteeing the deposited shape's accuracy. Researchers have linked yield stress to the material's capacity to maintain a desired structure under external forces, such as those induced by deposited geopolymer-based composite in subsequent layers or gravitational forces [51]. A material with a high yield stress exhibits a greater capacity to retain the desired shape after deposition.

The B-M1200s exhibit a distinctive behavior, particularly evident in amplitude oscillation data analysis at a 10% shear strain in Figure 7. Both the storage modulus G' and loss modulus G'' undergo a notable sequence of a sudden decrease followed by an increase. This phenomenon is attributed to the developing geopolymer network contributing to the elastic component, while stresses surpassing the elastic limit gradually disrupt the microstructural arrangement, facilitating the flow process. The observed response suggests a potential structural reorganization of the material.

It is noteworthy that the difference between the values of G' and G'' for the B-M88 is significantly higher than for the other binders, reaching a value of 26.38 Pa. This substantial difference may indicate the presence of a strongly crosslinked structure, similar to what occurs in a strongly bound dispersion. Furthermore, it may also suggest a low molecular weight distribution of crosslinked geopolymer chains. Consequently, this observation can be associated with a transition to a rubbery plateau in rheological behavior, highlighting a unique and strongly crosslinked geopolymer structure.

From a rheological perspective, B-M88 offers distinct performances. Viscosity tests reveal a rapid transition from the viscous to the liquid state, indicating enhanced pumpability of this binder. In amplitude sweep tests, B-M88 stands out with a high yield stress point, corresponding to a particularly broad linear viscoelastic region plateau. This signifies increased stability and the ability to maintain constant mechanical properties over an extensive range of stresses. The moderate viscosity of B-M88 compared to that of B-M1200s allows for a higher filling capacity, which is the primary focus of this study. By increasing the loading capacity, we aim to enhance mechanical properties and reduce shrinkage. This approach leverages the versatility of B-M88 in accommodating higher filler content, contributing to the overall improvement of the material's performance and its suitability for 3D printing. These overall findings lead to the conclusion that the B-M88 is most suitable for ensuring good quality in the context of 3D printing. Consequently, for the remainder of this article, we will rely on M88 as the binder of choice and seek to optimize the filler ratio to maximize its performance.

3.1.2. Filler Optimization and Effect of Adding Gelling Agent: Rheological Behavior Evaluation

Introducing fillers to the geopolymer binder is essential for optimizing the 3D printing process, enabling precise control of rheology.

In this section, our aim is to optimize the filler ratio to incorporate into the B-M88. To achieve this goal, we conducted a comparative analysis of four different formulations of geopolymer-based composite paste, as detailed in Table 2, outlining the respective proportions.

Viscosity Shear Tests

In the investigation of the initial formula GP-F, the incorporation of feldspar particles into B-M88 demonstrates a significant impact on both viscosity and shear stress, as evidenced by Figure 8a,b. A substantial disparity is evident between the formulations B-M88 and GP-F, with viscosity increasing by 68%, from an initial 180 Pa.s to 566 Pa.s. Additionally, the yield stress, defined as the point where shear stress sharply increases relative to shear rate, marking the onset of substantial material flow [52], has risen by 77%, climbing from 2 Pa to 9 Pa. This effect is primarily attributed to the inclusion of feldspar particles, with a size of 20 μm, as depicted in Figure 3. The size of these feldspar particles further contributes to the increase in viscosity, with larger particles hindering molecular movement to a greater extent. Moreover, the angular morphology of the feldspar particles, as shown in Figure 2a, introduces geometric complexity into the binder matrix, impeding the movement of binder molecules and consequently elevating material viscosity. Simultaneously, molecular interactions, particularly hydrogen bonds, between binder components and feldspar particles enhance the material's structure, leading to an increase in yield stress [53,54]. Consequently, the addition of feldspar fillers not only alters particle morphology and size but also influences molecular interactions, resulting in a simultaneous rise in viscosity and shear stress.

The disparity in viscosity and shear stress is further accentuated with the introduction of wollastonite GP-FW formulation compared to GP-F, as demonstrated in Figure 8a,b. With the addition of wollastonite, the viscosity reaches 1001 Pa.s, surpassing the one without wollastonite by 43%, while the yield stress rises to 14.32 Pa, marking a 37% increase. The incorporation of wollastonite into the GP-F enables the modulation of B-M88 viscosity through multiple mechanisms. Wollastonite, known for its fibrous morphology characterized by elongated needle-like particles, as depicted in Figure 2b, acts as a physical barrier, hindering molecular mobility within the binder, with a d50 around 10 μm (Table 2). These obstacles disrupt fluidity, creating zones of resistance that need increased energy for binder molecules to navigate [55]. Moreover, wollastonite particles can interact with binder molecules, forming physical bonds such as van der Waals forces and chemical interactions such as hydrogen bonding. These interactions increase the material's structure, enhancing molecular cohesion [56]. Simultaneously, wollastonite has the capability to alter the molecular network structure within the binder by intertwining with binder molecules, resulting in a more complex structure and contributing to increased viscosity.

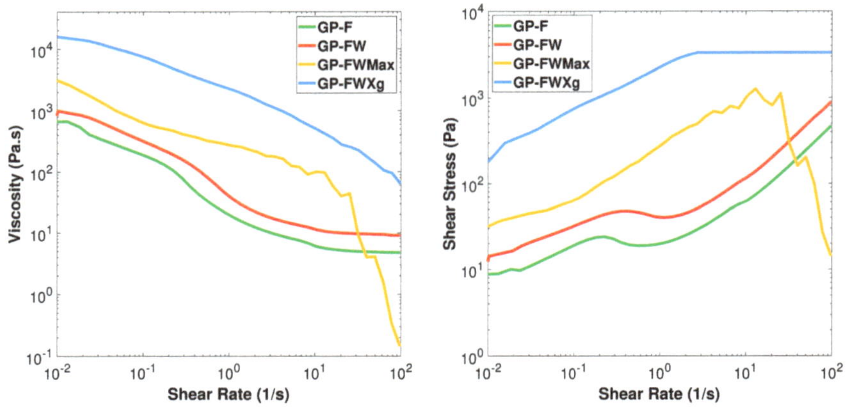

(**a**) (**b**)

Figure 8. (**a**) Viscosity test; (**b**) shear stress test.

The GP-FWMax formula, containing 55% more fillers compared to GP-FW, exhibited a significant increase in viscosity, almost tripling to 2620 Pa.s and doubling in yield stress to 36.83 Pa. The heightened filler concentration led to a denser particle arrangement, impeding molecular movement and causing a surge in viscosity. At high shear rates, fluctuations in values were observed due to internal friction from irregular particles colliding, leading to interference, agglomeration, and increased internal friction. This instability resulted in non-constant viscosity and shear stress values, disrupting molecular movement consistency within the material.

Finally, the formulation incorporating the Xanthan gum additive GP-FWXg exhibited the most significant increase, registering an almost 90 times higher viscosity compared to B-M88, setting a record at 15,970 Pa·s, with a yield stress value of 181.4 Pa, according to Figure 8a,b. The addition of Xanthan gum to a geopolymer-based composite results in increased viscosity for various reasons. Xanthan gum, a polysaccharide extracted from bacteria, has the ability to form network structures in liquids. When introduced into a geopolymer, it acts as a thickening agent by creating a three-dimensional network [57,58]. This network hinders molecular movement, generating resistance to flow and increasing viscosity. Moreover, Xanthan gum is hygroscopic, and by absorbing water present in the geopolymer, it swells, causing an increase in volume and contributing to the viscosity rise. In summary, the thickening effect of Xanthan gum results from network formation and its water-absorbing ability, limiting the fluidity of the geopolymer and increasing viscosity [59]. Furthermore, Xanthan gum chains can interact with geopolymer molecules through hydrogen bonding [60]. These interactions strengthen the overall material structure, enhancing molecular cohesion, and consequently, increasing shear stress.

Table 4 displays the peak rheological values associated with the inclusion of B-M88, as extracted from Figure 8a,b. The Herschel–Bulkley model offers a more comprehensive description of the flow dynamics of non-Newtonian fluids, especially when the material exhibits yield stress behavior. This model is governed by three key parameters: the yield stress τ_0, the consistency index K, and the flow index n. The yield stress τ_0 represents the stress that must be exceeded for the material to begin flowing, while K indicates the material's viscosity once the flow has started, and n describes the degree of shear-thinning or thickening behavior. A higher value of K reflects increased viscosity, and an n value of less than 1 indicates shear-thinning behavior [61]. To further investigate the non-Newtonian flow behavior, the Herschel–Bulkley model, as shown in Equation (1), was applied for modeling purposes [23], with R^2 representing the coefficient of determination. By fitting the experimental data using nonlinear regression, the parameters τ_0, K, and n were identified, offering a precise quantification of the flow behavior across the observed shear rate range.

$$\tau = \tau_0 + K\dot{\gamma}^n \quad (1)$$

Table 4. Shear thinning coefficients for geopolymer formulas.

Geopolymer Formulas	Viscosity (Pa.s)	τ_0 (Pa)	K (Pa.s)	n	R^2
GP-F	566	9.01	4.624	0.912	0.993
GP-FW	1001	14.32	8.883	0.898	0.998
GP-FWMax	2620	36.83	277.223	0.588	0.988
GP-FWXg	15,970	181.41	2110.793	0.413	0.976

In the context of 3D printing, the manifestation of shear thinning, as indicated in Table 4, is evident when $n < 1$. Weak shear-thinning behavior is observed at n values close to 1 but still below it, typically indicating a moderate reduction in viscosity under shear. In contrast, strongly pronounced shear-thinning occurs at much lower n values, generally less than 0.5, where the material exhibits a significant reduction in viscosity when forced through the nozzle, facilitating flow [62,63]. Illustrated in Figure 8, all formulations display distinctive shear-thinning behavior, characterized by decreasing viscosity with escalating shear rate. This property signifies the fluid's ability to flow more readily under increased shear forces, which is crucial for successful extrusion in 3D printing. Table 4 further highlights that formulas with fillers GP-F and GP-FW exhibit n values of 0.912 and 0.898, respectively, indicating relatively weak shear-thinning behavior. Conversely, GP-FWMax and GP-FWXg display notably lower n values of 0.588 and 0.413, respectively, suggesting a more pronounced shear-thinning behavior, particularly in the case of GP-FWXg.

Three-interval thixotropy test

An insightful rheological analysis is facilitated by the three-interval thixotropy test (3ITT), applicable in either rotational or oscillatory modes. This test emulates extrusion-based printing through three distinct steps, as shown in Figure 9a,b: (1) a small shear rate in the linear viscoelastic region (LVR), mimicking the paste at rest in the 3D printing cartridge; (2) a high shear rate surpassing the flow point to replicate extrusion through the nozzle; and (3) a return to a small shear rate resembling the geopolymer paste's rest state after deposition.

The main goal is to evaluate the rapidity of solid-like behavior recovery, ensuring nozzle shape maintenance and sufficient modulus for precise printing and self-supporting structures. The second interval probes time-dependent viscosity, revealing thixotropic effects that may necessitate adjustments in printing parameters for consistent material flow [64]. For a geopolymer-based composite material, a satisfactory recovery parameter is defined as achieving 85% structural reformation. This level ensures that the layer stacking ability remains significantly dependent on the material's initial mechanical properties [65].

Analysis of Figure 10 and the data presented in Table 4 reveals a distinct trend in the thixotropic behavior across various geopolymer formulations. In the case of GP-F, the initial recovery response exhibits characteristic features: a notable peak viscosity of 924.1 Pa.s at 900 s is followed by a sharp decline to 88.51 Pa.s at 920 s, further decreasing to 24.18 Pa.s at 1000 s. Subsequently, there is a gradual increase, reaching 119.8 Pa.s at 1500 s. Drawing on observations by Mewis and Wagner [66], this sequence suggests a viscoelastic nature, with the initial peak indicating immediate elasticity, the rapid decrease signifying a transition to viscosity, and the subsequent rise, implying a return to elasticity, resulting in a 25.02% recovery rate.

In contrast to GP-F, the GP-FW formulation, incorporating wollastonite with a needle-like morphology, exhibits a distinct thixotropic behavior, as shown in Figure 10. Initially, viscosity sharply decreases to 36.92 Pa.s at 900 s, followed by an abrupt drop to 18.59 Pa.s at 920 s, and then a substantial increase to 69.56 Pa.s at 1000 s (Table 5). This pattern aligns with the ideal thixotropy classification [66], indicating initial thixotropy, followed by reversibility and subsequent inelastic behavior. The observed phenomena are attributed to complex interactions between wollastonite and feldspar particles during recovery, leading to flocculation or agglomeration [67,68].

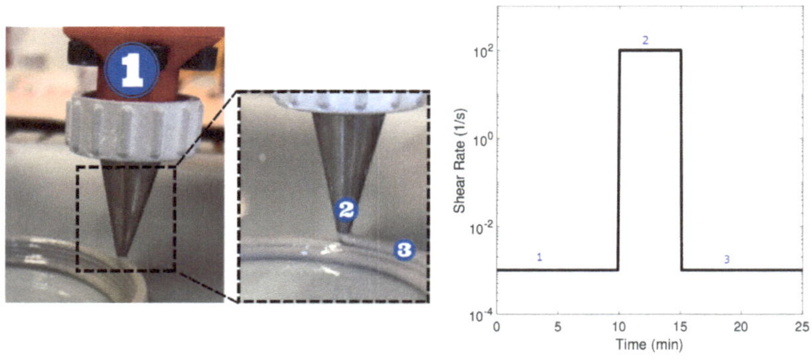

Figure 9. Correlation between thixotropy test and extrusion process: (**a**) extrusion process; (**b**) three intervals reproducing extrusion process.

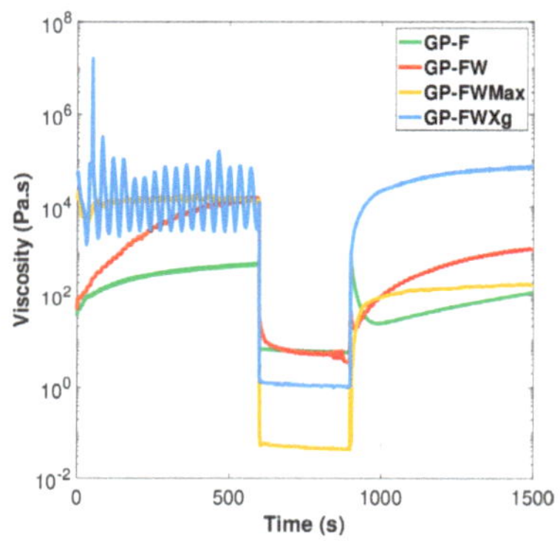

Figure 10. Three-step shear rate test.

Table 5. Three-step shear rate data.

Formulas	Viscosity (Pa.s)						Recovery (%)
GP-F	478.8	924.1	88.51	26.25	24.18	119.8	25.02
GP-FW	1.38×10^{4}	36.92	18.59	46.26	69.56	1.24×10^{3}	8.98
GP-FWMax	1.54×10^{4}	7.22	24.6	68	93	179	1.16
GP-FWXg	5.95×10^{4}	914	3.82×10^{3}	6.64×10^{3}	9.78×10^{3}	5.31×10^{4}	89.24
Time measurement (s)	500	900	920	950	1000	1500	

The GP-FWMax formula, reinforced with 55% of a mix of feldspar and wollastonite, results in a significant drop in the recovery rate to 1.16%, emphasizing the sensitivity of the material to specific filler combinations: a very high density can trap the molecules of the geopolymer, making their movement difficult and decreasing the material's ability to regain its initial viscosity after being subjected to stress or deformation. Therefore, a low recovery rate can be directly attributed to a very high density resulting from a high filling ratio in the formulation. Figure 10 indicates a combination of viscoelastic and thixotropic responses, reinforcing the nuanced nature of the material's behavior [69,70].

In the formula GP-FWXg, a distinctive observation emerges at the test's initiation with a very low shear rate of 0.001 s^{-1}, revealing a sinusoidal viscosity pattern in Figure 10, indicative of substantial flocculation. This complex behavior arises from Xanthan gum's ability to form a three-dimensional network, creating a gel-like structure in liquids [71]. At extremely low shear rates during testing, Xanthan gum promotes molecular bonding, leading to the formation of denser aggregates or network structures [57]. These structures may demonstrate a sinusoidal response as they pass through the rheometer, reflecting variations in viscosity or deformation specific to this phase of the test. However, it is also possible that the sinusoidal pattern observed is partially due to noise from the rheometer itself, highlighting the need for careful interpretation of such behaviors at very low shear rates.

To precisely characterize the recovery rate, we calculated the average of all values collected during the initial test phase, revealing an initial viscosity of 5.95×10^4 Pa.s, as shown in Table 5. In the post-thixotropy test at 1500 s, a viscosity of 5.31×10^4 Pa.s was attained, resulting in an outstanding recovery rate of 89.24%, the highest among all formulations. This notable improvement can be attributed to Xanthan gum's incorporation, which is crucial for maintaining a uniform particle distribution in the geopolymer-based composite. By minimizing zones with low local concentrations, Xanthan gum enhances material consistency, especially during deformation and recovery phases [72]. Moreover, Xanthan gum's network structure creation when hydrated acts as a physical barrier between particles, preventing their proximity and floc formation [73].

Oscillation shear tests

The observations from Figure 11a,b indicate that the GP-F exhibits a liquid-like behavior throughout the entire range, with the viscous component G'' around 23.98 Pa exceeding the elastic component G' with a value of 14.78 Pa. This behavior is attributed to several factors related to the material's internal structure and component segregation [74]. Shear-induced phase transitions, triggered by high stresses, lead to molecular rearrangements and alterations in the geopolymer's microstructure [75]. Filler segregation, especially in feldspar, significantly impacts rheological properties, with clustered fillers hindering energy storage and increasing viscosity. This uneven distribution of fillers can impede the material's elastic properties, leading to energy dissipation and a rise in the viscous component. In applications such as 3D printing, where precise structural integrity is crucial, such behavior poses challenges.

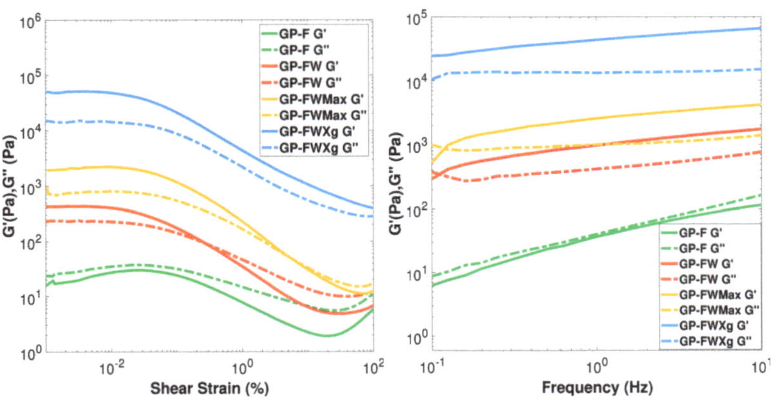

(**a**) (**b**)

Figure 11. Oscillation tests: (**a**) amplitude sweeps; (**b**) frequency sweeps.

The addition of wollastonite GP-FW leads to significant changes, as evident in Figure 11a,b. A notable shift in the G''/G' ratio, causing curves to intersect in Figure 11a, indicates complex interactions and microstructural evolution. This effect is also observed in the frequency oscillation shown in Figure 11b. Initially, feldspar acts as a filler, inducing a viscous response in the geopolymer-based composite. However, introducing wollastonite needles brings substantial microstructural modifications. These needles reinforce the structure, promoting an elastic response through stronger bonds and a resilient three-dimensional network. This alteration greatly influences the mechanical behavior, increasing elasticity with a yield stress of 96.54 Pa at 0.25% shear strain. The geopolymer's ability to recover its shape post-deformation is attributed to the robustness of needle interlacing. Consequently, the material exhibits elastic behavior under mechanical stress, highlighting its long-term structural stability, which is crucial for applications requiring resistance to repeated stresses.

A high amount of fillers in GP-FWMax results in more pronounced plateaus of G' and G'', indicating a significant expansion in the material's linear viscoelastic behavior (LVE) range. The material now shows a broader zone of proportional and elastic response to applied shear stresses. At the same time, there is a notable increase in yield stress, rising from 0.25% to 62% in shear strain, as shown in Figure 11. This higher yield stress suggests increased resistance to permanent deformations or a greater stress threshold required to induce lasting changes in the material's structure [53]. These changes in rheological properties, observed with the addition of 55% feldspar and wollastonite, (Table 2), are attributed to their significant influence on the material's structure and behavior. The combined effect of these fillers enhances the geopolymer's resistance, leading to more pronounced plateaus and an elevated yield stress. These characteristics are essential in various applications where structural stability and mechanical resilience against permanent deformations are critical.

The dominance of G' over G'' in rheological results after adding Xanthan gum, GP-FWXg, is due to its unique molecular structure. Xanthan gum consists of linear polysaccharide chains with branched side chains, forming a robust three-dimensional network in water. These side chains act as junction points, creating a strong polymeric matrix [57]. This structural strength results in a prevalent elastic component G' in the rheological response. Under stress, the matrix undergoes elastic deformation, returning to its original shape afterward. Xanthan gum also influences the microstructure of the geopolymer-based composite, organizing particle dispersion in geopolymer composite matrix and enhancing their connection. Its thickening properties and gel-forming ability contribute to a cohesive three-dimensional structure, reinforcing resistance to stress. Intermolecular forces from its branched side chains establish a structured network, further enhancing elasticity. Intramolecular forces, such as hydrogen bonding, reinforce molecular cohesion [58]. Overall, Xanthan gum enhances the elastic properties of the geopolymer-based composite material, primarily due to its structured network and molecular cohesion.

Concluding the rheological study, the GP-FWXg appears to be the most suitable for 3D printing, as it exhibits the highest viscosity and a recovery rate exceeding 85%. It displays a gel-like behavior and demonstrates good dispersion. This combination of characteristics suggests that GP-FWXg possesses the necessary flow properties to ensure proper material deposition during printing while also maintaining structural integrity.

3.2. Printable Geopolymer-Based Composite Paste

The two formulations most suitable for 3D printing are GP-FXMax and GP-FWXg, owing to their high viscosities and substantial LVE plateaus. Their high viscosity allows for better control of material flow during printing, while their extended LVE plateaus indicate a broader zone of proportional and elastic response to applied stresses. The first step in the printing protocol involved printing a four-layer wall to assess the sagging of the initial layer, as shown in Figure 12.

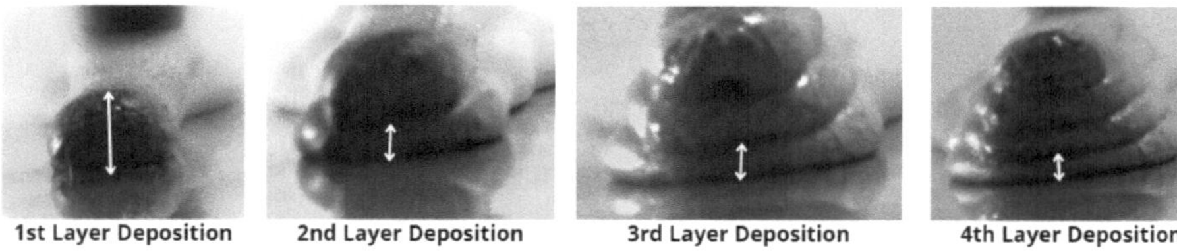

Figure 12. Sagging evolution of the 1st layer.

3.2.1. Three-Dimensional Printing of Geopolymer-Based Composite Formula "GP-FWMax"

Figure 13 quantitatively reveals a significant sagging in the first layer, originally designed as a circular shape with a height of 9.5 mm. It depicts the sagging rate of the first layer under its own weight, labeled "Layer1", and the evolution of this rate under the deposition of subsequent layers, labeled "D.Layer n.", (Figure 12).

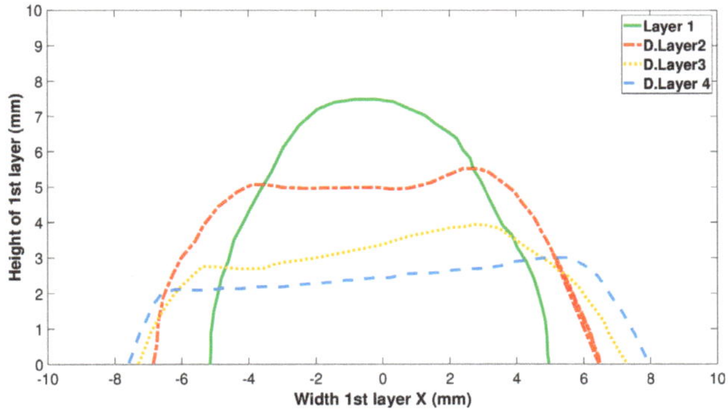

Figure 13. Shape of the first layer as a function of the deposition of subsequent layers.

The average height for the first layer measures 7.5 mm, a phenomenon explained by various physical and rheological mechanisms. The surface tension during extrusion plays a crucial role in material interaction with solid guide surfaces, influencing adhesion to the nozzle and mold walls. In instances of low surface tension and insufficient viscosity (2620 Pa.s in our case), poor adhesion may occur, leading to undesirable deformations. The material tends to spread beyond the intended shape, reducing the effective height of the cord. Post-extrusion, the geopolymer-based composite fails to fully regain its original shape, resulting in a permanent deformation and a 2 mm reduction in effective height. Additionally, continuous lateral flow contributes to a 10 mm increase in cord width. Gravity, notably pronounced due to our formula's high loading, exerts a significant downward force on the material, as visible in Figure 13. This gravitational impact intensifies sagging rates, which reach 35% after the second layer deposition, then an average of 55% after the third layer deposition, and finally 70% after the fourth layer deposition. These findings underscore the material's weight and gravitational force in cord sagging, attributed to a low recovery rate of approximately 1.16%, as shown in Figure 10. The geopolymer's elastic nature, combined with nozzle extrusion, induces deformation under pressure, further reducing cord height compared to its non-elastic counterpart.

During the 3D printing of a cylinder with GP-FWMax, we observed various undesirable phenomena that had a significant impact on the extrusion process. One of these phenomena was the sporadic occurrence of air bubble explosions, as illustrated in Figure 14, specifically in the area-a of the piece. These air bubble explosions were detrimental to the smooth progress of the printing process and the quality of the finished object. Two major factors were identified as the root cause of this phenomenon. First and foremost, inadequate extrusion pressure was found to be a decisive factor. If the extrusion pressure is not properly adjusted, it can lead to fluctuations in the material flow as it is being extruded. In particular, excessive extrusion pressures can exert force on the air bubbles trapped within the material, causing them to be abruptly expelled and leading to bubble explosions within the geopolymer composite being extruded. Secondly, a lack of proper degassing during the printing process was also identified as a contributing factor. Air bubbles can form during printing if the material has not been adequately degassed beforehand. This degassing deficiency can result from difficulties in removing the trapped air within the material or from an inadequate chemical reaction that releases air during the extrusion.

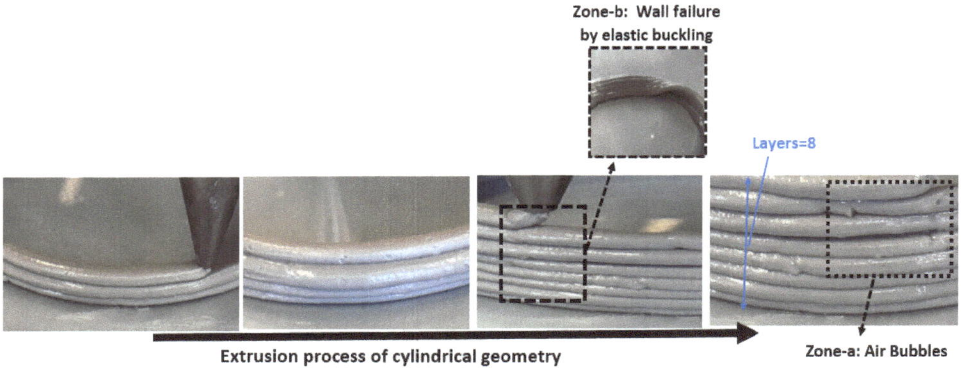

Figure 14. Critical issues for the extrusion process.

The presence of air bubbles during printing not only disrupted the workflow but also compromised the part's quality, resulting in surface defects such as visible cavities and pores. Addressing air bubble formation is essential for ensuring high-quality 3D printing outcomes and maintaining the prototype's intended functionality, particularly in critical roles such as sealing and mechanical strength.

While printing the final layer of the extrusion process, notably during the eighth and final layer, a mechanical phenomenon emerged (Figure 14, zone-b), characterized by instability and eventual structural collapse known as elastic buckling, a deformation occurring in thin structures under compressive loads [76]. This issue observed during the 3D printing of our geopolymer paste may be attributed to its low viscosity, approximately 2620 Pa·s, making the material excessively fluid and prone to difficulties in maintaining structural integrity. This fluidity, especially in vertical or overhanging areas, can result in sagging, deformation, and buckling, ultimately leading to unwanted distortions in the printed structure. The material's reduced resistance to compression is a key factor contributing to this elastic buckling phenomenon.

The challenges associated with the recovery rate and post-printing deformation, arising from the remarkably low recovery rate of approximately 1% in our geopolymer-based composite, can significantly impact the 3D printing process. Adequate adhesion between successive layers is imperative for ensuring structural integrity. Inadequate adhesion, especially in vertical or overhanging regions, can lead to surface irregularities and defects (as illustrated in Figure 15, zone-a). Furthermore, such deficiencies in adhesion may culminate in the premature termination of the extrusion process due to a loss of stability. Specifically, this low recovery rate can result in vulnerable areas within the printed part, particularly susceptible to adhesion issues during the printing process (Figure 15, zone-b). Thus, addressing these issues is paramount for achieving high-quality and reliable 3D-printed complex design prototypes, requiring careful consideration and optimization of the printing parameters and material composition.

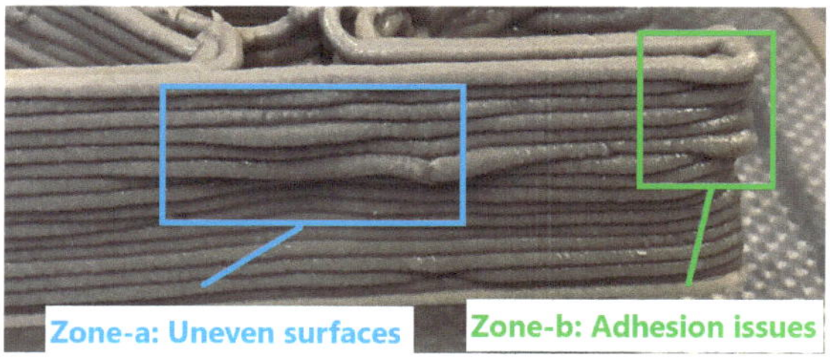

Figure 15. Formula printing issues.

3.2.2. Three-Dimensional Printing of Geopolymer-Based Composite Formula "GP-FWXg"

In comparison with the previous formula GP-FWMax, tests were conducted using the formulation containing Xanthan gum GP-FWXg, employing a 9.5 mm diameter nozzle to print a four-layer wall structure design analyzed through image sequences (Figure 12). Immediate improvements were evident, particularly in cord quality, with a significant reduction in sagging observed. While the previous formula exhibited a total sagging of 70%, the formula with Xanthan gum reduced this to approximately 40%. During the printing of the first cord, an average height of 8.5 mm was measured (Figure 16), resulting in a recovery rate of 90%, consistent with thixotropy tests (Table 4). The real revelation occurred during the printing of the second layer, where the influence of weight and gravity caused the first cord to deform, reducing its height to an average of 7.75 mm with a sagging rate of only 20%, significantly lower than observed previously.

This notable difference can be attributed to the extremely high viscosity of the Xanthan gum-containing formulation, promoting better-organized and more rigid interparticle interactions. These observations were further confirmed when adding the third and fourth layers, where we observed respective average sagging rates of 35% and 40% in height, as well as lateral expansion of 35% and 40% compared to the first layer, as indicated in Figure 16. These results clearly demonstrate the beneficial impact of adding Xanthan gum to the formulation, enhancing material cohesion, and significantly improving the stability of 3D-printed structures.

The second part of our experiments focuses on the printing of a cylinder under similar conditions to the previous section, using a 2 mm nozzle (Figure 17a). It is noteworthy that we have observed a marked improvement in cord quality, with a significant reduction in air bubble explosions on the surface, resulting in a visually more appealing texture. We printed the number of layers of the previous cylinder with GP-FWMax, and there was no buckling phenomenon, thanks to the solid viscoelastic behavior of the formulation.

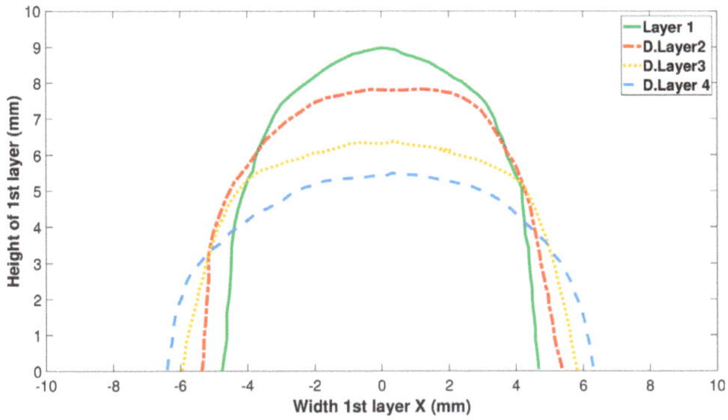

Figure 16. Shape of the first layer as a function of the deposition of subsequent layers.

(**a**) (**b**)

Figure 17. Printing of cylindrical geometry: (**a**) printing with a nozzle of 2 mm; (**b**) printing with a nozzle of 0.6 mm.

Subsequently, we aimed to push the boundaries of our tests to determine the minimum nozzle diameter for printing. The cylinder design presented in Figure 17b was printed using a 0.6 mm nozzle under similar environmental conditions. We successfully printed up to 12 layers, utilizing two adjacent cords to enhance stability.

Even though we have achieved the correct formulation based on rheology, challenges remain, as precise parameter control is crucial, especially when dealing with complex geometries. This has been evident from the beginning with the phenomenon of zigzag patterns, also called "wavy patterns", observed in Figure 18a and properly corrected in Figure 18b. In reality, an incorrect Z offset, i.e., the distance between the print nozzle and the print bed, can lead to an uneven printing surface, often manifested by irregular lines or ripples on the surface of the printed part. Thus, precise adjustment of the Z offset is essential to ensure high-quality 3D printing and minimize these visual defects.

(a) (b)

Figure 18. Complex shape printing: (**a**) wavy patterns; (**b**) conforming part.

The SEM analysis, as shown in Figure 19, shows a cross-section of the 3D-printed geopolymer composite part (GP-FWXg), using a 0.6 mm diameter nozzle, as depicted in Figure 18b, which includes metakaolin Tempozz M88, potassium silicate, feldspar (d50 20 µm), wollastonite (d50 10 µm), and xanthan gum. The SEM images at varying scales (500 µm, 50 µm, and 10 µm) provide detailed insights into the microstructural integrity and performance of the printed material.

(a) (b) (c)

Figure 19. SEM of 3D-printed geopolymer composite part: (**a**) at 500 µm, (**b**) at 50 µm, and (**c**) at 10 µm.

At the 500 µm scale, Figure 19a reveals a generally homogeneous distribution of constituents throughout the composite. This uniformity indicates efficient mixing during the composite preparation, which is essential to prevent cold joints between the printed layers. The observed homogeneity suggests that the interaction between the geopolymer composite matrix and fillers during printing ensures continuous bonding, thus maintaining structural strength and coherence. The feldspar and wollastonite fillers are well dispersed, indicating an effective mixing process prior to printing. Additionally, the SEM images reveal that both the shape and the diameter of the strand were maintained accurately, indicating that the 3D printing process successfully adhered to the intended design specifications. The absence of cold joints highlights excellent cohesion between the printed layers, ensured by effective interaction between the binder and fillers during the printing process.

At the 50 µm scale, as shown in Figure 19b, the detailed distribution of feldspar and wollastonite fillers becomes apparent. Feldspar particles, with a median size of 20 µm, and wollastonite fibers, with a median size of 10 μm, are uniformly dispersed within the matrix. The directional flow of material through the 0.6 mm nozzle during printing aligns the wollastonite fibers parallel to the print direction. This alignment is advantageous as it enhances the mechanical properties of the composite along the direction of printing. The alignment is particularly beneficial for applications requiring anisotropic mechanical properties, where strength in a specific direction is crucial.

The 10 μm scale image, as shown in Figure 19c, reveals well-defined interfaces with excellent adhesion between the geopolymer matrix and the fillers. The distinct morphology of the wollastonite fibers, as evident at this scale, further confirms their alignment and interaction with the matrix. The fibers' needle-like structure provides reinforcement, improving the composite's resistance to cracking and deformation under load.

Xanthan gum plays a vital role in stabilizing the suspension of particles within the geopolymer composite matrix. Its presence ensures a uniform distribution of fillers, preventing the sedimentation and agglomeration that could lead to heterogeneous regions within the printed structure. This stabilization is indirectly observed through the consistent microstructural features across different scales, highlighting xanthan gum's efficacy in maintaining a stable mixture during the printing process.

4. Conclusions

In conclusion, our study highlights the pivotal role of particle size and morphology in shaping the rheological properties of geopolymer-based composites, particularly for 3D-printed parts with complex designs. The use of finer metakaolin emerged as a critical factor in increasing material viscosity, which guided the selection of optimal binders like B-M88. This binder is characterized by its high viscosity and notable linear viscoelastic behavior, which is crucial for maintaining cord quality, stability, and precision during printing. The strong viscoelasticity in the formulation preserves shape integrity, while reducing air bubble formation enhances surface smoothness—both essential for fine, stable printing. Additionally, the incorporation of a high content of feldspar and wollastonite as fillers, up to 55%, significantly enhanced the mechanical properties and dimensional stability of the geopolymer composites, underscoring their importance in achieving high-performance results. The incorporation of Xanthan gum into the geopolymer paste induced significant improvements in recovery rates and viscoelastic behavior, and these effects were specifically observed under specific stress conditions relevant to extrusion.

However, some limitations of this study must be acknowledged. The long-term stability of these formulations, particularly under varying environmental conditions (e.g., humidity or temperature), remains to be fully explored. Moreover, while the rheological enhancements facilitated the printing of complex geometries, further work is needed to fully characterize the limits of print resolution and structural fidelity in intricate designs, especially at smaller scales.

Future research will focus on addressing these limitations and expanding the scope of this study. Characterizing the thermal properties of the printed parts is essential, particularly for our industrial applications involving mold inserts, which requires high-temperature resistance, up to 800 °C, and mechanical stability. Preliminary findings in this area have shown promise, but further investigation is necessary to confirm these results across a broader range of formulations and printing conditions. Additionally, we plan to leverage the rheological data obtained in computational fluid dynamics (CFDs) simulations to optimize the 3D printing process. These simulations will serve as predictive tools for anticipating material behavior post-extrusion and identifying potential issues early on, thus streamlining the printing workflow.

Author Contributions: Conceptualization, H.T.; methodology, A.G. and R.D.; validation, C.P. and M.G.; formal analysis, A.G. and F.J.; resources, H.T.; data curation, A.G. and F.J.; writing—original draft, A.G.; writing—review and editing, C.P., R.D. and M.G.; supervision, C.P., R.D. and M.G. All authors have read and agreed to the published version of the manuscript.

Funding: This work was realized under the GEOFAB project, funded by the region Hauts-de-France.

Institutional Review Board Statement: Not applicable.

Informed Consent Statement: Not applicable.

Data Availability Statement: Data are contained within the article.

Conflicts of Interest: The authors declare no conflicts of interest.

References

1. Hodžić, S.; Šikić, T.F.; Dogan, E. Green environment in the EU countries: The role of financial inclusion, natural resources and energy intensity. *Resour. Policy* **2023**, *82*, 103476. [CrossRef]
2. Satone, S.R.; Parbat, D.K.; Badar, A.M.; Varghese, V.P.; Satone, D.S.; Kawalkar, M.A. Application of green material on durability behaviour of green concrete. *Mater. Today Proc.* **2023**. . [CrossRef]
3. Kowalski, P.; Legendre, C. *Raw Materials Critical for the Green Transition: Production, International Trade and Export Restrictions*; OECD: Paris, France, 2023.
4. Mahmoodi, O.; Siad, H.; Lachemi, M.; Dadsetan, S.; Şahmaran, M. Optimized application of ternary brick, ceramic and concrete wastes in sustainable high strength geopolymers. *J. Clean. Prod.* **2022**, *338*, 130650. [CrossRef]
5. Lu, C.; Zhang, Z.; Shi, C.; Li, N.; Jiao, D.; Yuan, Q. Rheology of alkali-activated materials: A review. *Cem. Concr. Compos.* **2021**, *121*, 104061. [CrossRef]
6. Davidovits, J. *Geopolymer Chemistry and Applications*; Institute Geopolymer: Saint-Quentin, France, 2008.
7. Shi, C.; Jiménez, A.F.; Palomo, A. New cements for the 21st century: The pursuit of an alternative to Portland cement. *Cem. Concr. Res.* **2011**, *41*, 750–763. [CrossRef]
8. Davidovits, J. Geopolymers: Ceramic-like inorganic polymers. *J. Ceram. Sci. Technol.* **2017**, *8*, 335–350.
9. Hegab, H.; Khanna, N.; Monib, N.; Salem, A. Design for sustainable additive manufacturing: A review. *Sustain. Mater. Technol.* **2023**, *35*, e00576. [CrossRef]
10. Dey, D.; Srinivas, D.; Panda, B.; Suraneni, P.; Sitharam, T. Use of industrial waste materials for 3D printing of sustainable concrete: A review. *J. Clean. Prod.* **2022**, *340*, 130749. [CrossRef]
11. Xiong, Y.; Tang, Y.; Kim, S.; Rosen, D.W. Human-machine collaborative additive manufacturing. *J. Manuf. Syst.* **2023**, *66*, 82–91. [CrossRef]
12. Roussel, N. Rheological requirements for printable concretes. *Cem. Concr. Res.* **2018**, *112*, 76–85. [CrossRef]
13. Botti, R.; Innocentini, M.; Faleiros, T.; Mello, M.; Flumignan, D.; Santos, L.; Franchin, G.; Colombo, P. Additively manufactured geopolymer structured heterogeneous catalysts for biodiesel production. *Appl. Mater. Today* **2021**, *23*, 101022. [CrossRef]
14. Zoude, C.; Prud'homme, E.; Johannes, K.; Gremillard, L. The Mechanical Properties of Geopolymers as a Function of Their Shaping and Curing Parameters. *Ceramics* **2024**, *7*, 873–892. [CrossRef]
15. Franchin, G.; Scanferla, P.; Zeffiro, L.; Elsayed, H.; Baliello, A.; Giacomello, G.; Pasetto, M.; Colombo, P. Direct ink writing of geopolymeric inks. *J. Eur. Ceram. Soc.* **2017**, *37*, 2481–2489. [CrossRef]
16. Ma, S.; Fu, S.; Zhao, S.; He, P.; Ma, G.; Wang, M.; Jia, D.; Zhou, Y. Direct ink writing of geopolymer with high spatial resolution and tunable mechanical properties. *Addit. Manuf.* **2021**, *46*, 102202. [CrossRef]
17. Pramanik, A.; Subhani, T.; Yun, T.S.; Nagesh, R.; Ramakrishna, S. Exploring the Structural Performance of 3D-Printed Geopolymer-based Architectural Components. *Mater. Today Commun.* **2021**, *27*, 102382.
18. Ricciotti, L.; Apicella, A.; Perrotta, V.; Aversa, R. Geopolymer materials for extrusion-based 3D-printing: A review. *Polymers* **2023**, *15*, 4688. [CrossRef]
19. Jia, D.; He, P.; Wang, M.; Yan, S.; Jia, D.; He, P.; Wang, M.; Yan, S. Geopolymers and Their Matrix Composites: A State-of-the-Art Review. *Geopolymer Geopolymer Matrix Compos.* **2020**, *311*, 7–34.
20. Panda, B.; Singh, G.V.P.B.P.; Unluer, C.; Tan, M.J. Synthesis and characterization of one-part geopolymers for extrusion based 3D concrete printing. *J. Clean. Prod.* **2019**, *220*, 610–619.
21. Kovářík, T.; Hájek, J.; Hervert, T.; Deshmukh, K.; Pola, M.; Jansa, Z.; Beneš, J.; Svoboda, M. Silica-based geopolymer spherical beads: Influence of viscosity on porosity architecture. *Cem. Concr. Compos.* **2021**, *124*, 104261. [CrossRef]
22. Amorim, P.; d'Ávila, M.; Anand, R.; Moldenaers, P.; Van Puyvelde, P.; Bloemen, V. Insights on shear rheology of inks for extrusion-based 3D bioprinting. *Bioprinting* **2021**, *22*, e00129. [CrossRef]
23. Ma, S.; Fu, S.; Yang, T.; Li, K.; Chen, G.; Dong, Q.; He, P.; Sun, Z.; Duan, X.; Jia, D.; et al. Unveiling the critical role of rheology modifiers in additive manufacturing of geopolymers and their mechanical properties. *Addit. Manuf.* **2023**, *78*, 103826. [CrossRef]
24. Ilcan, H.; Sahin, O.; Kul, A.; Ozcelikci, E.; Sahmaran, M. Rheological property and extrudability performance assessment of construction and demolition waste-based geopolymer mortars with varied testing protocols. *Cem. Concr. Compos.* **2023**, *136*, 104891. [CrossRef]
25. Chen, Y.; Zhang, M.; Sun, Y.; Phuhongsung, P. Improving 3D/4D printing characteristics of natural food gels by novel additives: A review. *Food Hydrocoll.* **2022**, *123*, 107160. [CrossRef]
26. Cofelice, M.; Messia, M.C.; Marconi, E.; Cuomo, F.; Lopez, F. Effect of the xanthan gum on the rheological properties of alginate hydrogels. *Food Hydrocoll.* **2023**, *142*, 108768. [CrossRef]

27. Gonçalves, N.P.; Olhero, S.M.; Labrincha, J.A.; Novais, R.M. 3D-printed red mud/metakaolin-based geopolymers as water pollutant sorbents of methylene blue. *J. Clean. Prod.* **2023**, *383*, 135315. [CrossRef]
28. Chen, W.; Pan, J.; Zhu, B.; Ma, X.; Zhang, Y.; Chen, Y.; Li, X.; Meng, L.; Cai, J. Improving mechanical properties of 3D printable 'one-part'geopolymer concrete with steel fiber reinforcement. *J. Build. Eng.* **2023**, *75*, 107077. [CrossRef]
29. Asghar, R.; Khan, M.A.; Alyousef, R.; Javed, M.F.; Ali, M. Promoting the green construction: Scientometric review on the mechanical and structural performance of geopolymer concrete. *Constr. Build. Mater.* **2023**, *368*, 130502. [CrossRef]
30. Shilar, F.A.; Ganachari, S.V.; Patil, V.B.; Bhojaraja, B.; Khan, T.Y.; Almakayeel, N. A review of 3D printing of geopolymer composites for structural and functional applications. *Constr. Build. Mater.* **2023**, *400*, 132869. [CrossRef]
31. Arunothayan, A.R.; Nematollahi, B.; Khayat, K.H.; Ramesh, A.; Sanjayan, J.G. Rheological characterization of ultra-high performance concrete for 3D printing. *Cem. Concr. Compos.* **2023**, *136*, 104854. [CrossRef]
32. Davidovits, J. *Geopolymer Chemistry and Applications*, 5th ed.; Institut Géopolymère, Geopolymer Institute: Saint-Quentin, France, 2020; Chapter 21.
33. Sabbatini, A.; Vidal, L.; Pettinari, C.; Sobrados, I.; Rossignol, S. Control of shaping and thermal resistance of metakaolin-based geopolymers. *Mater. Des.* **2017**, *116*, 374–385. [CrossRef]
34. Favier, A. Mécanisme de Prise et Rhéologie de Liants Géopolymères Modèles. Ph.D. Thesis, Université Paris-Est, Paris, France, 2013.
35. Keskin-Topan, Y.; Bessaies-Bey, H.; Petit, L.; Tran, N.C.; d'Espinose de Lacaillerie, J.B.; Rossignol, S.; Roussel, N. Effect of maximum packing fraction of powders on the rheology of metakaolin-based geopolymer pastes. *Cem. Concr. Res.* **2024**, *179*, 107482.
36. Zhang, J.; Tuohey, J.; Amini, N.; Morton, D.A.; Hapgood, K.P. 3D printing of customised particles for powder rheology, mixing and segregation study. *Powder Technol.* **2023**, *425*, 118576. [CrossRef]
37. Dávila, J.; d'Ávila, M. Rheological evaluation of Laponite/alginate inks for 3D extrusion-based printing. *Int. J. Adv. Manuf. Technol.* **2019**, *101*, 675–686.
38. Yan, D.; Chen, S.; Zeng, Q.; Xu, S.; Li, H. Correlating the elastic properties of metakaolin-based geopolymer with its composition. *Mater. Des.* **2016**, *95*, 306–318. [CrossRef]
39. Hass, L.; Bos, F. Bending and pull-out tests on a novel screw type reinforcement for extrusion-based 3D printed concrete. In Proceedings of the Second RILEM International Conference on Concrete and Digital Fabrication: Digital Concrete 2020, Eindhoven, The Netherlands, 6–8 July 2020; Springer: Berlin/Heidelberg, Germany, 2020; pp. 632–645.
40. Sanjayan, J.; Jayathilakage, R.; Rajeev, P. Vibration induced active rheology control for 3D concrete printing. *Cem. Concr. Res.* **2021**, *140*, 106293. [CrossRef]
41. Tanner, R.I. *Engineering Rheology*; Oxford University Press: Oxford, UK, 2000; Volume 52.
42. Graessley, W.W. The entanglement concept in polymer rheology. In *The Entanglement Concept in Polymer Rheology*; Springer: Berlin/Heidelberg, Germany, 2005; pp. 1–179.
43. Barnes, H.A.; Hutton, J.F.; Walters, K. *An Introduction to Rheology*; Elsevier: Amsterdam, The Netherlands, 1989; Volume 3.
44. Rovnaník, P.; Rovnanikova, P.; Vyšvařil, M.; Grzeszczyk, S.; Janowska-Renkas, E. Rheological properties and microstructure of binary waste red brick powder/metakaolin geopolymer. *Constr. Build. Mater.* **2018**, *188*, 924–933. [CrossRef]
45. Romagnoli, M.; Leonelli, C.; Kamse, E.; Gualtieri, M. Rheology of geopolymer by DOE approach. *Constr. Build. Mater.* **2012**, *36*, 251–258.
46. Ewoldt, R.H.; Johnston, M.T.; Caretta, L.M. Experimental challenges of shear rheology: How to avoid bad data. In *Complex Fluids in Biological Systems: Experiment, Theory, and Computation*; Springer: Berlin/Heidelberg, Germany, 2015; pp. 207–241.
47. Guy, C.; Jean-Louis, G.; Nicolas, H. *Initiation à la Rhéologie (4 Éd.): Bases Théoriques et Applications Expérimentales*; Lavoisier: Paris, France, 2014.
48. de Andrade, R.; Paim, T.C.; Bertaco, I.; Naasani, L.S.; Buchner, S.; Kovářík, T.; Hájek, J.; Wink, M.R. Hierarchically porous bioceramics based on geopolymer-hydroxyapatite composite as a novel biomaterial: Structure, mechanical properties and biocompatibility evaluation. *Appl. Mater. Today* **2023**, *33*, 101875. [CrossRef]
49. Maldonado-Rosas, R.; Tejada-Ortigoza, V.; Cuan-Urquizo, E.; Mendoza-Cachú, D.; Morales-de La Pena, M.; Alvarado-Orozco, J.M.; Campanella, O.H. Evaluation of rheology and printability of 3D printing nutritious food with complex formulations. *Addit. Manuf.* **2022**, *58*, 103030. [CrossRef]
50. Malkin, A.; Kulichikhin, V.; Ilyin, S. A modern look on yield stress fluids. *Rheol. Acta* **2017**, *56*, 177–188. [CrossRef]
51. Liu, Z.; Bhandari, B.; Prakash, S.; Mantihal, S.; Zhang, M. Linking rheology and printability of a multicomponent gel system of carrageenan-xanthan-starch in extrusion based additive manufacturing. *Food Hydrocoll.* **2019**, *87*, 413–424. [CrossRef]
52. Nguyen, Q.D.; Akroyd, T.; Kee, D.C.D.; Zhu, L. Yield stress measurements in suspensions: An inter-laboratory study. *Korea-Aust. Rheol. J.* **2006**, *18*, 15–24.
53. Kolli, R. Experimental investigation on mechanical properties of fly ash-GGBFS based GO-geopolymer concrete using mineral sand (Quartz-Feldspar) as fine aggregate. *Mater. Today Proc.* **2022**, *60*, 40–45.
54. Tian, L.; Feng, W.; Ma, H.; Zhang, S.; Shi, H. Investigation on the microstructure and mechanism of geopolymer with different proportion of quartz and K-feldspar. *Constr. Build. Mater.* **2017**, *147*, 543–549. [CrossRef]
55. Bong, S.H.; Nematollahi, B.; Xia, M.; Nazari, A.; Sanjayan, J. Properties of one-part geopolymer incorporating wollastonite as partial replacement of geopolymer precursor or sand. *Mater. Lett.* **2020**, *263*, 127236. [CrossRef]

56. Archez, J.; Texier-Mandoki, N.; Bourbon, X.; Caron, J.F.; Rossignol, S. Influence of the wollastonite and glass fibers on geopolymer composites workability and mechanical properties. *Constr. Build. Mater.* **2020**, *257*, 119511. [CrossRef]
57. Dick, A.; Bhandari, B.; Dong, X.; Prakash, S. Feasibility study of hydrocolloid incorporated 3D printed pork as dysphagia food. *Food Hydrocoll.* **2020**, *107*, 105940. [CrossRef]
58. Zhang, C.; Wang, C.S.; Therriault, D.; Heuzey, M.C. Development of aqueous protein/polysaccharide mixture-based inks for 3D printing towards food applications. *Food Hydrocoll.* **2022**, *131*, 107742. [CrossRef]
59. Aboulayt, A.; Jaafri, R.; Samouh, H.; El Idrissi, A.C.; Roziere, E.; Moussa, R.; Loukili, A. Stability of a new geopolymer grout: Rheological and mechanical performances of metakaolin-fly ash binary mixtures. *Constr. Build. Mater.* **2018**, *181*, 420–436. [CrossRef]
60. Chang, I.; Im, J.; Prasidhi, A.K.; Cho, G.C. Effects of Xanthan gum biopolymer on soil strengthening. *Constr. Build. Mater.* **2015**, *74*, 65–72. [CrossRef]
61. Gao, H.; Qing, L.; Ma, G.; Zhang, D.; Wei, C. Numerical investigation into effects of rheological properties on grout flow in rock fracture using Herschel-Bulkley model. *Eng. Geol.* **2024**, *329*, 107402. [CrossRef]
62. Mizuno, H.; Ikeda, A.; Kawasaki, T.; Miyazaki, K. Universal mechanism of shear thinning in supercooled liquids. *Commun. Phys.* **2024**, *7*, 199. [CrossRef]
63. Rosti, M.E.; Takagi, S. Shear-thinning and shear-thickening emulsions in shear flows. *Phys. Fluids* **2021**, *33*, 083319. [CrossRef]
64. Schwab, A.; Levato, R.; D'Este, M.; Piluso, S.; Eglin, D.; Malda, J. Printability and shape fidelity of bioinks in 3D bioprinting. *Chem. Rev.* **2020**, *120*, 11028–11055. [CrossRef] [PubMed]
65. Panda, B.; Unluer, C.; Tan, M.J. Extrusion and rheology characterization of geopolymer nanocomposites used in 3D printing. *Compos. Part B Eng.* **2019**, *176*, 107290. [CrossRef]
66. Mewis, J.; Wagner, N.J. Thixotropy. *Adv. Colloid Interface Sci.* **2009**, *147*, 214–227. [CrossRef]
67. Prabhakar, S.; Rao, K.H.; Forsling, W. Dissolution of wollastonite and its flotation and surface interactions with tallow-1, 3-diaminopropane (duomeen T). *Miner. Eng.* **2005**, *18*, 691–700. [CrossRef]
68. Dong, H.; Aziz, N.A.; Shafri, H.Z.M.; Ahmad, K.A.B. Computational fluid dynamics study on cemented paste backfill slurry. *Constr. Build. Mater.* **2023**, *369*, 130558. [CrossRef]
69. Brady, J.F. The rheological behavior of concentrated colloidal dispersions. *J. Chem. Phys.* **1993**, *99*, 567–581. [CrossRef]
70. Brader, J.M. Nonlinear rheology of colloidal dispersions. *J. Phys. Condens. Matter* **2010**, *22*, 363101. [CrossRef]
71. Gholamipour-Shirazi, A.; Norton, I.T.; Mills, T. Designing hydrocolloid based food-ink formulations for extrusion 3D printing. *Food Hydrocoll.* **2019**, *95*, 161–167. [CrossRef]
72. Baniasadi, H.; Kimiaei, E.; Polez, R.T.; Ajdary, R.; Rojas, O.J.; Österberg, M.; Seppälä, J. High-resolution 3D printing of xanthan gum/nanocellulose bio-inks. *Int. J. Biol. Macromol.* **2022**, *209*, 2020–2031. [CrossRef] [PubMed]
73. Avery, M.P.; Klein, S.; Richardson, R.; Bartlett, P.; Adams, G.; Dickin, F.; Simske, S. The rheology of dense colloidal pastes used in 3D-printing. In Proceedings of the Nip & Digital Fabrication Conference, Philadelphia, PA, USA, 7–11 September 2014; Society for Imaging Science and Technology: Springfield, IL, USA, 2014; Volume 1, pp. 140–145.
74. Quemada, D.; Berli, C.L.A. *Interactions Dans Les Dispersions*; Stabilisation; Lavoisier: Paris, France, 2006.
75. Kröger, M.; Vermant, J. *The Structure and Rheology of Complex Fluids*; Oxford University Press: Oxford, UK, 2000.
76. Suiker, A.S.; Wolfs, R.J.; Lucas, S.M.; Salet, T.A. Elastic buckling and plastic collapse during 3D concrete printing. *Cem. Concr. Res.* **2020**, *135*, 106016. [CrossRef]

ceramics

MDPI

Article

Characterization of the Evolution with Temperature of the Structure and Properties of Geopolymer-Cordierite Composites

Franklin Casarrubios [1,*], Alexandre Marlier [1], Charlotte Lang [2], Sandra Abdelouhab [2], Isabella Mastroianni [2], Geoffroy Bister [2] and Maurice-François Gonon [1,*]

1 Materials Institute, University of Mons (UMONS), 7000 Mons, Belgium; alexandre.marlier@student.umons.ac.be
2 Belgian Ceramic Research Centre (BCRC), 7000 Mons, Belgium; c.lang@bcrc.be (C.L.); s.abdelouhab@bcrc.be (S.A.); i.mastroianni@bcrc.be (I.M.); g.bister@bcrc.be (G.B.)
* Correspondence: franklin.casarrubios@umons.ac.be (F.C.); mauricefrancois.gonon@umons.ac.be (M.-F.G.)

Abstract: This work is part of a research project aimed at producing ceramic-like materials, without the need for an initial sintering, for potential applications in catalysis or filtration at temperatures up to 1000 °C. In that context, cordierite-derived materials were prepared from recycled cordierite powder (automotive industry waste) bonded with metakaolin-potassium silicate geopolymer. The principle is that these materials, prepared at temperatures below 100 °C, acquire their final properties during the high-temperature commissioning. The focus is on the influence of the K/Al ratio and cordierite fraction on the stability of the dimensions and porosity during heating at 1000 °C, and on the final Young's modulus and coefficient of thermal expansion. Conventional and high-temperature XRD evidenced the absence of crystallization of the geopolymer binder and interaction with the cordierite filler during the heating stage when K/Al = 1 or 0.75. By contrast, crystallization of kalsilite and leucite, and diffusion of potassium ions in the structure of cordierite is evidenced for K/Al = 1.5 and 2.3. These differences strongly influence the shrinkage due to sintering and the final properties. It is shown that a K/Al ratio of 0.75 or 1 is favorable to the stability of the porosity, around 25 to 30%. Moreover, a low coefficient of thermal expansion of 4 to 4.5 $\times 10^{-6}$ K^{-1} and a Young's modulus of 40 to 45 GPa is obtained.

Keywords: cordierite; geopolymer; Young's modulus; coefficient of thermal expansion

1. Introduction

Cordierite $Mg_2Al_4Si_5O_{18}$ (or $2MgO{\cdot}2Al_2O_3{\cdot}5SiO_2$) is one of the main compounds of the ternary system MgO-SiO_2-Al_2O_3. Although there are different ways to synthesize cordierite, the most common is by reactive sintering, at a temperature of around 1400 °C, of mixtures of raw materials containing the required oxides, such as talc, clay and alumina [1]. Besides their interesting dielectric properties [2], these materials have received a great deal of attention due to their very low coefficients of thermal expansion (<3 $\times 10^{-6}$ K^{-1}) and thermal conductivity (2–3 W m^{-1} K^{-1}) [3,4]. They are thus widely used in high-temperature applications, such as catalyst substrates and filtration [5–8]. For the automotive industry, most of the catalytic converters are produced by the extrusion of a cordierite honeycomb substrate coated with a "washcoat", consisting of precious metals (Pt, Pd, ...) [9,10].

Considering the huge number of motor vehicles produced in the world, recycling of end-of-life converters is of growing interest from economic and environmental points of view. Studies are carried out on the recovery of the precious metals of the washcoat by pyrometallurgy and hydrometallurgy. Furthermore, new approaches are considered to obtain zero waste by recycling the ceramic substrate [11]. On the other hand, a non-negligible quantity of non-conform fired cordierite substrates (without washcoat) are produced by the industry. The reuse of these production wastes for the manufacture of high-added value products is, therefore, of high interest and would be a first step to demonstrate the recyclability of the cordierite.

However, the use, as raw material, of a powder of milled fired substrates, and, therefore, mineralogically containing cordierite as nearly single phase, fundamentally changes the sintering mechanism. Indeed, the absence of physicochemical transformations and reactions between the different raw materials of a mixture (e.g., talc, kaolin, ...) consequently leads to sintering essentially based on diffusion mechanisms in the solid state [12,13]. In this case, the suitability of the powder edifice for consolidation and densification is strongly conditioned to the use of a sub-micron size powder and a high sintering temperature. Using dilatometry curves, Camerucci et al. evidenced the solid-state sintering of cordierite between 850 °C and 1350 °C. Above 1350 °C, shrinkage is accelerated by the occurrence of a liquid phase. Their study also highlights the influence of particle size distribution on the densification. Firstly, using a monomodal distribution, the finer the particle size, the greater the densification. Secondly, using a bimodal distribution, with an optimum ratio, allows an increase in the degree of densification [12]. However, obtaining a powder from downgraded parts requires several grinding steps, and the energy cost and process duration increase sharply with the desired fineness. An alternative solution is the use of a sintering additive to lower the sintering temperature. Ogiwara et al. studied the influence of the addition of 3%wt of $Li_2O–Bi_2O_3$. They succeeded in achieving a densification of 97% at 1050 °C for 2 h. However, the mechanical properties were degraded, for example, flexural strength going from 243 MPa to 120 MPa [13]. Even if it is considered that extensive densification is not useful for the intended application, sufficient consolidation to ensure mechanical strength is essential. It is, therefore, of great interest to find an alternative to conventional sintering to produce consolidated and geometrically stable parts at the lowest possible energy cost.

For several years now, geopolymer materials have been attracting growing interest from scientists and industrialists alike. Without the need for energy-intensive heat treatments, the geopolymerization process makes it possible to form silicate inorganic materials with good performance in terms of mechanical properties, chemical resistance and durability [14–16]. Geopolymers are obtained by reaction between alkaline silicate, or more rarely acidic reagent, and amorphous aluminosilicate powders, at room or low temperatures. The resulting solid has a three-dimensional amorphous structure that contains mostly covalent bonds. A large body of research focuses on the formulation of geopolymers for construction applications [17,18]. However, more and more studies are also focusing on the development of geopolymers for applications that usually use ceramic materials [19,20].

Unfortunately, one of the downsides of geopolymers is their poor mechanical properties at high temperatures. During a rise in temperature, dehydration and dihydroxylation of the inorganic network occur, possibly followed by its crystallization and other phenomena. These transformations induce strong dimensional variations and cracking [21,22]. To overcome this issue, the geopolymer paste can be loaded with mineral filler to produce a "composite" material [19,23,24]. By doing so, the geopolymer allows a "cold" consolidation of the powder edifice and the reaction between the geopolymer and the filler during a rise in temperature can possibly stabilize the structure of the material and enhance its final properties.

The type of alkali and the alkali/Al ratio have a strong influence on the properties of the geopolymer materials after curing but also at high temperatures. Bernal et al. studied the influence of alkali over the evolution of shrinkage with temperature until 1000 °C on samples cured at ambient temperature for 7 days and 5 years. For alkali/Al = 1, Na-based geopolymers show a first stage of shrinkage of 7.5% at 250 °C related to dehydration and dihydroxylation, and a second shrinkage around 750 °C related to viscous sintering. The final shrinkage is 23%. For K-geopolymers, the first shrinkage is lower (5%) and the viscous sintering begins at 900 °C; the final shrinkage is 10%. Finally, Rb- and Cs-geopolymers show the same behavior with a smoother shrinkage for the removal of water, and no sintering is observable. The final shrinkage is around 7.5%. After 5 years of curing,

Na-geopolymers show the same behavior but the shrinkage related to sintering is more drastic and the final shrinkage is 27.5%. Finally, K-geopolymers show the same behavior as the Rb-geopolymers, and the curing time has no effect over the evolution of shrinkage with temperature [25]. The potassium-based geopolymers show better resistance to high temperatures compared to sodium-based geopolymer pastes [23]. A study was undertaken by Kohout et al. [26] about the effect of the K/Al molar ratio on the thermomechanical properties of metakaolinite-based geopolymer composites. In their study, a chamotte was used as a filler for the geopolymer matrix. From the structural point of view, it was shown that the pure geopolymer stays amorphous until 800 °C. For higher temperatures, the crystallization of leucite ($KAlSi_2O_6$) and kalsilite ($KAlSiO_4$) happens depending on the K/Al ratio; the higher the ratio, the more phases appear at lower temperatures. It was also highlighted that the geopolymer in the composites shows the same transformations as the pure geopolymer. Regarding the mechanical properties, a high K/Al molar ratio tends to bring down the mechanical properties of the samples. For unfired samples, the maximal value of the Young's modulus is obtained with a ratio near 1.0. Additionally, the modulus tends towards lower values when the sample is fired at temperatures up to 1000 °C. On the other hand, in their work, K. Hemra and P. Aungkavattana demonstrated that using cordierite as filler leads to a strong increase in the compressive strength after heat treatment at 800 °C/2 h compared to the geopolymer alone [24]. They also showed that no new crystalline compounds are formed during the heat treatment.

Cesium-based geopolymers reinforced with 5–35 wt% cordierite were synthesized by Wei Chen et al. [27]. The main crystalline phases in the product, obtained after thermal treatment at 1400 °C, are cordierite and cubic pollucite, $Cs_2O{\cdot}Al_2O_3{\cdot}4SiO_2$, which is formed by crystallization of the geopolymer matrix over 700 °C. The cordierite filler does not react with the geopolymer. The shrinkage resulting from the sintering is less than 5% and occurs over 800 °C. The lowest coefficient of thermal expansion results from the highest cordierite load and is 1.89×10^{-6} K^{-1}, which is unexpectedly lower than the CTE of pure cordierite. In their work, Chengying Bai et al. [28] investigated the synthesis by a replica route of cordierite-based geopolymer foams that have potential applications such as porous supports or high-temperature filter components. The geopolymers were prepared by incorporating MgO as an additional magnesium source in metakaolin and fly ash reacted with NaOH solution and sodium silicate solution. Sintering at 1200 °C for 2 h led to porous cordierite-based ceramics. However, an extensive structural characterization showed that the transformation of cordierite after the thermal treatment was incomplete.

Finally, some recent works demonstrate that it is possible to obtain geopolymer composite parts by additive manufacturing. Gasmi et al. studied the production of metakaolin-based geopolymer and composites with feldspar and wollastonite by robocasting [29]. Additive manufacturing techniques appear to be highly promising for producing substrates or filters with optimized geometries. Kovacev et al. used digital-light processing (DLP) to produce substrates with two designs of diamond unit cells. For the same inlet gas temperature, it was found that the conversion efficiency was higher for the substrates produced by DLP compared to traditional honeycomb substrates [30].

In this context, we investigate the manufacture of cordierite–geopolymer composites. As previously explained, the first motivations for the choice of cordierite as the mineral filler are its intrinsic properties and the reuse of industrial wastes. Besides this, it is well known that the crystal structure of the cordierite can accept the insertion of potassium atoms [31,32]. This can possibly lead to interactions between the potassium silicate-based geopolymer and the cordierite filler during the rise in temperature, and then affect the dimension stability and the final properties of the composite. More precisely, our interest is in the influence of the K/Al ratio and the cordierite fraction on the stability of the dimensions and porosity during heating at 1000 °C, and on the final Young's modulus and coefficient of thermal expansion. The final goal of this work is to select the most promising compositions for a further study that will focus on their use in additive manufacturing.

2. Materials and Methods

2.1. Materials

The cordierite powder used in this work comes from non-conform (over tolerance limits or presence of flaws) honeycomb substrates for catalytic converters that were supplied by an industrial partner. These substrates are first crushed with a pestle into a coarse powder, which is further wet-milled using a Turbula shaker (Willy A. Bachofen AG, Basel, Switzerland) at a speed of 46 RPM with alumina media, and finally dried to obtain the final powder. The parameters for the milling are the following:

- Weight ratio cordierite/water = 5:7
- Weight ratio cordierite/balls = 1:1
- Shaking time = 7.5 h

The particle size distribution of the final cordierite powder was characterized by using a Malvern Mastersizer 2000 laser granulometer (Malvern Panalytical, Malvern, UK) (Table 1). The median size of the particles d_{50} is 26.9 μm. An X-ray diffraction (XRD) analysis was performed on the powder using a Panalytical Empyrean diffractometer (Malvern Panalytical, Malvern, UK) with a Cu Kα radiation source. The X-ray pattern shows the diffraction peaks of cordierite (JCPDS #00-013-0294), and a small amount of mullite (JCPDS #00-015-0776) (Figure 1). An XRF analysis was performed with Horiba XGT 9000 equipment (Horiba, Kyoto, Japan) to ascertain the composition of the powder and compare it to the stoichiometric cordierite (Table 2).

Table 1. Particle size distribution of the cordierite and metakaolin powders.

	D_{10} (μm)	D_{50} (μm)	D_{90} (μm)
Cordierite	3.2	26.9	61.2
Metakaolin	1.2	4.4	18.9

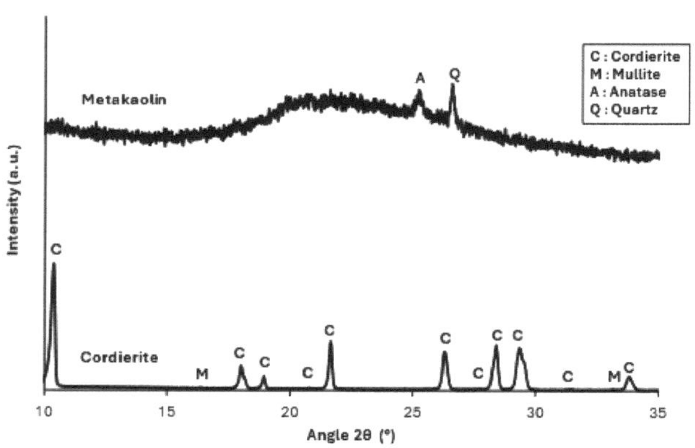

Figure 1. XRD patterns of the cordierite powder and metakaolin powders.

Table 2. Equivalent oxide composition (%mol) of the cordierite and metakaolin powders.

	MgO	SiO_2	Al_2O_3	Fe_2O_3	TiO_2	CaO	Other
Cordierite powder	21.7	55.7	21.4	0.3	0.3	0.1	0.5
Theoretical cordierite	22	56	22	/	/	/	/
Metakaolin powder	0.1	65.3	32.4	0.3	1.2	0.5	0.3

The metakaolin ($Al_2O_3{\cdot}2SiO_2$) used for this work comes from Temcon-Solutions GmbH (Winnweiler, Germany) under the name Tempozz® M88. Its particle size distribution shows a median particle size d_{50} = 4.4 μm (Table 1). Its composition measured from XRF analysis (Table 2) indicates a ratio SiO_2/Al_2O_3 = 2.01, which is very close to 2.00, corresponding to pure kaolinite. The XRD analysis (Figure 1) shows the characteristic background bump of an amorphous material, and two weak intensity peaks that can be attributed to the presence of small amounts of quartz (JCPDS #00-033-1161) and anatase (JCPDS #00-021-1272).

The potassium silicate used in this work comes from Wöllner GmbH (Ludwigshafen, Germany) and is named Geosil® 14517. The characteristics of the potassium silicate solution are listed in Table 3.

Table 3. Characteristics of the potassium silicate solution at 20 °C.

	SiO_2/K_2O (Molar Ratio)	Water Fraction (wt%)	Density (g/cm^3)	pH	Viscosity (mPa·s)
Geosil 14517	1.7	55	1.5	12.5	20

2.2. Preparation of Geopolymer Composites

The process used for the synthesis of the geopolymer composites is shown in Figure 2.

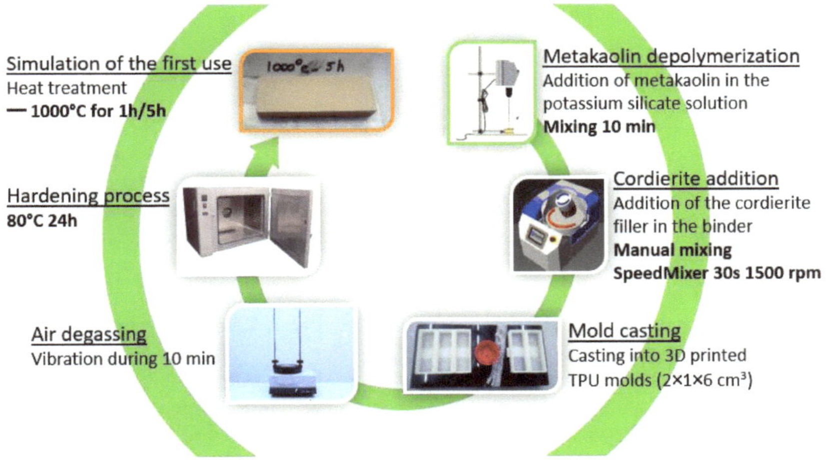

Figure 2. Process used for the synthesis of the geopolymer composites.

- Geopolymer paste:

The preparation of the geopolymer binder starts with the potassium silicate solution and the metakaolin. The metakaolin is added into the silicate solution and is mixed with a laboratory mixer for 10 min. During this stage, the depolymerization of the metakaolin takes place and there is formation of oligomers. The mixture is very fluid and easy to mix. For some samples of 100% geopolymer, no cordierite filler is added, and step 2 is skipped.

- Cordierite addition:

When the geopolymer is ready, it is added to the cordierite powder and mixed by hand. Depending on the percentage of filler, the mixture is more difficult to mix, this is why a SpeedMixer at 1500 rpm for 30 s is used to obtain good homogeneity of the paste and avoid lumps. The geopolymer paste flows easily under agitation.

- Mold casting

The mixture is then cast into 3D-printed TPU molds of dimensions 2 × 1 × 6 cm. The 3D-printed TPU molds are flexible and make it easier to remove the samples. They also limit the adhesion of the geopolymer paste from the walls.

- Air degassing:

Due to its high viscosity, the geopolymer composite can capture air, which makes large pores. To prevent air in the samples, the filled molds are vibrated on a vibrating table for 10 min.

- Hardening:

The samples are sealed in plastic bags with a water recipient to ensure constant humidity and to prevent water from evaporating. Then, the samples are kept in an oven at 80 °C for 24 h. Once the samples are cured, they are unmolded and polished at grade 120. Next, the characterization of the samples is performed.

- Heat treatment:

To investigate the evolution of the properties of the composites during the first use at high temperature, some samples undergo heat treatment with the following conditions:

- Heating ramp: 5 °C/min
- Plateau: 1000 °C for 5 h
- Cooling ramp: 5 °C/min

2.3. Investigated Compositions and Characterization Methods

The purpose of this work is to investigate the influences of (i) the K/Al ratio of the geopolymer binder and (ii) the cordierite fraction on the stability of the dimensions and porosity during heating at 1000 °C, and on the final Young's modulus and coefficient of thermal expansion. Table 4 presents the compositions that were prepared using K/Al ratios of 0.75, 1, 1.5 and 2.3, and cordierite weight fractions of 0, 53%, 60% and 65%.

After curing, five samples of each composition are characterized in terms of:

- Geometric apparent density d_a;
- Absolute density d using a Micromeritics AccuPyc II 1340 device (Micromeritics, Norcross, GA, USA) (after milling);
- Total porosity p calculated from the equation $p = 1 - d_a/d$;
- Crystalline and amorphous phases by means of a Panalytical Empyrean diffractometer with a Cu kα radiation source with a range of 2θ between 9° and 35°. The detector is a PIXcel1D (RTMS detector) with an active length of 3.3473°. The step size and the counting time are 0.0131° and 8.67 s, respectively
- Room temperature Young's modulus E using the Impulse Excitation Technique (IET) with a GrindoSonic MK3 materials tester (Grindosonic, Heverlee, Belgium).

For each composition, the same characterization was also performed on one sample that underwent heat treatment at 1000 °C for 5 h.

For selected compositions, additional characterization techniques were applied on cured samples to follow the evolution of their properties during heat treatment at 1000 °C, with a soaking time of 1 h and heating ramp of 5 °C/min as follows:

- Crystallographic changes by high-temperature diffraction (HT-XRD) on powdered samples by means of a Panalytical Empyrean (HTK-2000N) diffractometer with a Cu Kα radiation source with a range of 2θ between 9° and 35°. The detector is a 1Der detector (RTMS detector) with an active length of 2.1223°. The step size and the counting time are 0.0167° and 10.16 s, respectively
- Dimension changes by dilatometry analysis on 5 × 5 × 20 mm^3 samples using a Dil 402 Expedis dilatometer (Netzsch, Selb, Germany);
- Evolution of the Young's modulus E using a GrindoSonic MK7 non-destructive materials tester coupled with a Nabertherm furnace HT-1600 °C.

Table 4. The different compositions investigated, and the characterization techniques performed.

		wt% Cordierite			
		0	**53**	**60**	**65**
K/Al	0.75		P + D + I		
	1	P + D + X	P + D + I	P + D + X + I	
	1.5		P + D + I		
	2.3	(*) X	P + D	P + D + X	P

P: Density/porosity/Young's modulus after curing and after heat treatment; D = Dimension changes by dilatometry; X = HT-XRD; I = Measurement of E by IET in temperature; (*) In the case of the ratio K/Al = 2.3, without cordierite, the samples obtained after curing were seriously cracked, and it was not possible to perform most of the characterizations.

3. Results

3.1. Characterization of the Samples after Curing and after Heat Treatment

Figures 3 and 4 show the pictures of the samples after curing. The top side of some samples clearly shows spherical pores that are the result of the trapping of air bubbles during the casting due to high viscosity of the paste. This is particularly the case when the K/Al ratio is low, and the cordierite fraction high. These pores influence the apparent density of the samples, and consequently their total porosity. Therefore, in this work, we will focus on the evolution of the porosity after thermal treatment, and not on the comparison of the absolute total porosity between the different compositions.

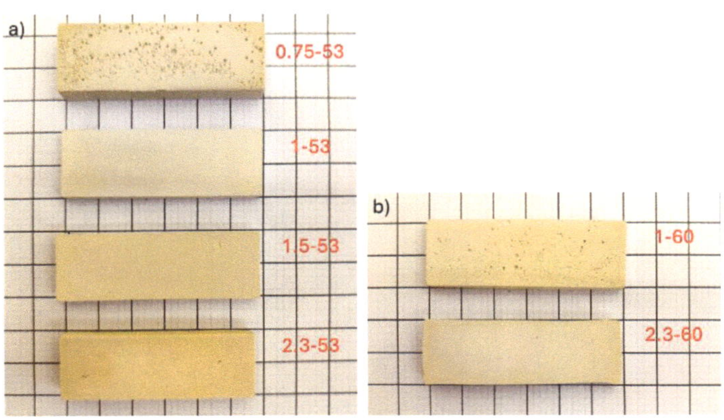

Figure 3. Pictures of samples with 53%wt (**a**) and 60%wt (**b**) with different K/Al ratio.

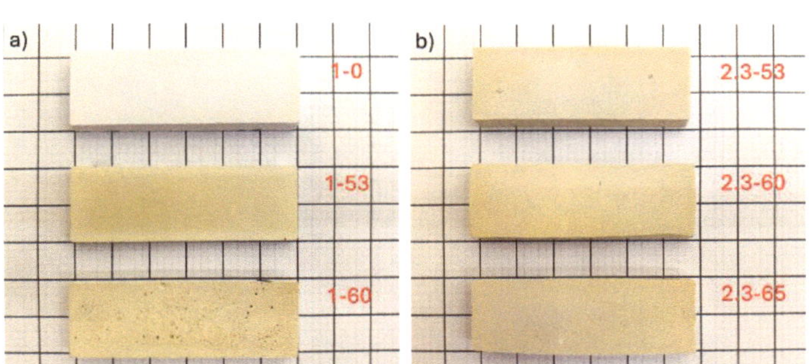

Figure 4. Pictures of samples with K/Al ratio of 1 (**a**) and 2.3 (**b**) with different amounts in cordierite.

Table 5 shows the characteristics of the sample after curing (C) and after treatment (T). Due to the higher value of the absolute density of the cordierite (i.e., 2.55 g/cm^3, experimentally measured by He pycnometry) by comparison to that of the geopolymer binder, the cured composites show an absolute density increasing with the cordierite load. We also note that the absolute density trend is of a decrease with increase in the K/Al ratio. After heat treatment, the differences are less, and all composites show a significant increase in the density ranging from 2.55 to 2.60 g/cm^3. This difference in density influences the evolution of the porosity during the thermal treatment. The increase in the porosity is low for the composites with K/Al = 0.75 and 1, and large for K/Al = 1.5 and 2.3. Concerning the Young's modulus, the composites with the lowest ratios (i.e., 0.75 and 1) show a strong increase in the Young's modulus after heat treatment.

Table 5. Characteristics of the samples before (C) and after (T) heat treatment.

Sample	Apparent Density (g/cm^3)		Absolute Density (g/cm^3)		Porosity (%)		Young's Modulus (GPa)	
	C	T	C	T	C	T	C	T
0–1	1.39	2.20	2.11	2.48	34	12	8	15
0–2.30	-	-	1.92	-	-	-	-	-
53–0.75	1.80	1.89	2.35	2.57	23	26	20	32
53–1.00	1.77	1.89	2.36	2.53	25	25	21	46
53–1.50	1.67	1.61	2.34	2.57	28	37	14	16
53–2.30	1.92	1.56	2.24	2.58	14	39	9	13
60–1.00	1.78	1.81	2.36	2.54	24	29	23	37
60–2.30	1.88	1.62	2.27	2.59	18	38	11	14
65–2.30	1.87	1.68	2.31	2.60	19	36	16	17

Figure 5 shows the XRD pattern of the composites with 53 wt% cordierite and K/Al = 0.75; 1.0; 1.5 and 2.3. After curing, only the diffraction peaks of cordierite are observed. After heat treatment of 5 h at 1000 °C, all the diffraction patterns highlight the presence of leucite $KAlSi_2O_6$ (JCPDS #00-038-1423), and the possibility of kalsilite $KAlSiO_4$ (JCPDS #00-011-0579) when K/Al = 1.5 and 2.3.

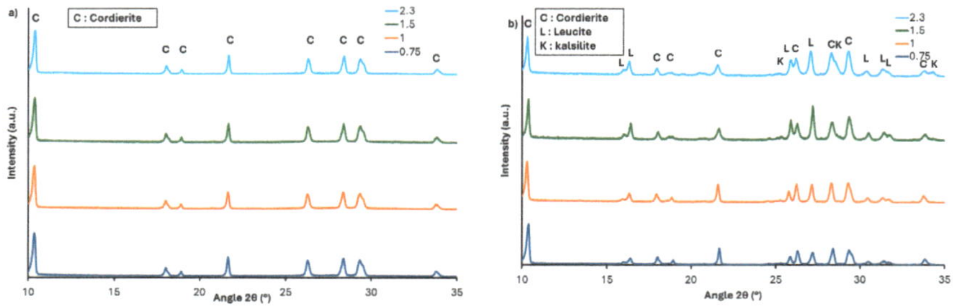

Figure 5. XRD patterns of the composites with 53 wt% cordierite and K/Al = 0.75, 1.0, 1.5 and 2.3 after curing (**a**) and after heat treatment (**b**).

The XRD patterns of the samples corresponding to K/Al = 1 and cordierite loads of 0, 53 and 60 wt% (Figure 6) evidence that the occurrence of the leucite is due to the crystallization of the geopolymer binder, whereas the presence of kalsilite after heat treatment is uncertain.

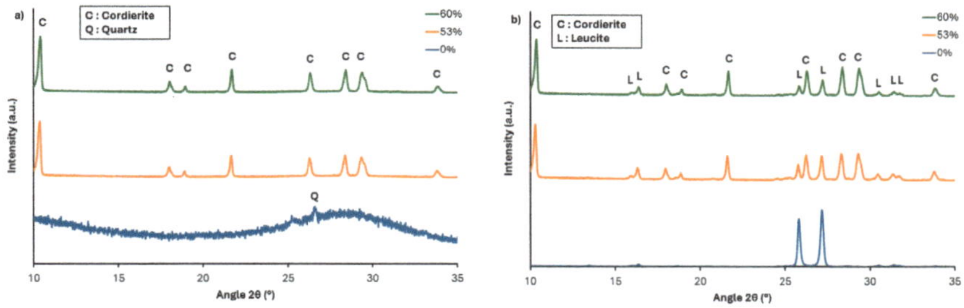

Figure 6. XRD patterns of the composite with K/Al = 1 and cordierite loads of 0, 53 and 60 wt% after curing (**a**) and after heat treatment (**b**).

When K/Al = 2.3, the presence of kalsilite in addition to leucite is clear on the XRD patterns of the heat-treated geopolymer and of the composite loaded with 53 wt% cordierite (Figure 7). The composites loaded with 60 wt% and 65 wt% cordierite seem to contain only leucite.

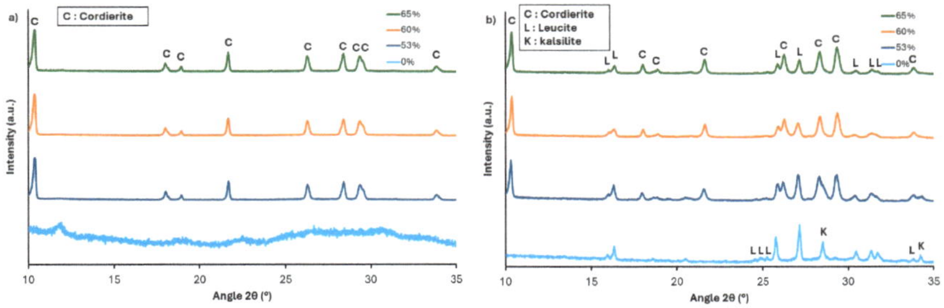

Figure 7. XRD patterns of the composites with K/Al = 2.3 and cordierite loads of 0, 53, 60 and 65 wt% after curing (**a**) and after heat treatment (**b**).

3.2. HT-XRD

a. Phases analysis

For a better understanding of the crystallographic changes occurring during the heat treatment, high-temperature XRD analysis was performed on the geopolymers and composites with 60 wt% cordierite with K/Al = 1 (Figure 8) and 2.3 (Figure 9).

For the geopolymer with K/Al = 1, the crystallization of leucite and kalsilite occurs between 900 °C and 1000 °C on heating. During the plateau, the intensities of the peaks of kalsilite decrease whereas those of leucite increase. The XRD patterns collected during the cooling to RT between 600 °C and 400 °C evidence the crystal symmetry transformation of leucite from the high-temperature cubic form (JCPDS #00-031-0967) to the low-temperature tetragonal form [33]. In the case of the composite with 60 wt% cordierite, the occurrence of kalsilite or leucite is not detected during the thermal cycle. It should be noted that the main diffraction peaks of kalsilite are convoluted with high intensity peaks of cordierite, so it is not possible to conclude with certitude the absence of a weak amount of this phase.

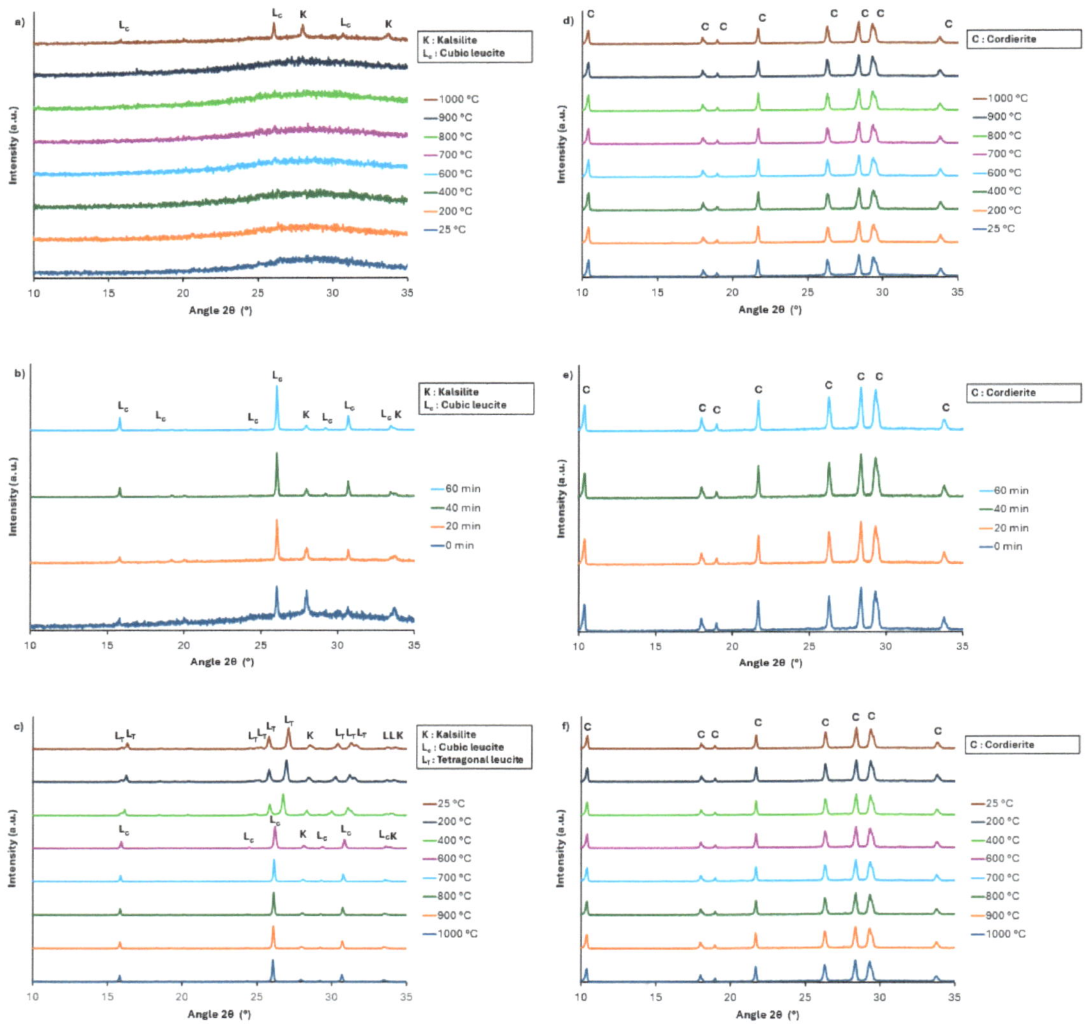

Figure 8. HT-XRD analysis of the geopolymer (**a–c**) and composite (**d–f**) with K/Al =1 with heating until 1000 °C (**a**,**d**), plateau at 1000 °C (**b**,**e**) and cooling to RT (**c**,**f**).

The HT-XRD analysis performed on the geopolymer with K/Al = 2.3 also shows the crystallization of kalsilite and leucite, but with the occurrence of kalsilite at a slightly lower temperature during heating (between 800 °C and 900 °C). As for the other geopolymer, the intensity of the peaks of leucite increases during the plateau at 1000 °C, and allotropic transformation of leucite is observed during cooling. For the composite with 60 wt% cordierite, the presence of leucite is also detected at 1000 °C during heating. The peaks of kalsilite are not observed but, as stated previously, this is possibly due to their convolution with strong peaks of cordierite.

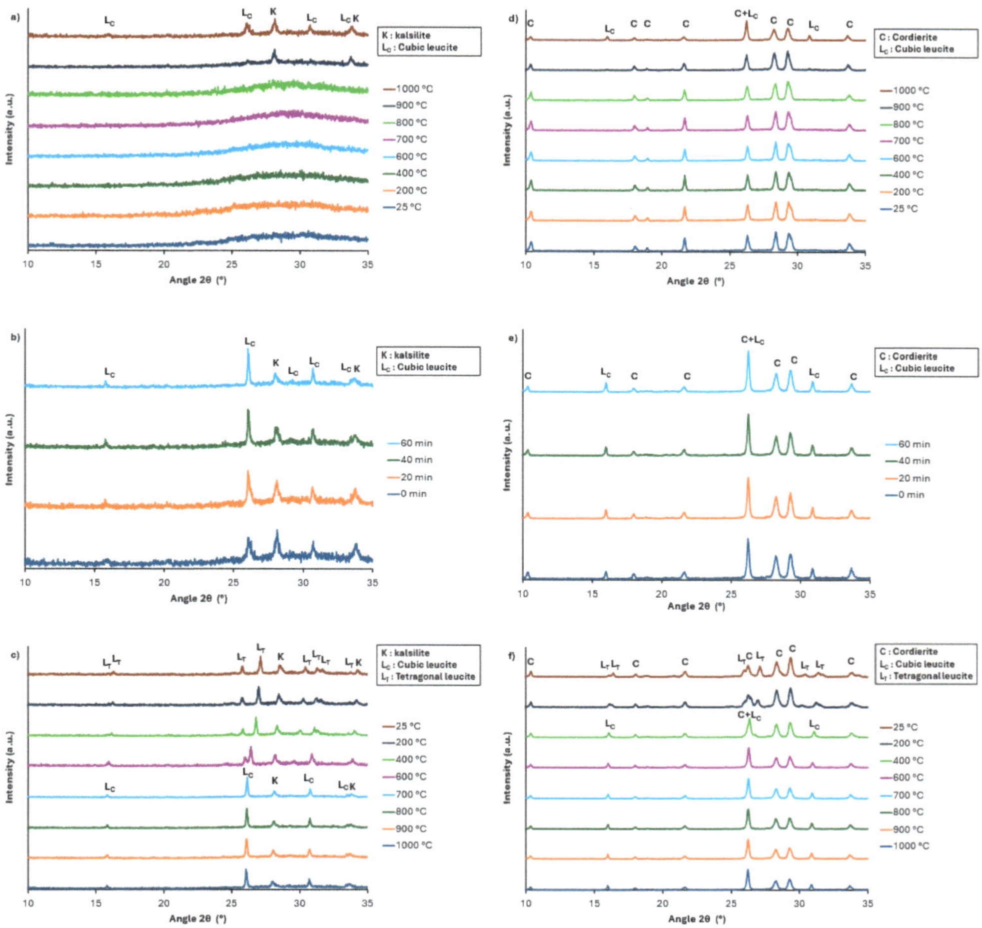

Figure 9. HT-XRD analysis of the geopolymer (**a–c**) and composite (**d–f**) with K/Al = 2.3 with heating until 1000 °C (**a**,**d**), plateau at 1000 °C (**b**,**e**) and cooling to RT (**c**,**f**).

b. Evolution of the size of the cordierite unit cell

To evidence possible change in the unit cell parameters of the cordierite, Rietveld refinement was performed on all the HT-XRD patterns using the MAUD software (Version 2.996). Figure 10 and Tables 6 and 7 show the evolution of the parameters a, b and c with temperature on heating and on cooling, for the cordierite powder, and for the composites with K/Al = 1 and 2.3 with 60 wt% cordierite. Tables 8 and 9 give the values of the profile residuals calculated by the MAUD software. The graphs of Figure 10 highlight weak differences between the cordierite powder and the cordierite in the composite when the K/Al ratio of the binder is 1. By contrast, with the binder with K/Al = 2.3, all the unit cell parameters significantly increase when the temperature goes over 700 °C on heating. Moreover, the contraction of the cell during the cooling stage is slightly larger, especially along the c-axis.

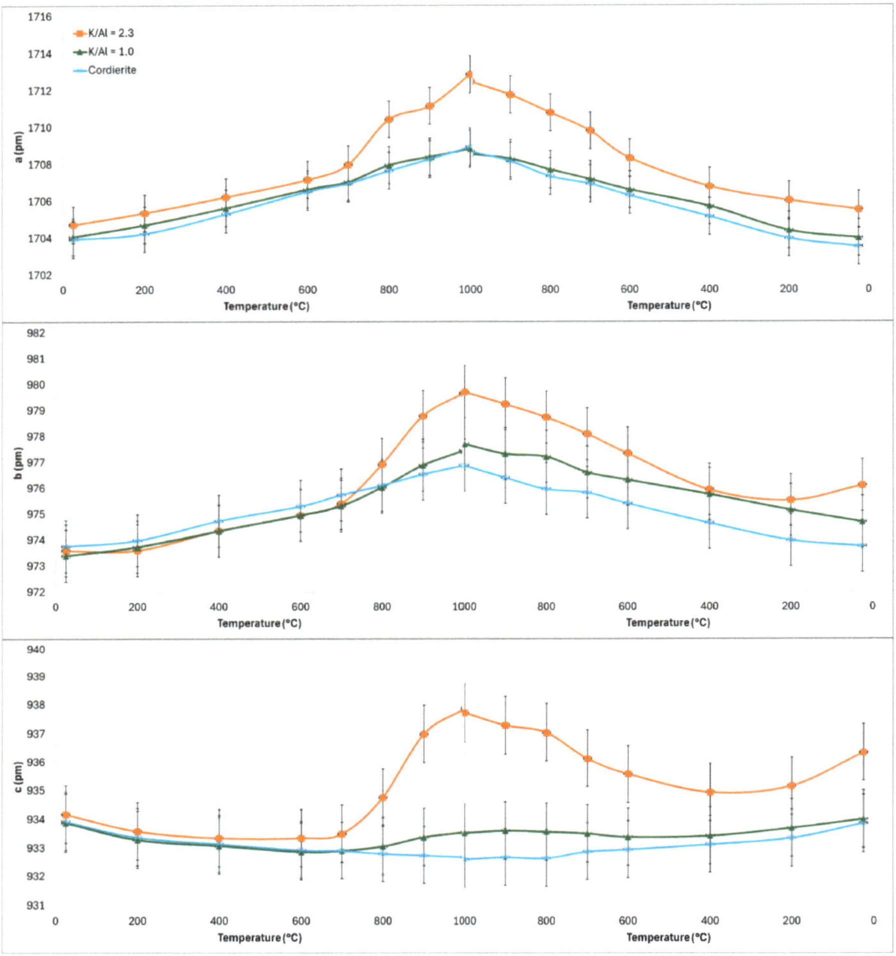

Figure 10. Evolution of the unit cell parameters (**a**–**c**) of cordierite on heating (left) and on cooling (right) for the cordierite powder, and for the composites with K/Al = 1 and 2.3 with 60 wt% cordierite.

Table 6. Unit cell parameters of the cordierite measured by Rietveld refinement on the HT-XRD patterns of the cordierite powder and the composites 1–60 and 2.3–60 during heating.

		Temperature Heating (°C)							
		25	**200**	**400**	**600**	**700**	**800**	**900**	**1000**
Cordierite	a (pm)	1703.9	1704.2	1705.3	1706.5	1707.0	1707.7	1708.3	1708.9
	b (pm)	973.8	974.0	974.7	975.3	975.7	976.1	976.5	976.9
	c (pm)	933.9	933.4	933.2	932.9	932.9	932.8	932.8	932.7
1–60	a (pm)	1704.1	1704.7	1705.6	1706.6	1707.1	1708.0	1708.4	1708.8
	b (pm)	973.3	973.7	974.3	975.0	975.3	976.0	976.9	977.4
	c (pm)	933.6	933.3	933.1	932.9	932.9	933.1	933.4	933.6
2.3–60	a (pm)	1704.8	1705.3	1706.1	1707.2	1708.0	1710.4	1711.2	1712.8
	b (pm)	973.5	973.5	974.3	975.0	975.3	976.9	978.7	979.6
	c (pm)	934.2	933.6	933.4	933.4	933.5	934.8	937.0	937.8

Table 7. Unit cell parameters of the cordierite measured by Rietveld refinement on the HT-XRD patterns of the cordierite powder and the composites 1–60 and 2.3–60 during cooling.

		Temperature Cooling (°C)							
		1000	**900**	**800**	**700**	**600**	**400**	**200**	**25**
Cordierite	a (pm)	1708.7	1708.2	1707.4	1706.9	1706.3	1705.2	1704.0	1703.5
	b (pm)	976.9	976.4	976.0	975.8	975.4	974.7	974.0	973.8
	c (pm)	932.6	932.7	932.7	932.9	933.0	933.1	933.4	933.9
1–60	a (pm)	1708.6	1708.3	1707.7	1707.2	1706.6	1705.7	1704.4	1704.0
	b (pm)	977.7	977.3	977.2	976.6	976.3	975.8	975.2	974.7
	c (pm)	933.5	933.6	933.6	933.5	933.4	933.4	933.7	934.0
2.3–60	a (pm)	1712.6	1711.8	1710.8	1709.9	1708.3	1706.9	1706.0	1705.5
	b (pm)	979.6	979.2	978.8	978.1	977.4	976.0	975.6	976.0
	c (pm)	937.7	937.3	937.0	936.1	935.6	935.0	935.1	936.4

Table 8. Weighted profile residual Rwp, Bragg residual Rb, and expected profile residual Rexp of the Rietveld refinement performed on the XT-XRD patterns of the cordierite powder and of the composites 10–60 and 23–60 during heating.

		Temperature Heating (°C)							
		25	**200**	**400**	**600**	**700**	**800**	**900**	**1000**
Cordierite	Rwp (%)	17.37	18.10	18.34	18.59	18.99	18.33	18.63	18.67
	Rb (%)	11.75	12.12	12.48	12.30	12.77	12.29	12.31	12.54
	Rexp (%)	10.40	10.53	10.54	10.62	10.57	10.59	10.59	10.54
1–60	Rwp (%)	12.28	13.23	13.93	13.85	13.70	13.98	14.34	14.68
	Rb (%)	9.07	9.63	10.08	10.08	10.05	10.29	10.80	11.16
	Rexp (%)	9.92	10.12	10.28	10.39	10.40	10.39	10.40	10.43
2.3–60	Rwp (%)	13.82	14.56	14.78	15.61	15.05	15.63	15.70	17.35
	Rb (%)	10.03	10.45	10.54	11.20	10.61	10.82	11.44	12.53
	Rexp (%)	11.82	12.04	12.24	12.34	12.32	12.35	12.35	12.42

Table 9. Weighted profile residual Rwp, Bragg residual Rb, and expected profile residual Rexp of the Rietveld refinement performed on the XT-XRD patterns of the cordierite powder and of the composites 10–60 and 23–60 during cooling.

		Temperature Cooling (°C)							
		1000	**900**	**800**	**700**	**600**	**400**	**200**	**25**
Cordierite	Rwp (%)	19.30	18.64	19.08	18.92	18.99	18.32	18.23	17.32
	Rb (%)	12.86	12.39	12.98	12.71	12.84	12.16	12.27	11.81
	Rexp (%)	10.55	10.49	10.47	10.45	10.38	10.34	10.25	10.02
1–60	Rwp (%)	14.68	14.61	14.72	14.24	14.24	14.09	13.38	13.02
	Rb (%)	11.20	11.21	11.14	10.75	10.75	10.63	10.15	9.89
	Rexp (%)	10.44	10.40	10.39	10.35	10.35	10.24	10.15	9.90
2.3–60	Rwp (%)	17.56	16.91	17.42	16.45	17.18	16.82	15.74	13.64
	Rb (%)	12.48	12.03	12.56	12.93	12.35	12.74	11.76	10.14
	Rexp (%)	12.34	12.23	12.15	12.12	12.09	11.93	11.77	11.50

3.3. Dimension Changes During the Thermal Cycle

a. Effect of the K/Al ratio

Figure 11 shows the dilatometry curves recorded during heating up to 1000 °C, during the plateau of 1 h at 1000 °C, and on cooling to RT, for the composition with 53 wt% of cordierite and different K/Al ratios. When this ratio is 0.75 and 1, at low temperatures, the curves show a first stage corresponding to a moderate shrinkage (<2%), slowing down when the temperature increases. Next, when the temperature goes over 800 °C, a second stage leads to a rapid shrinkage greater than 2%. The plateau of 1 h at 1000 °C leads to a stabilization of the shrinkage. The curves recorded during the cooling are linear. The final shrinkage is around 5% for both compositions. When K/Al = 1.5 or 2.3, the second stage of the shrinkage starts at lower temperatures (i.e., 600 °C for K/Al = 2.3 and 700 °C for K/Al = 1.5) but stops soon after to be replaced by an expansion at higher temperatures. During the plateau, the expansion still slightly increases. During cooling, a rapid contraction is observed around 500 °C.

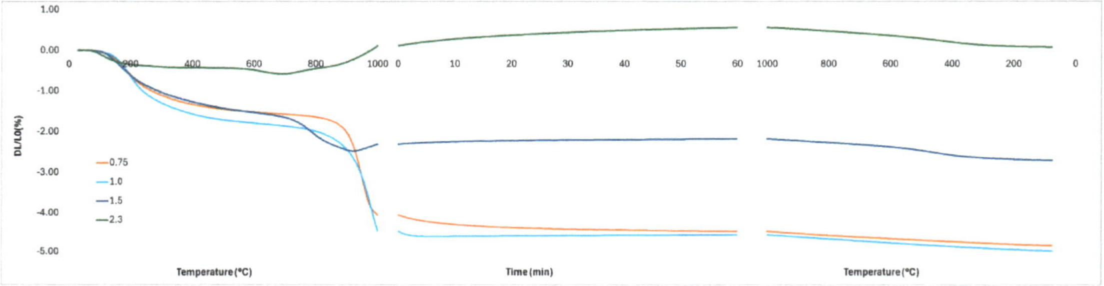

Figure 11. Dilatometry curves during heating up to 1000 °C, during the plateau at 1000 °C for 1 h, and on cooling to RT for the compositions with 53 wt% of cordierite and K/Al = 0.75, 1, 1.5 and 2.3.

b. Effect of cordierite load

Figure 12 shows the dilatometry curves recorded during heating up to 1000 °C, during the plateau of 1 h at 1000 °C, and on cooling to RT for the compositions with a ratio K/Al = 1.0 and cordierite loads of 0, 53 and 60 wt%. It is observed that the cordierite load seriously limits the shrinkage by comparison to the geopolymer alone. Moreover, the contraction observed around 500 °C for the geopolymer binder is not present for the composites.

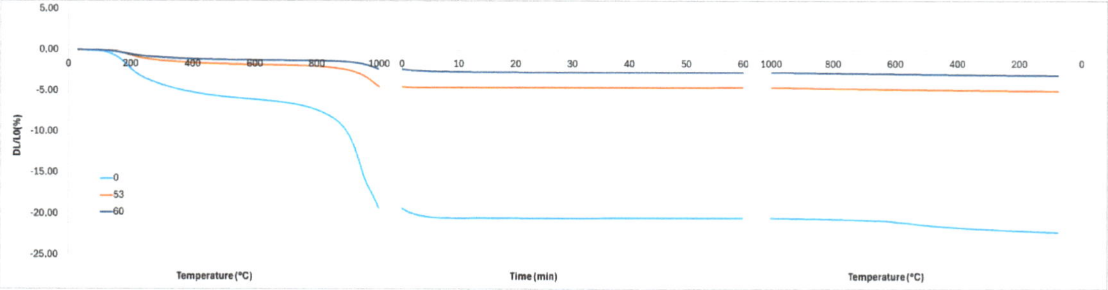

Figure 12. Dilatometry curves recorded during heating up to 1000 °C, during the plateau at 1000 °C for 1 h, and on cooling to RT for the compositions with a ratio K/Al = 1 and cordierite loads of 0, 53 and 60 wt%.

When the ratio K/Al is 2.3, the increase in the cordierite load has no significant effect during the heating stage, but lowers the expansion observed during the soaking stage (Figure 13). On cooling, a rapid contraction around 400 °C is observed for both cordierite loads.

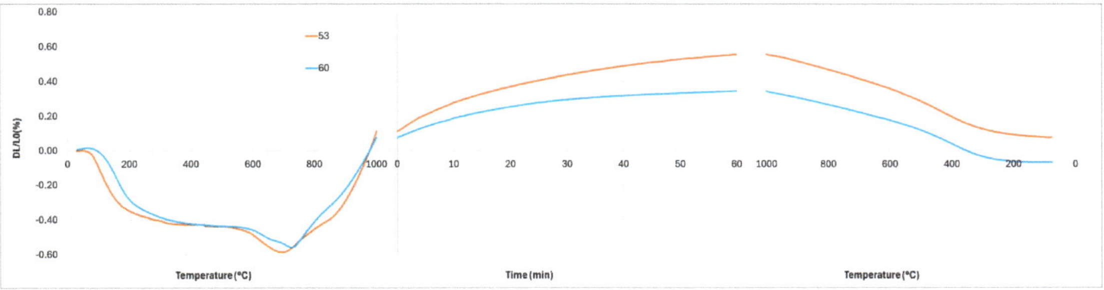

Figure 13. Dilatometry curves recorded during heating up to 1000 °C, during the plateau at 1000 °C for 1 h, and on cooling to RT for the compositions with a ratio K/Al = 2.3 and cordierite loads of 53 and 60 wt%.

The coefficients of thermal expansion (CTEs) calculated from the slope of the curves recorded on cooling are given in Table 10. One notes the very high value (27.8×10^{-6} K^{-1}) obtained for the pure geopolymer with K/Al = 1. The addition of the cordierite filler drastically lowers the CTE. The lowest value is obtained for the lowest K/Al ratio, whereas the maximum value is for K/Al = 1.5.

Table 10. Average coefficient of thermal expansion (CTE) calculated between 1000 °C and RT during cooling after thermal treatment at 1000 °C for 1 h.

CTE (10^{-6} K^{-1})		wt% Cordierite		
		0	53	60
K/Al	0.75	-	4.0	-
	1	27.8	4.6	4.1
	1.5	-	6.7	-
	2.3	-	5.9	5.2

3.4. Evolution of the Young's Modulus with Temperature

The evolution of the Young's modulus E with temperature was characterized using an impulse excitation technique for compositions of different ratio K/Al and a cordierite load of 53 wt%. Figure 14 shows the relative evolution of the Young's modulus E with respect to its initial value E_0 measured at room temperature. During the heating, we observe first a slow increase in E for all compositions followed by a more rapid increase when the temperature reaches 900 °C for compositions with K/Al = 0.75 and 1. Moreover, for those two compositions, the value of E continuously rises until reaching a plateau during the soaking stage at 1000 °C and remains almost stable during cooling. For the composition with K/Al = 1.5, the value of E remains low during the plateau and rapidly decreases to its initial value on cooling.

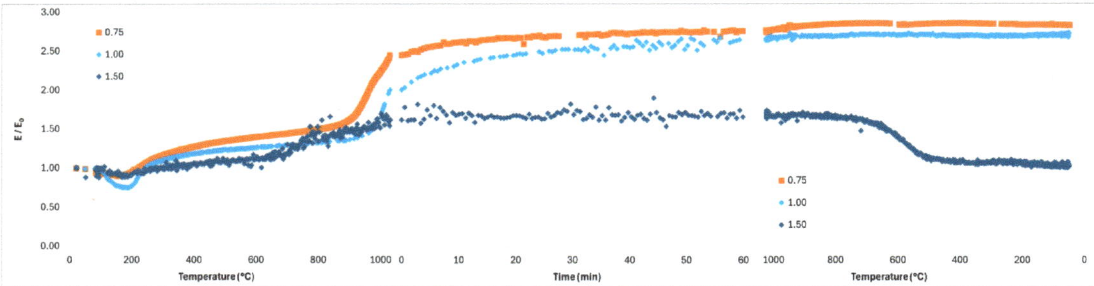

Figure 14. Evolution of the Young's modulus E during heating up to 1000 °C, during the plateau at 1000 °C for 1 h, and during the cooling to RT for the compositions with 53 wt% of cordierite and K/Al = 0.75, 1 and 1.5.

For the binder K/Al =1, Figure 15 shows that the cordierite load has almost no influence on the evolution of the Young's modulus during heating until the temperature reaches 1000 °C. However, during the plateau at 1000 °C, the increase in the modulus is larger for the lowest load.

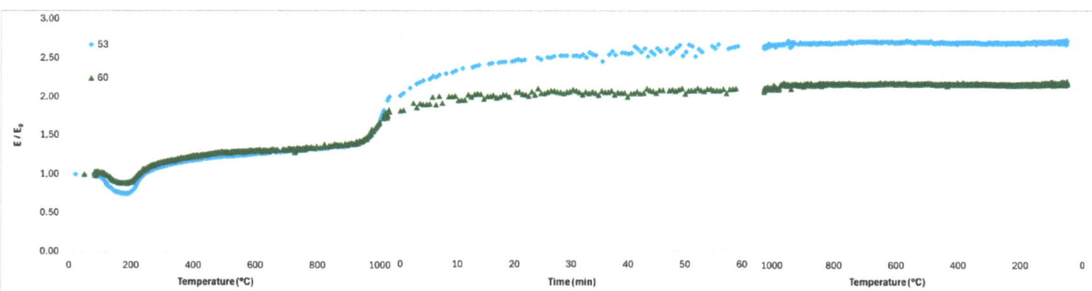

Figure 15. Effect of the cordierite load on the evolution of the Young's modulus E during heating up to 1000 °C, during the plateau at 1000 °C for 1 h, and during cooling at RT for compositions with K/Al = 1 and a load of cordierite of 53 and 60 wt%.

4. Discussion

The HT-XRD analysis shows that for the geopolymer with K/Al = 1 (Figure 8), the crystallization of the leucite and kalsilite occurs during heating between 900 and 1000 °C. During the plateau at 1000 °C for 1 h, the amount of kalsilite decreases and that of leucite increases, thereby indicating the possible reaction (1):

$$KAlSiO_4 + SiO_2 \rightarrow KAlSi_2O_6 \quad (1)$$

The XRD pattern of the geopolymer heat-treated at 1000 °C for 5 h (Figure 6) shows leucite as the single crystalline phase. For the composites synthesized with this geopolymer and 60 wt% cordierite, neither leucite nor kalsilite is detected during the HT-XRD analysis. When comparing the HT-XRD analysis to those performed at RT after heat treatment of 5 h (Figures 5 and 6), which show the presence of weak peaks of leucite, we can conclude that the addition of cordierite to the geopolymer delays the crystallization of leucite. When K/Al = 2.3, the HT-XRD analysis evidences the crystallization of kalsilite between 800 °C and 900 °C, and leucite between 900 °C and 1000 °C during the heating stage (Figure 9). As for the geopolymer with K/Al = 1, a transformation of kalsilite to leucite occurs during the plateau of 1 h at 1000 °C. However, due to the higher potassium content, the XRD pattern of this geopolymer heat-treated during 5 h (Figure 7) still shows the presence of kalsilite. For the composites synthesized with this geopolymer and 60 wt% cordierite, the HT-XRD analysis also shows the crystallization of leucite at 1000 °C but does not allow the presence of kalsilite to be inferred. However, this latter is present on the XRD pattern of the composite heat-treated for 5 h (Figure 7). When comparing the results of the XRD analysis, according to the K/Al ratio, it is concluded that the addition of cordierite has a stronger influence on the crystallization of the geopolymer when K/Al is low. More specifically, the crystallization does not take place during the heating stage up to 1000 °C when K/Al = 1, in contrast to what happens when K/Al = 2.3. On the other hand, the calculation by Rietveld refinement of the unit cell parameters on the cordierite highlights the expansion of the cell when the temperature goes over 700 °C for the composite with K/Al = 2.3 and 60 wt% cordierite (Figure 10). This can be explained by the insertion of potassium ions in the channels of the structure of the cordierite [31,32]. This mechanism also involves the substitution of Si atoms by Al atoms in the tetrahedral sites, and possibly ends by the formation of the compound $K_{0.5}Mg_2Al_{4.5}Si_{4.5}O_{18}$, which has a hexagonal symmetry (indialite-like):

$$2\,Mg_2Al_4Si_5O_{18} + \tfrac{1}{2}K_2O + \tfrac{1}{2}Al_2O_3 \rightarrow 2K_{0.5}Mg_2Al_{4.5}Si_{4.5}O_{18} + SiO_2 \quad (2)$$

The consequence is a change in the composition of the geopolymer that can modify its behavior at sintering and crystallization. This phenomenon is not observed when K/Al = 1.

The crystallization of the geopolymer and the interaction between the geopolymer and the cordierite can explain the differences observed on the dilatometry curves (Figures 11–13). Firstly, the first stage of shrinkage, from room temperature to 300 °C, is related to dehydration and dihydroxylation [25]. Secondly, the increase in the amount of K_2O lowers the temperature of the beginning of the second stage of the shrinkage that can be associated with sintering of the material. This can be explained by the amorphous nature of the geopolymer that probably induces a viscous flow sintering mechanism [34,35]. On the other hand, the shrinkage due to sintering stops around 700 °C when K/Al = 2.3. This temperature corresponds to the beginning of the expansion of the unit cell of the cordierite, resulting in the diffusion of potassium ions in the structure. This phenomenon, together with the crystallization of kalsilite, lowers the K_2O content in the geopolymer and thereby stops the viscous flow mechanism. Moreover, the transformation of kalsilite to leucite (Equation (1)) leads to a volume expansion due to the significantly lowest absolute density of leucite (2.47 g/cm^3 against 2.60 g/cm^3). The magnitude of this expansion depends on the density of the silica involved in the reaction, and consequently its state. This reaction may explain the dilatation observed on the dilatometric curves when the temperature is over 900 °C and during the plateau at 1000 °C, with dilatation enhanced by the geopolymer fraction (Figure 13). Finally, the presence of a significant amount of leucite after 1 h at 1000 °C in the composite when K/Al = 2.3 explains the non-linear contraction during the cooling by the allotropic transformation of cubic-leucite to tetragonal-leucite. It is noted that this transformation is also present for the composite with K/Al =1.5. By contrast, no diffusion of potassium ions in the cordierite nor crystallization of the geopolymer take place during the heating stage when K/Al = 1, and probably 0.75. So, a significant shrinkage due to sintering is observed between 800 °C and 1000 °C. Sintering still takes place during the plateau of 1 h at 1000 °C as the crystallization of leucite occurs later, as previously explained. To investigate the effect of this crystallization on the dimensions of the composite, a new dilatometry analysis, with a plateau of 10 h at 1000 °C, was performed. Figure 16 shows the relative variation in dimensions during the plateau. This new analysis confirms the shrinkage observed with the first analysis with a plateau of 1 h, but it also highlights a weak expansion after 1 h. If we attribute this expansion to the transformation of kalsilite to leucite, as explained previously, this leads to the inference of prior crystallization of a low amount of kalsilite.

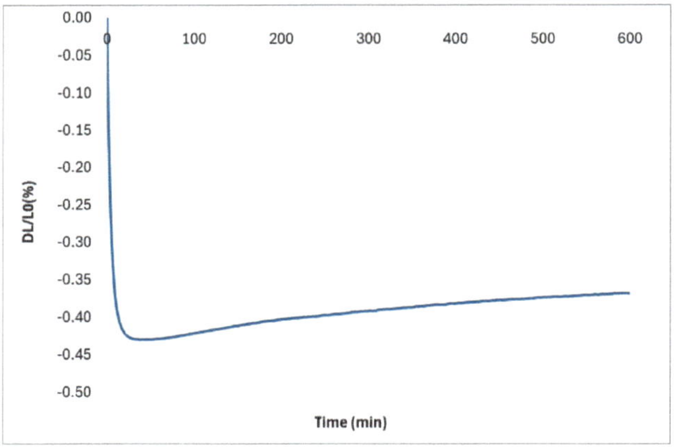

Figure 16. Relative variation in dimension of the composite with K/Al = 1 and 53 wt% cordierite during a plateau of 10 h at 1000 °C.

In terms of absolute density (Table 5), after curing, the density decreases when the K/Al ratio increases because of the larger amount of potassium silicate used, which has a lower density (1.50 g/cm^3) than the metakaolin (2.70 g/cm^3). For a given K/Al ratio, the absolute density of the composite increases with the amount of cordierite, as the absolute density of cordierite is higher (2.55 g/cm^3) than the geopolymer alone. After heat treatment, the absolute density of the composites increases due to the partial crystallization of the geopolymer in leucite (2.47 g/cm^3) and/or kalsilite (2.60 g/cm^3). Consequently, the variation in absolute density is larger when K/Al increases because of the lowest initial value, but also related to the presence of kalsilite in the heat-treated samples.

The differences in shrinkage, together with the differences in the evolution of the absolute density, explain why the porosity of the composite remains nearly constant when K/Al is low whereas it strongly increases when it is high (Table 5).

The fractions of leucite, kalsilite and cordierite in the composites play a key role in the coefficient of thermal expansion (CTE) (Table 10). A high fraction of cordierite and low fractions of leucite and kalsilite are favorable to a low value due to the very high CTEs of the two potassium aluminum silicates—20×10^{-6} K^{-1} for kalsilite, 22×10^{-6} K^{-1} for tetragonal leucite (above 600 °C) and 3×10^{-6} K^{-1} for cubic leucite (under 600 °C) [33,36]. Moreover, the CTEs given in the present work are the average values calculated between 1000 °C and RT, so they include the contraction due to the allotropic transformation of leucite. This explains the very high value measured for the pure geopolymer, but also the much lower CTEs of the composite with K/Al = 0.75 and 1 by comparison to those of the composites with K/Al = 1.5 and 2.3. The decrease in the CTE when K/Al increases from 1.5 to 2.3 is probably due to the crystallization of less leucite due to the insertion of a part of the potassium in the structure of the cordierite. This is also supported by the presence of kalsilite in the composite heat-treated during 5 h at 1000 °C, which evidences the non-complete transformation of kalsilite in leucite. One must also consider the effect of the insertion of the potassium ions and the aluminum/silicon substitution in the structure of the cordierite that possibly lowers the CTE when stabilizing the hexagonal symmetry [32]. However, the measurement of the unit cell parameters of orthorhombic cordierite derived from the HT-XRD patterns seems, by contrast, to highlight an increase in the variation with temperature (Figure 10). This point must be clarified in future work as the heating conditions could influence the relative importance of the diffusion and stabilization of the potassium ions in the cordierite structure, and their contribution to the crystallization of kalsilite and leucite.

Finally, the strong increase in the Young's modulus of the composite with K/Al = 0.75 and 1 when the temperature goes over 900–950 °C during the heating stage of the thermal treatment (Figures 14 and 15) evidences a consolidation mechanism correlated to the shrinkage due to sintering. The rapid decrease in the Young's modulus during the cooling stage when K/Al = 1.5 occurs at the same temperature as the contraction observed on the dilatometry curve (Figure 11). This can be explained by damage due to internal stresses generated by the rapid volume change of 1.2% induced by the allotropic transformation of leucite [33].

5. Conclusions

Geopolymer composites containing 53 to 65 wt% of a recycled cordierite powder as filler were synthesized.

- The K/Al ratio of the geopolymer plays a key role in the evolution of the dimensions and properties during thermal treatment at 1000 °C. A low K/Al ratio of 0.75 or 1 leads to shrinkage of around 5% due to sintering of the composite.
- A higher ratio induces a reaction between the geopolymer and the cordierite filler that inhibits sintering, and greater crystallization of kalsilite and leucite. This influences the evolution of the properties.
- A K/Al ratio of 0.75 or 1 is favorable to the stability of the porosity around 25 to 30%. Moreover, the lower amount of leucite is beneficial to a low coefficient of thermal expansion of 4 to 4.5 $\times$ 10^{-6} K^{-1}. The stronger consolidation during sintering leads to a significant increase in the Young's modulus and values of 40 to 45 GPa are obtained after heat treatment.

Author Contributions: Methodology, M.-F.G., C.L. and G.B.; investigation, F.C., A.M., S.A. and I.M.; writing—original draft preparation, F.C. and M.-F.G.; writing—review and editing, F.C., G.B. and M.-F.G.; supervision, M.-F.G.; project administration, M.-F.G.; funding acquisition, M.-F.G. All authors have read and agreed to the published version of the manuscript.

Funding: This research was partially funded by the Wallonia region and the E.U. within the FEDER-FTJ 21-27 project ECOMAT-UMONS.

Institutional Review Board Statement: Not applicable.

Informed Consent Statement: Not applicable.

Data Availability Statement: The original contributions presented in the study are included in the article, further inquiries can be directed to the corresponding authors.

Conflicts of Interest: The authors declare no conflicts of interest.

References

1. Wang, H.; Wang, S.; Meng, Z.; Chen, Z.; Liu, L.; Wang, X.; Qian, D.; Xing, Y. Mechanism of Cordierite Formation Obtained by High Temperature Sintering Technique. *Ceram. Int.* **2023**, *49*, 20544–20555. [CrossRef]
2. Camerucci, M.A.; Urretavizcaya, G.; Castro, M.S.; Cavalieri, A.L. Electrical Properties and Thermal Expansion of Cordierite and Cordierite-Mullite Materials. *J. Eur. Ceram. Soc.* **2001**, *21*, 2917–2923. [CrossRef]
3. Evans, D.L.; Fischer, G.R.; Geiger, J.E.; Martin, F.W. Thermal Expansions and Chemical Modifications of Cordierite. *J. Am. Ceram. Soc.* **1980**, *63*, 629–634. [CrossRef]
4. Rohan, P.; Neufuss, K.; Matějíček, J.; Dubský, J.; Prchlík, L.; Holzgartner, C. Thermal and Mechanical Properties of Cordierite, Mullite and Steatite Produced by Plasma Spraying. *Ceram. Int.* **2004**, *30*, 597–603. [CrossRef]
5. Hosseini, S.; Moghaddas, H.; Masoudi Soltani, S.; Kheawhom, S. Technological Applications of Honeycomb Monoliths in Environmental Processes: A Review. *Process. Saf. Environ. Prot.* **2020**, *133*, 286–300. [CrossRef]
6. Zhou, T.; Li, L.; Cheng, J.; Hao, Z. Preparation of Binary Washcoat Deposited on Cordierite Substrate for Catalytic Applications. *Ceram. Int.* **2010**, *36*, 529–534. [CrossRef]
7. Fuji, M.; Shiroki, Y.; Menchavez, R.L.; Takegami, H.; Takahashi, M.; Suzuki, H.; Izuhara, S.; Yokoyama, T. Fabrication of Cordierite Filter by In-Situ Solidification for High Temperature Dust Collection. *Powder Technol.* **2007**, *172*, 57–62. [CrossRef]
8. Nazarova, E.A.; Alimova, D.S.; Mikhaylov, V.I.; Krivoshapkina, E.F.; Krivoshapkin, P.V. Macroporous Ceramic Filters from Mineral Raw Materials for Machine Oils Filtration. *Ceram. Int.* **2019**, *45 Pt A*, 8767–8773. [CrossRef]
9. Robles-Lorite, L.; Dorado-Vicente, R.; Torres-Jiménez, E.; Bombek, G.; Lešnik, L. Recent Advances in the Development of Automotive Catalytic Converters: A Systematic Review. *Energies* **2023**, *16*, 6425. [CrossRef]
10. Williams, J.L. Monolith Structures, Materials, Properties and Uses. *Catal. Today* **2001**, *69*, 3–9. [CrossRef]

11. Trinh, H.B.; Lee, J.; Suh, Y.; Lee, J. A Review on the Recycling Processes of Spent Auto-Catalysts: Towards the Development of Sustainable Metallurgy. *Waste Manag.* **2020**, *114*, 148–165. [CrossRef] [PubMed]
12. Camerucci, M.A.; Urretavizcaya, G.; Cavalieri, A.L. Sintering of Cordierite Based Materials. *Ceram. Int.* **2003**, *29*, 159–168. [CrossRef]
13. Ogiwara, T.; Noda, Y.; Shoji, K.; Kimura, O. Low Temperature Sintering of α-Cordierite Ceramics with Low Thermal Expansion Using Li2O–Bi2O3 as a Sintering Additive. *J. Ceram. Soc. Jpn.* **2011**, *119*, 557–562. [CrossRef]
14. Duxson, P.; Fernández-Jiménez, A.; Provis, J.L.; Lukey, G.C.; Palomo, A.; van Deventer, J.S.J. Geopolymer Technology: The Current State of the Art. *J. Mater. Sci.* **2007**, *42*, 2917–2933. [CrossRef]
15. Provis, J.L.; Van Deventer, J.S.J. 1—Introduction to Geopolymers. In *Geopolymers*; Provis, J.L., van Deventer, J.S.J., Eds.; Woodhead Publishing: Cambridge, UK, 2009; pp. 1–11. [CrossRef]
16. Davidovits, J. Geopolymers: Ceramic-like Inorganic Polymers. *J. Ceram. Sci. Technol.* **2017**, *8*, 335–350. [CrossRef]
17. Yuan, J.; Li, L.; He, P.; Chen, Z.; Lao, C.; Jia, D.; Zhou, Y. Effects of Kinds of Alkali-Activated Ions on Geopolymerization Process of Geopolymer Cement Pastes. *Constr. Build. Mater.* **2021**, *293*, 123536. [CrossRef]
18. Farooq, F.; Jin, X.; Faisal Javed, M.; Akbar, A.; Izhar Shah, M.; Aslam, F.; Alyousef, R. Geopolymer Concrete as Sustainable Material: A State of the Art Review. *Constr. Build. Mater.* **2021**, *306*, 124762. [CrossRef]
19. Djangang, C.N.; Tealdi, C.; Cattaneo, A.S.; Mustarelli, P.; Kamseu, E.; Leonelli, C. Cold-Setting Refractory Composites from Cordierite and Mullite–Cordierite Design with Geopolymer Paste as Binder: Thermal Behavior and Phase Evolution. *Mater. Chem. Phys.* **2015**, *154*, 66–77. [CrossRef]
20. Mohd Mortar, N.A.; Abdullah, M.M.; Abdul Razak, R.; Abd Rahim, S.Z.; Aziz, I.H.; Nabiałek, M.; Jaya, R.P.; Semenescu, A.; Mohamed, R.; Ghazali, M.F. Geopolymer Ceramic Application: A Review on Mix Design, Properties and Reinforcement Enhancement. *Materials* **2022**, *15*, 7567. [CrossRef]
21. Bernal, S.A.; Rodríguez, E.D.; Mejía de Gutiérrez, R.; Gordillo, M.; Provis, J.L. Mechanical and Thermal Characterisation of Geopolymers Based on Silicate-Activated Metakaolin/Slag Blends. *J. Mater. Sci.* **2011**, *46*, 5477–5486. [CrossRef]
22. Lahoti, M.; Wong, K.K.; Yang, E.-H.; Tan, K.H. Effects of Si/Al Molar Ratio on Strength Endurance and Volume Stability of Metakaolin Geopolymers Subject to Elevated Temperature. *Ceram. Int.* **2018**, *44*, 5726–5734. [CrossRef]
23. Kamseu, E.; Rizzuti, A.; Leonelli, C.; Perera, D. Enhanced Thermal Stability in K2O-Metakaolin-Based Geopolymer Concretes by Al2O3 and SiO2 Fillers Addition. *J. Mater. Sci.* **2010**, *45*, 1715–1724. [CrossRef]
24. Hemra, K.; Aungkavattana, P. Effect of Cordierite Addition on Compressive Strength and Thermal Stability of Metakaolin Based Geopolymer. *Adv. Powder Technol.* **2016**, *27*, 1021–1026. [CrossRef]
25. Bernal, S.; Deventer, J.; Provis, J. *What Happens to 5 Year Old Metakaolin Geopolymers? The Effect of Alkali Cation*; Springer: Dordrecht, The Netherlands, 2015; Volume 10. [CrossRef]
26. Kohout, J.; Koutník, P.; Hájková, P.; Kohoutová, E.; Soukup, A. Effect of K/Al Molar Ratio on the Thermo-Mechanical Properties of Metakaolinite-Based Geopolymer Composites. *Polymers* **2021**, *13*, 3754. [CrossRef]
27. Chen, W.; Garofalo, A.C.; Geng, H.; Liu, Y.; Wang, D.; Li, Q. Effect of High Temperature Heating on the Microstructure and Performance of Cesium-Based Geopolymer Reinforced by Cordierite. *Cem. Concr. Compos.* **2022**, *129*, 104474. [CrossRef]
28. Bai, C.; Li, B.; Ma, C.; Li, X.; Wang, X.; Wang, B.; Yang, K.; Colombo, P. Open-Cell Cordierite-Based Foams from Eco-Friendly Geopolymer Precursors via Replica Route. *Ceram. Int.* **2024**, *50 Pt A*, 15340–15347. [CrossRef]
29. Gasmi, A.; Pélegris, C.; Davidovits, R.; Guessasma, M.; Tortajada, H.; Jean, F. Advanced Refinement of Geopolymer Composites for Enhanced 3D Printing via In-Depth Rheological Insights. *Ceramics* **2024**, *7*, 1316–1339. [CrossRef]
30. Kovacev, N.; Doustdar, O.; Li, S.; Tsolakis, A.; Herreros, J.M.; Essa, K. The Synergy between Substrate Architecture of 3D-Printed Catalytic Converters and Hydrogen for Low-Temperature Aftertreatment Systems. *Chem. Eng. Sci.* **2023**, *269*, 118490. [CrossRef]
31. Mercurio, D.; Thomas, P.; Mercurio, J.P.; Frit, B.; Kim, Y.H.; Roult, G. Powder Neutron Diffraction Study of the Thermal Expansion of a K-Substituted Cordierite. *J. Mater. Sci.* **1989**, *24*, 3976–3983. [CrossRef]
32. Kim, Y.H.; Mercurio, D.; Mercurio, J.P.; Frit, B. Structural Study of a K-Substituted Synthetic Cordierite. *Mater. Res. Bull.* **1984**, *19*, 209–217. [CrossRef]
33. El-Meliegy, E.; van Noort, R. Leucite Glass-Ceramics. In *Glasses and Glass Ceramics for Medical Applications*; El-Meliegy, E., van Noort, R., Eds.; Springer: New York, NY, USA, 2012; pp. 167–192. [CrossRef]
34. Duxson, P.; Lukey, G.C.; van Deventer, J.S.J. Thermal Evolution of Metakaolin Geopolymers: Part 1—Physical Evolution. *J. Non-Cryst. Solids* **2006**, *352*, 5541–5555. [CrossRef]
35. Duxson, P.; Lukey, G.C.; van Deventer, J.S.J. The Thermal Evolution of Metakaolin Geopolymers: Part 2—Phase Stability and Structural Development. *J. Non-Cryst. Solids* **2007**, *353*, 2186–2200. [CrossRef]
36. Ota, T.; Takebayashi, T.; Takahashi, M.; Hikichi, Y.; Suzuki, H. High Thermal Expansion $KAlSiO_4$ Ceramic. *J. Mater. Sci.* **1996**, *31*, 1431–1433. [CrossRef]

ceramics

MDPI

Article

Mullite 3D Printed Humidity Sensors

Yurii Milovanov [1,2], Arianna Bertero [1], Bartolomeo Coppola [1], Paola Palmero [1] and Jean-Marc Tulliani [1,*]

1 INSTM R.U PoliTO-LINCE Laboratory, Department of Applied Science and Technology, Politecnico di Torino, Corso Duca degli Abruzzi 24, 10129 Turin, Italy; yurii.milovanov@polito.it or juri_milovanov @yahoo.com (Y.M.); arianna.bertero@polito.it (A.B.); bartolomeo.coppola@polito.it (B.C.); paola.palmero@polito.it (P.P.)

2 Institute of High Technologies, Taras Shevchenko National University of Kyiv, Volodymyrska Street 64, 01601 Kyiv, Ukraine

* Correspondence: jeanmarc.tulliani@polito.it; Tel.: +39-0110904700

Abstract: Mullite substrates with two different porosities were 3D printed, and tested as humidity sensors. To evaluate the effects of porosity on humidity sensitivity, the samples were sintered at 1400 °C (Sensor 1) and 1450 °C (Sensor 2). The sensors were tested in a range from 0% to 85% relative humidity (RH) at room temperature. When exposed to water vapor at room temperature, the impedance value dropped down from 155 MΩ under dry air to 480 kΩ under 85 RH% for Sensor 1 and from 115 MΩ under dry air to 410 kΩ for Sensor 2. In addition, response time and recovery time were below 2 min, whatever the firing temperature, when RH changed from 0% to 74%. Finally, tests carried out involving ammonia, methane, carbon dioxide and nitrogenous oxide, as well as ethanol and acetone, showed no interference.

Keywords: mullite; stereolithography; humidity sensor

1. Introduction

Over the last few decades, the use of humidity sensors to measure and control relative humidity (RH) has attracted significant interest in many industrial sectors, such as automation, food processing (e.g., microwave ovens), agriculture, health care (e.g., drug preparation, medical equipment, and air conditioners) and manufacturing (e.g., electronics and paper manufacturing) [1–4]. In addition, the need for humidity sensors able to operate in extreme conditions (high temperatures and corrosive atmospheres) is also constantly increasing. Thus, there is still an interest in studying humidity sensors today. These sensors can be classified as capacitive, resistive, and thermal conductive [5]. Their operating principle is based on the adsorption of water molecules, which changes the electrical properties, such as the capacitance or resistance, of the devices, or on the measurement of the difference between thermal conductivity in dry and in humid air [6–15]. Resistive sensors are low-cost and can be easily integrated in electronic circuits, making them ideal candidates for remote monitoring applications.

Among the resistive sensors, many sensing materials were investigated, such as metal oxides (Al_2O_3, Fe_2O_3, TiO_2, SiO_2, WO_3, ZnO, CuO and ZrO_2), perovskites ($ZrTiO_4$, $LaFeO_3$, $BaTiO_3$, $LiNbO_3$, $SmCrO_3$, etc.) and spinels ($ZnWoO_4$, $MnWO_4$, $NiWO_4$, $CoWO_4$, $MgCr_2O_4$, $ZnCr_2O_4$, $MgAl_2O_4$, Fe_3O_4, etc.) [16,17], as well as carbon-based materials (e.g., carbon nanotubes, graphene oxides and biochar) [18–24]. Clay minerals with layered structures have also been also investigated, due to their high specific surface area and their capacity for ion exchange and the hydration process [25–27]. More recently, because of new applications such as the Internet of Things (IoT) and human-body monitoring, new sensors based on different materials such as paper [28], sulfides [29], sulfates [30], metal–organic frameworks (MOFs) [31] and other materials [32] have also been proposed.

Gas and humidity resistive sensors are often manufactured by screen-printing or spin-coating, or by brushing the sensing layer onto inert rigid or soft substrates, such as ceramic or polymeric ones. However, in this case, only a limited volume of material is used for the detection of the target gas.

Additive manufacturing techniques allow the production of complex geometries. In this research, Digital Light Processing (DLP) was exploited for the direct preparation of active substrates able to support metallic electrodes and to act, at the same time, as sensing layers. Thus, preliminary printing tests were performed to produce planar substrates with thickness values that could allow them to survive the successive screen-printing process, while, at the same time, not being too thick. The obtained thickness value (1.07 mm, after firing) was a good compromise on this basis. In addition, typical tape-cast films are 100–300 μm thick [33], which would have required the pressing of several stacked green layers to obtain a substrate with the same thickness. In such a case, pore size and pore size distribution could be changed during this pressing step. Finally, uniaxially pressing pellets 1 mm thick was another alternative, even if handling them without any organic temporary binder could be challenging. For these reasons, stereolithography was chosen for the sensors' manufacturing, with the aim of producing more-complex porous geometries in the future.

Mullite ($3Al_2O_3{\cdot}2SiO_2$) is the only intermediate stable compound in the Al_2O_3-SiO_2 system, and it is known for its chemical inertness in harsh environments [34]. Thus, in this work, a commercial mullite powder was first characterized by X-ray diffraction (XRD) and laser granulometry. Then, a slurry based on a photocurable commercial resin with a dispersant, a sintering additive and the mullite powder was prepared, and planar substrates were printed by DLP. These substrates were partially sintered at two different temperatures in order to investigate the effects of residual porosity on humidity-sensing properties. Their electrical responses with respect to humidity and to interfering gases such as NH_3 (44 ppm in air), CH_4 (100 ppm in air), CO_2 (500 ppm in air) and NO_2 (2.5 ppm in air), as well as ethanol (68,800 ppm) and acetone (276,900 ppm), were studied for the first time.

2. Materials and Methods

2.1. Materials

A mullite commercial powder (SA 193 CR, from Baikowski SAS, Annecy, France) was used. Chemical characterization of the as-received powder was performed by XRD analysis using a Pananalytical X′Pert Pro instrument (Pananalytical, Eindhoven, The Netherlands) with CuKα radiation (0.154056 nm) in the 2θ range, 5–70°. Furthermore, the particle size distribution of the as-received powder was checked by laser granulometry (Mastersizer 3000, Malvern Panalytical, Malvern, Worcestershire, UK) in alcohol after 5 min sonication. A photocurable commercial resin (Admatec Europe BV, Alkmaar, The Netherlands) consisting of a liquid monomer resin system with a photo-initiator, diphenyl(2,4,6,-trimethylbenzoyl)phosphine oxide and acrylates was used to prepare the slurry. A dispersant (Disperbyk-103, BYK Chemie, Wesel, Germany) was added to enhance the stability and increase the solid loading of the suspension.

2.2. Slurry Preparation

The slurry was manufactured by mechanically mixing the resin and the dispersant with the ceramic powder in order to obtain a solid loading of 69 wt%. After preliminary tests, 5.0 wt% dispersant was added (with respect to the dry powder) to the monomer, under mechanical stirring with a helix for 15 min. Then, mullite powder was gradually introduced into the mixture under continuous stirring, accurately avoiding the formation of agglomerates. Magnesium carbonate was also mixed as a sintering aid (1 wt% of equivalent MgO with respect to mullite) [35]. The total mixing process lasted 45 min. The obtained slurry was then milled in agate jars with agate spheres (d = 10 mm) for 6 h at 350 rpm in a planetary miller (Fritsch Pulverisette 5, Fritsch GmbH, Oberstein, Germany). The slurry was finally degassed for 30 min under vacuum by means of a rotative pump in order to allow the entrapped gas to leave the mixture and to obtain a homogeneous slurry. A rotational rheometer (Kinexus Pro+, Malvern Panalytical, Malvern, Worcestershire,

UK) equipped with stainless steel parallel plates (20 mm diameter, 1 mm gap between plates) was used to assess the rheology of the slurry at 25 °C. Shear rates in the range of 0.1–1000 s^{-1} were applied.

2.3. Digital Light Processing (DLP) and Post-Processing

Planar-shaped specimens were designed using the AutoCAD 2018 software and printed using a DLP-based stereolithographic printer (ADMAFLEX 130, ADMATEC Europe BV, Alkmaar, The Netherlands) operating with a 405 nm wavelength UV light. Curing depths and curing degrees of the UV-curable suspensions were investigated through several tests. Curing parameters are of paramount importance for the integrity and quality of the printed samples. Thus, first, the best compromise between high resolution, good adhesion, and the uniformity of each layer was investigated. The most reliable parameters were the following: layer thickness, exposure time, and LED power; these were equal to 30 μm, 1 s and 13.93 mW/cm^3, respectively. During the printing process, a delay of 30 s before exposure was set to allow air bubbles to be expelled from the slurry. A 125 μm doctor blade was used. Slurries were exposed to the UV light in a chessboard configuration for different exposure times to determine the curing depth. The uncured slurry was then cleaned with paper and the thickness of the single layer was determined by means of a digital micrometer.

After printing, the samples were soaked in deionized water at approx. 40 °C for 6 h to enable researchers to easily remove with a brush the uncured slurry attached to the samples, promoting the removal of the contained water-soluble fraction. Indeed, the resin used (Blank resin C, water de-binding, ADMAFLEX, Alkmaar, The Netherlands) is specifically designed to avoid thermal de-binding defects during resin decomposition by creating a porous network (via water de-binding) through which decomposition gases can be removed without stressing the green samples. Then, the substrates were dried in an oven at 70 °C for 6 h. A thermal de-binding and sintering cycle (Carbolite 1800 electric furnace, Carbolite Gero GmbH, Neuhausen, Germany) (Figure 1) was set up. Specifically, the samples were slowly heated to 1000 °C to prevent cracking during resin decomposition and then were sintered up to 1400 °C (Sensor 1) and 1450 °C (Sensor 2) for 1 h, at a heating rate of 1 °C/min. These sintering temperatures were chosen based on preliminary tests: a too-high temperature guarantees higher mechanical properties but a lower porosity and sensor response. However, a minimum mechanical strength is needed because the substrates must withstand the screen-printing process for the metallic electrodes. The sintered density was determined according to the theoretical density (TD) of mullite (3.17 g/cm^3) and the microstructures of the films were observed by means of a field emission-scanning electron microscope (FE-SEM, Hitachi S3800, Tokyo, Japan).

2.4. Fabrication and Measurement of Gas Sensors

The platinum electrodes were screen-printed (Ferro 5545, King of Prussia, PA, USA; 2 successive prints) using an automatic machine with a 270-mesh steel screen. After drying overnight, the Pt ink was fired at 980 °C for 15 min to obtain a high-grade adhesion and optimize its electrical conductivity, as per the ink producer's indications. The electrodes had widths of 400 μm and were spaced 450 μm from each other.

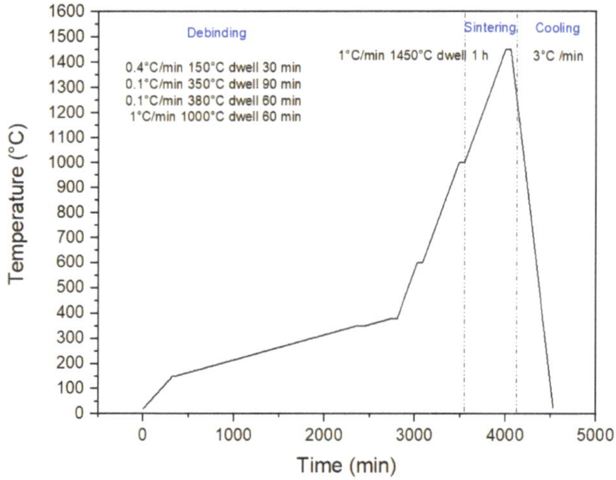

Figure 1. Schematic diagram of thermal de-binding and sintering cycles.

The sensor's humidity response in the 0–100% relative humidity (RH) range was studied by means of a laboratory system at room temperature, under an airflow rate of 150 mL/min. The RH was increased progressively by steps, each one 15 min. In this system, the airflow (Siad, Turin, Italy, research grade) was split into two flows and controlled by means of mass flows (MF302, Teledyne Hastings, Hampton, VA, USA; mass flow controller: Teledyne 4000): the first one was kept anhydrous, whereas the second one passed through a water bubbler, generating a saturated humid flow. Then, both flows were combined. A commercial probe for humidity and temperature measurements (Delta Ohm DO9406, Caselle di Selvazzano, PD, Italy; accuracy: ±0.1% in the 0–100% RH range from 50 to 250 °C) was used as reference for temperature, and RH values were determined inside the measurement chamber. During tests under a dynamic flow, the sensors' impedance and phase were measured by an LCR meter (Hioki 3533-01, Ueda, Nagano, Japan). The alternating voltage was 2 V at a frequency of 1 kHz. Additionally, tests were also carried out for different gases, like NH_3 (44 ppm in air), CH_4 (100 ppm in air), CO_2 (500 ppm in air) and NO_2 (2.5 ppm in air), all under the same flow rate (all gases were provided by Siad, Turin, Italy and were research grade). Then, a bubbler with pure ethanol and acetone (Sigma Aldrich, Milan, Italy, reagent grades) was added and the response of the sensor was investigated at 23 °C. The gas concentrations (in %) were calculated according to the respective vapor pressures (52.3 mmHg for ethanol and 210.4 mmHg for acetone [36]) and Equation (1) [37]:

$$\text{gas concentration (\%)} = 100 \times \text{vapor pressure of the liquid in mmHg}/760 \quad (1)$$

The response of the sensor (R) was calculated according to Equation (2)

$$R = Z_0/Z_g \quad (2)$$

where Z_0 and Z_g are the impedance values of the sensor under dry air and under humid air, respectively.

The sensitivity (S) of the sensor is the slope of the calibration curve, and it can be determined in accordance with Equation (3):

$$S(Z) = \Delta Z/\Delta RH \quad (3)$$

The response time (the time required by a sensor to achieve 90% of the total impedance change in the presence of humid air) as well as the recovery time (the time needed for a sensor to reach 90% of the total impedance variation during gas desorption) were calculated as well.

3. Results and Discussion

3.1. Powder Characterization

The as-received powder showed a multimodal distribution with a $\varnothing_{50}$ value close to ~29 μm (Table 1). After 5 min of sonication, the $\varnothing_{50}$ was equal to ~1.8 μm. Thus, sonication reduced the size of the coarser fractions. It can be concluded that the as-received powder presented soft agglomeration.

Table 1. Particle size (μm) corresponding to 10% (∅10), 50% (∅50) and 90% (∅90) of the cumulative distribution of the as-received and sonicated (5 min) mullite powders from laser granulometry.

	$\varnothing_{10}$ (μm)	$\varnothing_{50}$ (μm)	$\varnothing_{90}$ (μm)
As received	0.90	28.70	106.00
5 min sonication	0.58	1.76	4.36

The XRD pattern of the as-received mullite is reported in Figure 2; the pattern corresponds mainly to mullite (JCPDS card n○ 15-0776). Cristobalite (SiO_2, JCPD file 96-900-8230) and aluminum oxide (Al_2O_3, JCPFD files 96-900-7499 and 00-035-0121, for the hexagonal and the monoclinic phase, respectively) were also detected.

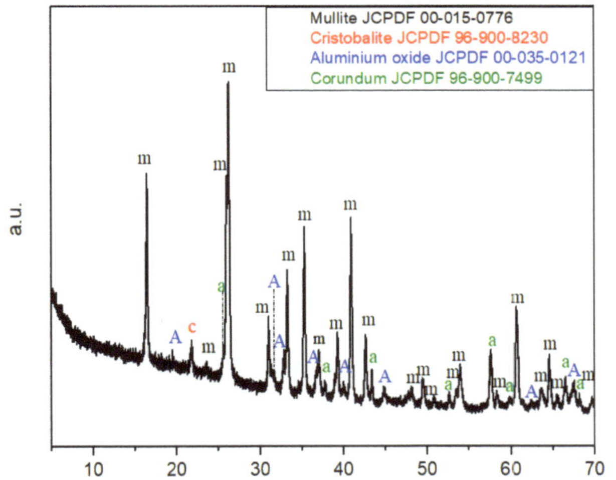

Figure 2. XRD patterns of the as-received mullite (m: mullite, c: cristobalite, A: aluminum oxide, a: corundum).

3.2. Sensor-Based Microstructural Characterization

The different sintering temperatures (1400 or 1450 °C, for 1h) produced a slight difference in substrate porosity: in fact, Sensor 1 samples were characterized by a porosity of 19% (determined according to the Archimedes method), and Sensor 2, of 17%. Before sintering, the average dimensions of the substrates were: 23.12 × 15.48 × 1.19 mm^3. After sintering, the average dimensions of the plates were: 21.56 × 14.4 × 1.07 mm^3 for Sensor 1 and 20.15 × 13.46 × 1.07 mm^3 for Sensor 2. Volume shrinkages of 20% for Sensor 1 and 32% for Sensor 2 were then calculated after sintering. The substrates were to be handled with care because of the rather low sintering temperature (with the same mullite powder and 1 wt% addition of MgO, full density can be reached after sintering at 1550 °C for 3 h [35], while the sensors were heat-treated at 1400 °C and 1450 °C for 1 h only). However, they were able to survive the screen-printing step, in which the steel mesh comes into contact with the sample and the squeegee forces the ink to pass through the former. When flat surfaces were obtained, no breaking of samples was observed. An image of the printed sensor with screen-printed electrodes is reported in Figure 3.

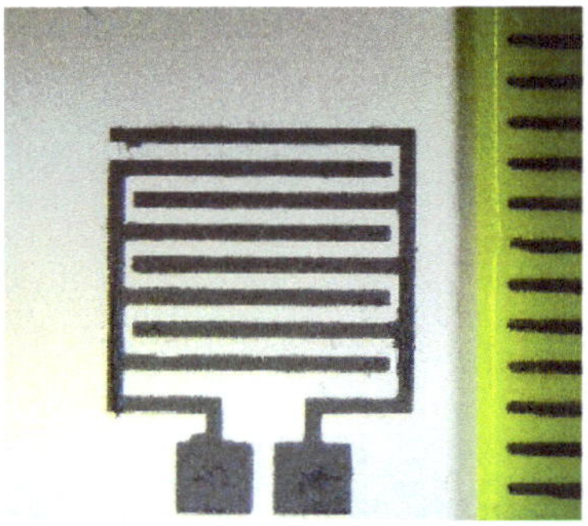

Figure 3. Optical micrograph of the 3D printed mullite sensor with screen-printed interdigitated electrodes, sintered at 1400 °C for 60 min.

The FE-SEM micrographs of the 3D printed mullite sintered at 1400 °C and 1450 °C are presented in Figure 4. The microstructures appear porous, and the lower annealing temperature (1400 °C) led to the creation of a spongier structure (Figure 4a,c), in comparison with the one obtained at a higher temperature (1450 °C) (Figure 4b,d), which was as expected, considering the respective densities of the sensors. Sensor 1 looks like a consolidated powder at the very beginning of the sintering process (Figure 4c), while on sensor 2 a coarsening of the grain powders is already visible (Figure 4d).

(**a**)

(**b**)

Figure 4. *Cont.*

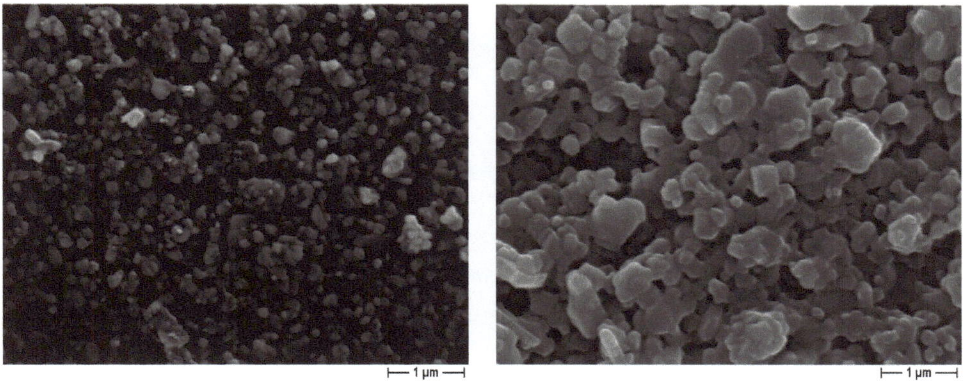

(**c**) (**d**)

Figure 4. FE-SEM micrographs of the 3D printed mullite sintered at 1400 °C (**a**,**c**) and 1450 °C (**b**,**d**) for 1 h.

3.3. Humidity-Sensing Properties

The sensor response, in terms of impedance and the phase variation of Sensors 1 and 2 are plotted in Figures 5 and 6. When in contact with humidity, the impedance value dropped from 155 MΩ under dry air to 480 kΩ under 85 RH% for Sensor 1, and from 115 MΩ under dry air to 410 kΩ for Sensor 2 (Figures 5a and 6a). The higher porosity of Sensor 1 can explain the higher initial (under dry air) impedance value.

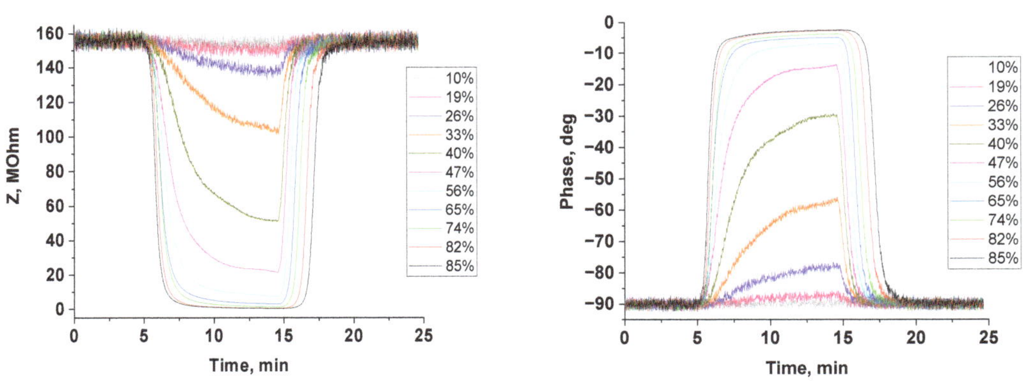

(**a**) (**b**)

Figure 5. Response as a function of relative humidity value for mullite Sensor 1, sintered at 1400 °C: (**a**) impedance variation and (**b**) phase variation.

Initially, at a low humidity level, the impedance value slightly changes, while above 19 RH%, the impedance of the sensor sharply drops down with the increase in humidity level (Figures 5a and 6a). At the same time, the phase increased continuously, from approximately −90° under dry air to approximately −3° under 85 RH% (Figures 5b and 6b), as already observed with humidity sensors based on rice husk ash [38].

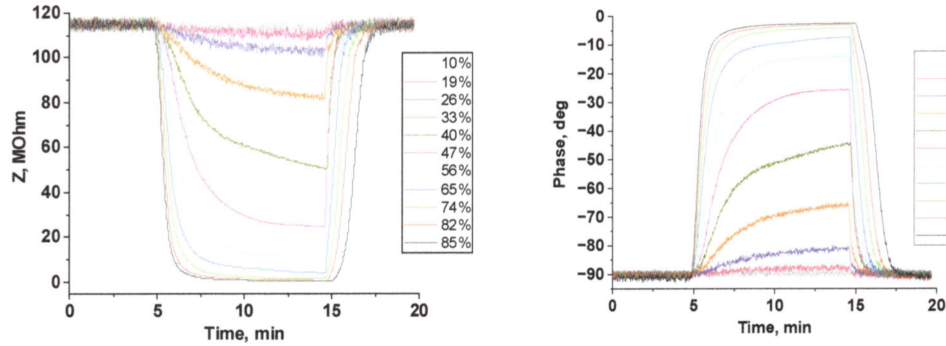

(a) (b)

Figure 6. Response as a function of relative humidity values for mullite Sensor 2, sintered at 1450 °C: (**a**) impedance variation and (**b**) phase variation.

Comparing the sensors' performances under 85% humidity (see Figure 7), Sensor 1 shows 13% higher sensor response as compared to Sensor 2 (322.9 and 280.5, respectively). The sensor also had negligible hysteresis within the measurement error.

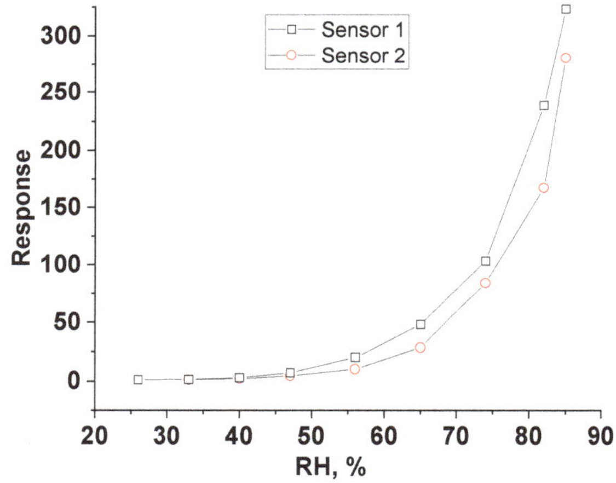

Figure 7. Responses of Sensors 1 and 2 at different RH values (in the range 26–85% RH).

Table 2 summarizes the sensors' response and response/recovery times at different humidity levels associated with mullite Sensors 1 and 2. The response and recovery times of Sensor 1 (91 and 167 s, respectively) are longer than those of Sensor 2 (64 and 119 s, respectively), probably because of a higher surface area, due to the lower sintering temperature.

Table 2. The sensors' response and response/recovery times at different humidity levels associated with mullite Sensors 1 and 2.

Humidity	Sensor Response ($R = Z_o/Z_g$)		Response Time, s		Recovery Time, s	
	Sensor 1	Sensor 2	Sensor 1	Sensor 2	Sensor 1	Sensor 2
19%	1.0	1.0				
26%	1.1	1.1				
33%	1.5	1.4	406	372	67	35
40%	3.0	2.3	350	382	66	36
47%	7.2	4.7	255	238	71	46
56%	20.2	10.4	174	174	87	56
65%	48.1	28.5	130	122	113	71
74%	103.3	83.9	117	97	122	81
82%	238.5	166.9	93	70	149	104
85%	322.9	280.5	91	64	167	119

In Figure 8, three consecutive measurements on mullite Sensors 1 and 2 under 85% RH are displayed. Both sensors show an excellent repeatability.

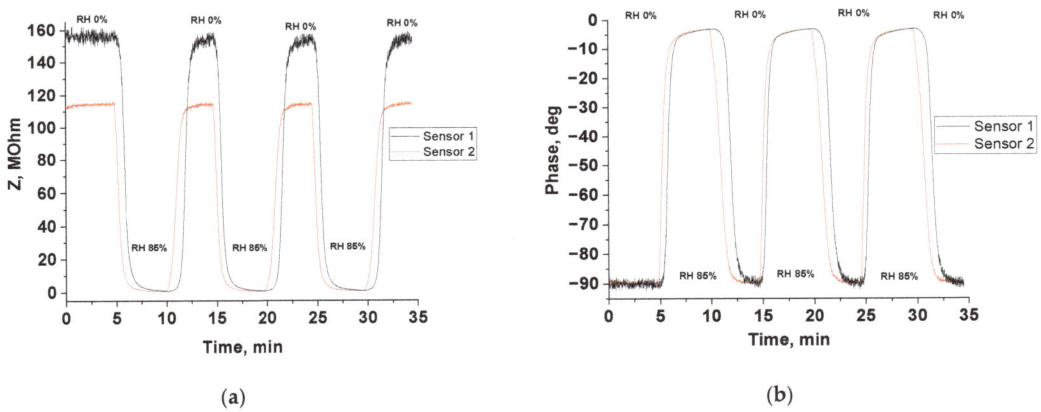

Figure 8. Repeatability measurements of mullite Sensors 1 and 2 under 85% RH: (**a**) impedance variation and (**b**) phase variation.

The calibration curves of the mullite sensors are illustrated in Figure 9. The slopes (sensor sensitivity) are 0.0434 RH^{-1} and 0.0446 RH^{-1} for Sensor 1 and Sensor 2, respectively, when log(R) is plotted as a function of the relative humidity value.

The humidity-sensing mechanism of the sintered mullite sensor is based on the protonic conduction on the surface. When exposed to the humidity of the environment, the impedance changes in the mullite sensors depend on the number of adsorbed water molecules. The RH level directly dictates the number of adsorbed water molecules on the mullite's surface. As the RH level increases, more water molecules are adsorbed, and this leads to a further decrease in the resistance value.

In Ref. [39], nanostructured ZnO, ZnO-TiO_2 humidity sensors were investigated. At the first stage of adsorption process, the negatively charged oxygen species were electrostatically attached to the positively charged metallic ions of the sensing material and formed a hydroxide layer. Thereby, the grain surfaces adjacent to the pores are covered by a chemisorbed monolayer. With the increase in relative humidity, the first physisorbed layer forms when single water molecules bind to two surface hydroxyls. A hydronium group H_3O^+ is then formed through those molecules in the second layer that are singly bound to the underlying layer by hydrogen bonds.

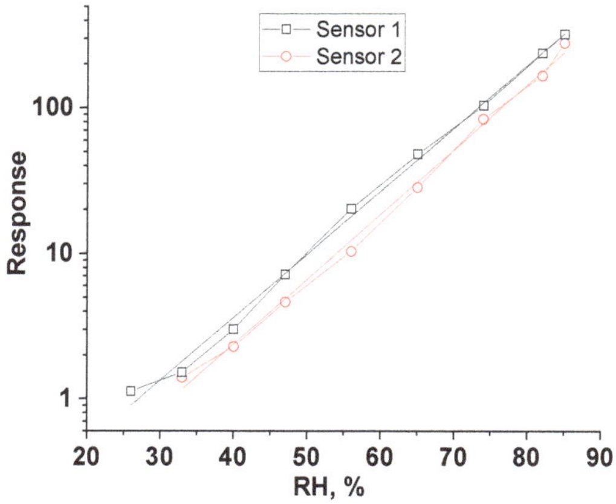

Figure 9. Calibration curves of mullite Sensor 1 and mullite Sensor 2 towards different RH values at room temperature.

A proton is thereby released to neighboring water molecules, which accept it while releasing another proton, and so on (this is the so-called Grotthuss chain-reaction mechanism) [17]. This proton is free to move along the water molecules and thus governs the sensor conductivity. At higher humidity levels, liquid water can also condense in pores, and electrolytic conduction can occur at the same time [16].

At low humidity levels, during the progressive formation of the first physisorbed layer of water molecules, the equivalent electrical circuit of oxidic humidity sensors is a parallel RC circuit [40,41]. The higher the degree of physical adsorption, the emerging ionic conduction and successive capillary water condensation, previously described, lead to a stronger decrease of the impedance of the film, as observed and supported by the phase change from almost -90° to about -3°. For these higher RH values, the equivalent circuit can be drawn by a number of (typically two) series-connected parallel RC circuits [40,41]. Then, a sweep in frequency (from 22 kHz to 100 kHz) was performed under 65% RH (Figure S7). The resistance and the capacitance of the two elements in parallel in the equivalent circuit were calculated and were equal to 60 MOhm and 4.94×10^{-14} F, respectively, at 53.7 kHz (at the apex of the Nyquist plot of the impedance diagram).

Tests were also performed at room temperature for ammonia (44 ppm), methane (100 ppm) and carbon dioxide (500 ppm), and nitrogenous oxide (2.5 ppm), as well as ethanol (68,800 ppm) and acetone (276,900 ppm) (Figures S1–S6, Supplementary Information). No interference towards the tested gases was observed, so the mullite sensors also exhibited a good selectivity with respect to humidity at room temperature.

Table 3 presents the results illustrated in this paper, and the recent literature data (of the last 5 years) on resistive sensors based on ceramic oxides; the obtained performances are rather promising. The response and recovery times can probably be further improved by printing more complex geometries to fully exploit the potentialities of DLP, as in Ref. [42].

Table 3. Comparison of mullite 3D printed sensor's performance with those in described in recent literature data on resistive sensors based on ceramic oxides.

Material	Sensor Response, $R = Z_0/Z_g$	Response Time, s	Recovery Time, s	Reference
Pt decorated MoS_2 nanoflakes	~4000 at 85% RH	92	154	[29]
ZnO/MoS_2	~301 at 85% RH	138	166	[43]
Porous aluminum-doped ZnO	733% at 90% RH	~238	~202	[44]
Copper ferrite-yttrium oxide nanocomposite	4895 at 97% RH	9	23	[45]
Titanium dioxide nanotubes	58.5 at 90% RH	NA	NA	[46]
$Cs_3Bi_2Br_9$ perovskite	987 at 90% RH	5.56	6.24	[47]
$SrTiO_3$ nanoparticles	1.12 at 85% RH	100	300	[48]
$GdAlO_3$	8000 at 97% RH	45	60	[49]
$Mn_{0.5}Zn_{0.5}Dy_xHo_yFe_{2-x}O_4$ (x = 0.005 to 0.03) nanoparticles	99% at 97% RH	90	18	[50]
Al–Sr and Al–Cd nano-materials	2.87 at 95% RH 3.19 at 95% RH	60 44	29 45	[51]
Reduced graphene oxide/zinc oxide nanostructured powder	172 at 90% RH	NA	NA	[52]
$Zn_{x-1}Al_2O_4(TiO_2)_x$	265 at 97% RH	195	28	[53]
Dy_2O_3 nanorods	15 at 97% RH	2	5	[54]
Ta-doped TiO_2/reduced graphene oxide	232% at 90% RH	4.2	3.3	[55]
Sr-doped $LaFeO_3$ nanofibers	60,597 at 90% RH	NA	NA	[56]
N-doped graphene oxide-WO_3	3427 at 98% RH	24	53	[57]
Nanosized α-Fe_2O_3 nanoparticles	48,569 at 95% RH	9	4	[58]
$(CaSO_4{\cdot}2H_2O)_{0.975}$-$(CuSO_4{\cdot}5H_2O)_{0.025}$	6.75 at 90% RH	5	3	[30]
SnO_2 thin film	3.1 at 95% RH	84	576	[59]
Porous SnO_2/MCM-48	10^5 at 98% RH	9	12	[60]
2D MoO_3	4024 at 75% RH	8	40	[61]
Perovskite $CsPbBr_3$-Fe quantum dots	1.1 at 70% RH	38	38	[62]
Mullite	322.9 at 85% RH	91	167	This work

4. Conclusions

In this work, 3D printed mullite substrates with two different porosities were fabricated to evaluate the effects of porosity and surface area on humidity sensitivity: one was sintered at 1400 °C, and the other one at 1450 °C. As a result of this difference, the resultant porosity of the samples affected the sensitivity of the sensors; this effect can be controlled by varying the heat treatment's temperature.

At room temperature, the impedance value dropped downward, from 155 MΩ under dry air to 480 kΩ under 85 RH% for Sensor 1, and from 115 MΩ under dry air to 410 kΩ for Sensor 2. At low humidity levels, the impedance change is small, while above 19 RH%, the impedance of the sensor decreases sharply with the increase in humidity content. When the RH level changed from 0% to 85%, the response times were equal to 91 s for Sensor 1 and 64 s for Sensor 2, whereas the recovery times were equal to 167 s for Sensor 1 and 119 s for Sensor 2. In addition, tests carried out for ammonia (44 ppm), methane (100 ppm), carbon dioxide (500 ppm), and nitrogenous oxide (2.5 ppm), as well as ethanol (68,800 ppm) and acetone (276,900 ppm) showed no interference.

This work shows that resistive ceramic humidity sensors can be easily produced by 3D printing techniques and that their use in harsh environments (in corrosive atmospheres, for example) can be envisaged.

Supplementary Materials: The following supporting information can be downloaded at: https://www.mdpi.com/article/10.3390/ceramics7020053/s1, Figure S1: Response of the Sensor S2 to 500 ppm CO_2 in air, Figure S2: Response of the Sensor S2 to 100 ppm CH_4 in air, Figure S3: Response of the Sensor S2 to 2.5 ppm NO_2 in air, Figure S4: Response of the Sensor S2 to 44 ppm NH_3 in air, Figure S5: Response of the Sensor S2 to 276,900 ppm (27.69%) acetone in air, Figure S6: Response of the Sensor S2 to 68,800 ppm (6.88%) ethanol in air, Figure S7: Impedance diagram for the sensor S2 under 65% RH.

Author Contributions: Conceptualization, J.-M.T. and P.P.; methodology, Y.M., B.C. and A.B.; software, Y.M. and A.B.; validation, J.-M.T., P.P. and B.C.; formal analysis, J.-M.T.; investigation, Y.M. and A.B.; data curation, Y.M. and A.B.; writing—original draft preparation, Y.M.; writing—review and editing, all authors; supervision, J.-M.T. All authors have read and agreed to the published version of the manuscript.

Funding: This research received no external funding.

Institutional Review Board Statement: Not applicable.

Informed Consent Statement: Not applicable.

Data Availability Statement: The data presented in this study are available on request from the corresponding author.

Acknowledgments: The Authors greatly acknowledge the Interdepartmental laboratory SISCON (Safety of Infrastructures and Constructions) from Politecnico di Torino, for the use of the rheometer. The Authors also acknowledge Politecnico di Torino, for the grant to carry out research activity "Development of new materials and devices for the detection of gaseous pollutants".

Conflicts of Interest: The authors declare no conflicts of interest.

References

1. Barmpakos, D.; Kaltsas, G.A. Review on Humidity, Temperature and Strain Printed Sensors—Current Trends and Future Perspectives. *Sensors* **2021**, *21*, 739. [CrossRef] [PubMed]
2. Esteban-Cubillo, A.; Tulliani, J.-M.; Pecharromán, C.; Moya, J.S. Iron-oxide nanoparticles supported on sepiolite as a novel humidity sensor. *J. Eur. Ceram. Soc.* **2007**, *27*, 1983–1989. [CrossRef]
3. Céline Laville, C.P.; Deletage, J.-Y. Humidity sensors for a pulmonary function diagnostic microsystem. *Sens. Actuators B Chem.* **2001**, *76*, 304–309. [CrossRef]
4. Tulliani, J.-M.; Baroni, C.; Zavattaro, L.; Grignani, C. Strontium-Doped Hematite as a Possible Humidity Sensing Material for Soil Water Content Determination. *Sensors* **2013**, *13*, 12070–12092. [CrossRef] [PubMed]
5. Duraia, E.M.; Beall, G.W. Humidity sensing properties of reduced humic acid. *Sens. Actuators B* **2015**, *220*, 22–26. [CrossRef]
6. Islam, T.; Nimal, A.T.; Mittal, U.; Sharma, M.U. A micro interdigitated thin film metal oxide capacitive sensor for measuring moisture in the range of 175–625 ppm. *Sens. Actuators B* **2015**, *221*, 357–364. [CrossRef]
7. Nanto, H.; Minami, T.; Takata, S. Zinc-oxide thin-film ammonia gas sensors with high sensitivity and excellent selectivity. *J. Appl. Phys.* **1986**, *60*, 482–484. [CrossRef]
8. Kim, Y.; Jung, B.; Lee, H.; Kim, H.; Lee, K.; Park, H. Capacitive humidity sensor design based on anodic aluminum oxide. *Sens. Actuators B Chem.* **2009**, *141*, 441–446. [CrossRef]
9. Balde, M.; Vena, A.; Sorli, B. Fabrication of porous anodic aluminium oxide layers on paper for humidity sensors. *Sens. Actuators B* **2015**, *220*, 829–839. [CrossRef]
10. Feng, X.; Chen, W.; Yan, L. Sens. and Actuators B: Chemical Free-standing dried foam films of graphene oxide for humidity sensing. *Sens. Actuators B Chem.* **2015**, *215*, 316–322. [CrossRef]
11. Fernández-Ramos, M.D.; Ordóñez, Y.F.; Capitán-Vallvey, L.F.; De Vargas-Sansalvador, I.M.P.; Ballesta-Claver, J. Optical humidity sensor using methylene blue immobilized on a hydrophilic polymer. *Sens. Actuators B Chem.* **2015**, *220*, 528–533. [CrossRef]
12. Jung, D.Y.; Yang, S.Y.; Park, H.; Shin, W.C.; Oh, J.G.; Cho, B.J.; Choi, S.Y. Interface engineering for high performance graphene electronic devices. *Nano Converg.* **2015**, *2*, 11. [CrossRef]
13. Phan, D.-T.; Chung, G.-S. Effects of rapid thermal annealing on humidity sensor based on graphene oxide thin films. *Sens. Actuators B Chem.* **2015**, *220*, 1050–1055. [CrossRef]
14. Su, P.-G.; Shiu, W.-L.; Tsai, M.-S. Flexible humidity sensor based on Au nanoparticles/graphene oxide/thiolated silica sol–gel film. *Sens. Actuators B Chem.* **2015**, *216*, 467–475. [CrossRef]
15. Su, P.-G.; Wang, C.S. Novel flexible resistive-type humidity sensor. *Sens. Actuators B Chem.* **2007**, *123*, 1071–1076. [CrossRef]
16. Traversa, E. Ceramic sensors for humidity detection: The state-of-the-art and future developments. *Sens. Actuators B Chem.* **1995**, *B23*, 135–156. [CrossRef]
17. Chen, Z.; Lu, C. Humidity Sensors: A Review of Materials and Mechanisms. *Sens. Lett.* **2005**, *3*, 274–295. [CrossRef]

18. Lukaszewicz, J.P. Carbon-film-based humidity sensor containing sodium or potassium. Recovery effect. *Sens. Actuators B Chem.* **1999**, *60*, 184–190. [CrossRef]
19. Varghese, O.K.; Kichambre, P.D.; Gong, D.; Ong, K.G.; Dickey, E.C.; Grimes, C.A. Gas sensing characteristics of multi-wall carbon nanotubes. *Sens. Actuators B Chem.* **2001**, *81*, 32–41. [CrossRef]
20. Yao, Y.; Chen, X.; Guo, H.; Wu, Z. Graphene oxide thin film coated quartz crystal microbalance for humidity detection. *Appl. Surf. Sci.* **2011**, *257*, 7778–7782. [CrossRef]
21. Chu, J.; Peng, X.; Feng, P.; Sheng, V.; Zhang, J. Study of humidity sensors based on nanostructured carbon films produced by physical vapor deposition. *Sens. Actuators B Chem.* **2013**, *178*, 508–513. [CrossRef]
22. Tulliani, J.-M.; Inserra, B.; Ziegler, D. Carbon-Based Materials for Humidity Sensing: A Short Review. *Micromachines* **2019**, *10*, 232. [CrossRef] [PubMed]
23. Afify, A.S.; Ahmad, S.; Khushnood, R.A.; Jagdale, P.; Tulliani, J.-M. Elaboration and characterization of novel humidity sensor based on micro-carbonized bamboo particles. *Sens. Actuators B Chem.* **2017**, *239*, 1251–1256. [CrossRef]
24. Ziegler, D.; Palmero, P.; Giorcelli, M.; Tagliaferro, A.; Tulliani, J.-M.; Ziegler, D.; Palmero, P.; Giorcelli, M.; Tagliaferro, A.; Tulliani, J.-M. Biochars as Innovative Humidity Sensing Materials. *Chemosensors* **2017**, *5*, 35. [CrossRef]
25. Konta, J. Clay and man: Clay raw materials in the service of man. *Appl. Clay Sci.* **1995**, *4*, 269–335. [CrossRef]
26. Murray, H.H. Traditional and new applications for kaolin, smectite, and palygorskite: A general overview. *Appl. Clay Sci.* **2000**, *17*, 207–221. [CrossRef]
27. Choy, J.-H.; Choi, S.-J.; Oh, J.-M.; Park, T. Clay minerals and layered double hydroxides for novel biological applications. *Appl. Clay Sci.* **2007**, *36*, 122–132. [CrossRef]
28. Andersson, H.; Manuilskiy, A.; Unander, T.; Lidenmark, C.; Forsberg, S.; Nilsson, H.-E. Inkjet printed silver nanoparticle humidity sensor with memory effect on paper. *IEEE Sens. J.* **2012**, *12*, 1901–1905. [CrossRef]
29. Burman, D.; Santra, S.; Pramank, P.; Guha, P. Pt decorated MoS_2 nanoflakes for ultrasensitive resistive humidity sensor. *Nanotechnology* **2018**, *29*, 115504. [CrossRef]
30. Nikulicheva, T.B.; Zakhvalinskii, V.S.; Pilyuk, E.A.; Nikulin, I.S.; Vyazmin, V.V.; Mishunin, M.V. New humidity sensor material $(CaSO_4{\cdot}2H_2O)_{0.975}$-$(CuSO_4{\cdot}5H_2O)_{0.025}$. *Materialia* **2023**, *27*, 101662. [CrossRef]
31. Wu, K.; Fei, T.; Zhang, T. Humidity Sensors Based on Metal–Organic Frameworks. *Nanomaterials* **2022**, *12*, 4208. [CrossRef] [PubMed]
32. Ku, C.-A.; Chung, C.-K. Advances in Humidity Nanosensors and Their Application: Review. *Sensors* **2023**, *23*, 2328. [CrossRef] [PubMed]
33. Hotza, D.; Greil, P. Review: Aqueous tape casting of ceramic powders. *Mater. Sci. Eng.* **1995**, *A202*, 206–217. [CrossRef]
34. Somiya, S.; Hirata, Y. Mullite Powder Technology and Applications in Japan. *Bull. Am. Ceram. Soc.* **1991**, *70*, 1624–1632.
35. Montanaro, L.; Tulliani, J.M.; Perrot, C.; Negro, A. Sintering of industrial mullites. *J. Eur. Ceram. Soc.* **1997**, *17*, 14, 1715–1723. [CrossRef]
36. Dortmund Data Bank (DDB). Available online: http://ddbonline.ddbst.com/antoinecalculation/antoinecalculationcgi.exe (accessed on 29 May 2024).
37. Firehouse. Available online: https://www.firehouse.com/rescue/article/10574732/hazmat-math-calculating-vapor-concentrations (accessed on 29 May 2024).
38. Ziegler, D.; Boschetto, F.; Marin, E.; Palmero, P.; Pezzotti, G.; Tulliani, J.M. Rice husk ash as a new humidity sensing material and its aging behavior. *Sens. Actuators B Chem.* **2021**, *328*, 129049. [CrossRef]
39. Srivastava, R.; Yadav, B. Nanostructured ZnO. ZnO-TiO_2 and ZnO-Nb_2O_5 as solid state humidity sensor. *Adv. Mater. Lett.* **2012**, *3*, 197–203. [CrossRef]
40. Zhang, Y.; Chen, Y.; Zhang, Y.; Cheng, X.; Feng, C.; Chen, L.; Zhou, J.; Ruan, S. A novel humidity sensor based on $NaTaO_3$ nanocrystalline. *Sens. Actuators B Chem.* **2012**, *174*, 485–489. [CrossRef]
41. Georgieva, B.; Nenova, Z.; Podolesheva, I.; Pirov, J.; Nenov, T. Investigation of humidity sensors based on Sn-O-Te films by impedance spectroscopy. *Bulg. Chem. Commun.* **2013**, *45*, 63–67.
42. Qi, G.; Weng, Y.; Chen, J. Preparation of porous SnO_2-based ceramics with lattice structure by DLP. *Ceram. Int.* **2022**, *48*, 14568–14577. [CrossRef]
43. Burman, D.; Choudhary, D.S.; Guha, P.K. ZnO/MoS_2-based enhanced humidity sensor prototype with android app interface for mobile platform. *IEEE Sens. J.* **2019**, *19*, 3993–3999. [CrossRef]
44. Kundu, S.; Majumder, R.; Ghosh, R.; Pal Chowdhury, M. Superior positive relative humidity sensing properties of porous nanostructured Al:ZnO thin films deposited by jet-atomizer spray pyrolysis technique. *J. Mater. Sci Mater. Electron.* **2019**, *30*, 4618–4625. [CrossRef]
45. Babu Reddy, L.P.; Megha, R.; Raj Prakash, H.G.; Ravikiran, Y.T.; Ramana, C.H.V.V.; Vijaya Kumari, S.C.; Kim, D. Copper ferrite-yttrium oxide (CFYO) nanocomposite as remarkable humidity sensor. *Inorg. Chem. Commun.* **2019**, *99*, 180–188. [CrossRef]
46. Manut, A.; Zoolfakar, A.S.; Mamat, M.H.; Ab Ghani, N.S.; Zolkapli, M. Characterization of Titanium Dioxide (TiO_2) Nanotubes for Resistive-type Humidity Sensor. In Proceedings of the IEEE International Conference on Semiconductor Electronics (ICSE), Kuala Lumpur, Malaysia, 28–29 July 2020; pp. 104–107.
47. Pi, C.; Chen, W.; Zhou, W.; Yan, S.; Liu, Z.; Wang, C.; Guo, Q.; Qiu, J.; Yu, X.; Liu, B.; et al. Highly stable humidity sensor based on lead-free $Cs_3Bi_2Br_9$ perovskite for breath monitoring. *J. Mater. Chem. C* **2021**, *9*, 11299–11305. [CrossRef]

48. Duy, L.T.; Baek, J.Y.; Mun, Y.J.; Seo, H. Patternable production of $SrTiO_3$ nanoparticles using 1-W laser directly on flexible humidity sensor platform based on $ITO/SrTiO_3/CNT$. *J. Mater. Sci. Technol.* **2021**, *71*, 186–194. [CrossRef]
49. Doroftei, C.; Leontie, L. Porous nanostructured gadolinium aluminate for high-sensitivity humidity sensors. *Materials* **2021**, *14*, 7102. [CrossRef] [PubMed]
50. El-Denglawey, A.; Manjunatha, K.; Vijay Sekhar, E.; Chethan, B.; Zhuang, J.; Jagadeesha Angadi, V. Rapid response in recovery time, humidity sensing behavior and magnetic properties of rare earth(Dy & Ho) doped Mn–Zn ceramics. *Ceram. Int.* **2021**, *47*, 28614–28622.
51. Shah, Z.; Shaheen, K.; Arshad, T.; Ahmad, B.; Khan, S.B. Al doped Sr and Cd metal oxide nanomaterials for resistive response of humidity sensing. *Mater. Chem. Phys.* **2022**, *290*, 126632. [CrossRef]
52. Subki, A.S.R.A.; Mamat, M.H.; Musa, M.Z.; Abdullah, M.H.; Shameem Banu, I.B.; Vasimalai, N.; Ahmad, M.K.; Nafarizal, N.; Suriani, A.B.; Mohamad, A.; et al. Effects of varying the amount of reduced graphene oxide loading on the humidity sensing performance of zinc oxide/reduced graphene oxide nanocomposites on cellulose filter paper. *J. Alloys Compd.* **2022**, *926*, 166728. [CrossRef]
53. Dubey, R.S.; Srilali, S.; Ravikiran, Y.T.; Babu, G.S.; Katta, K.V. Synthesis and characterization of $Zn_{x\text{-}1}Al_2O4(TiO_2)_x$ nanocomposite ceramics and their humidity sensing properties. *J. Mater. Sci.* **2022**, *57*, 2636–2649. [CrossRef]
54. Yasin, E.; Javed, Y.; Imran, Z.; Anwar, H.; Shahid, M. Exploration of dielectric and humidity sensing properties of dysprosium oxide nanorods. *Eur. Phys. J. Plus* **2023**, *138*, 1050. [CrossRef]
55. Mohamed Zahidi, M.; Mamat, M.H.; Subki, A.S.R.A.; Abdullah, M.H.; Hassan, H.; Ahmad, M.K.; Bakar, S.A.; Mohamed, A.; Ohtani, B. Formation of a nanorod-assembled TiO_2 actinomorphic-flower-like microsphere film via Ta doping using a facile solution immersion method for humidity sensing. *Nanomaterials* **2023**, *13*, 256. [CrossRef] [PubMed]
56. Zhang, Z.; Li, F.; Zheng, Y. Highly sensitive resistive humidity sensor based on strontium-doped lanthanum ferrite nanofibers. *Sens. Actuators A Phys.* **2023**, *358*, 114435. [CrossRef]
57. Ravichandran, R.; Quine, S.D.; Arularasu, M.V. Humidity sensing performance of nitrogen doped reduced graphene oxide-WO_3 composite. *BioNanoSci* **2023**, *13*, 2205–2214. [CrossRef]
58. Dhariwal, N.; Yadav, P.; Kumari, M.; Jain, P.; Sanger, A.; Kumar, V.; Thakur, O.P. Iron oxide-based nanoparticles for fast-response humidity sensing, real-time respiration monitoring, and noncontact sensing. *IEEE Sens. J.* **2023**, *23*, 22217–22224. [CrossRef]
59. Kumar, A.; Kumari, P.; Kumar, M.S.; Gupta, G.; Shivagan, D.D.; Bapna, K. SnO_2 nanostructured thin film as humidity sensor and its application in breath monitoring. *Ceram. Int.* **2023**, *49*, 24911–24921. [CrossRef]
60. Malik, P.; Duhan, S.; Malik, R. A high-performance humidity sensor based on 3D porous SnO_2-encapsulated MCM-48 for real-time breath monitoring and contactless gesture detection. *Mater. Adv.* **2024**, *5*, 2510–2525. [CrossRef]
61. Jiang, W.; Su, M.; Zheng, Y.; Fei, T. Efficient electron transfer through interfacial water molecules across two-dimensional MoO_3 for humidity sensing. *ACS Appl. Mater. Interfaces* **2024**, *16*, 7406–7414. [CrossRef]
62. Huang, S.; Shan, H.; Zhao, Y.; Ren, Y.; Gu, X. Preparation of humidity sensors based on $CsPbBr_3$ quantum dots for applications in microcrack detection. *Chin. J. Inorg. Chem.* **2024**, *40*, 383–393.

ceramics

MDPI

Review

An Overview on the Manufacture and Properties of Clay-Based Porous Ceramics for Water Filtration

Iffat Qoudsiyyah Maury Njoya [1], Gisèle Laure Lecomte-Nana [1,*], Kassoum Barry [1,2], Dayirou Njoya [3], Youssef El Hafiane [1] and Claire Peyratout [1]

1 Institute of Research for Ceramics (IRCER), European Ceramic Centre (CEC), University of Limoges, 12 Atlantis Street, 87068 Limoges, France; iffat_qoudsiyyah.maury_njoya@unilim.fr (I.Q.M.N.)
2 Laboratory of Molecular Chemistry and Materials (L2CM), University Joseph KI-ZERBO, Ouagadougou 03 BP 702, Burkina Faso
3 Applied Inorganic Chemistry Laboratory, Department of Inorganic Chemistry, Faculty of Sciences, University of Yaounde 1, 3.315 Street of University, P.O. Box 812, Yaounde 00000, Cameroon
* Correspondence: gisele.lecomte@unilim.fr

Abstract: This study explores the different techniques used to manufacture porous clay-based ceramics, examining their properties such as porosity, strength, permeability and filtration efficiency. Different techniques are discussed in this review, with additive manufacturing being one of the most innovative techniques for manufacturing porous ceramics. Porous ceramics have their applications in numerous domains. Such ceramic filters have the advantages of retaining heavy materials, suspended particles, bacteria, viruses and, water turbidity. Thus, the choice of the technique and propriety is a crucial step in obtaining a porous ceramic with the best performance. Barry et al. prepared porous phyllosilicate-based ceramics by freeze-tape casting on four samples and obtained porosity values in the range of 67–79% and diametrical compressive strength in the range of 3–7 MPa. Manni et al. prepared porous red ceramics from Moroccan clay and coffee waste (10, 20 and 30 wt.%) via uniaxial pressing and sintering at 1150 °C. They obtained porosities ranging from 30.2 to 63.8% and flexural strength values from 1.8 to 19.5 MPa. Medri et al. prepared ZrB2-based porous bodies with the use of sponges and polyurethane foams as templates via the replica method and obtained high porosity over 80% and compressive strength up to 4.8 MPa. The use of clay and peanut shell mixtures was used in preparing porous silicate ceramics after unidirectional pressing and sintering at 1100 °C. These samples included 25 mass% of peanut shells, and exhibited porosity in the range of 40 to 60% and diametrical compressive strength in the range of 1–6 MPa. Such properties are suitable for domestic use of these types of clay-based ceramic filters. Moreover, the permeability values and removal of some pollutants, like arsenic, have been satisfactory for the first set of samples.

Keywords: clay materials; porous ceramics; processing; properties; water filtration

1. Introduction

The World Water Council estimates that by 2030, about 3.9 billion people will live in regions characterized as "water scarce". According to the World Health Organization, diarrheal diseases cause 1.5 million deaths every year, and it is estimated that more than half of these deaths can be attributed to a lack of sanitation and hygiene, as well as to the consumption of contaminated water [1]. In order to satisfy the lack of good-quality drinking water, the majority of the population in under-developed countries makes use of well, river or even drilling water (Figure 1). Water from these sources is not suitable for human

 https://doi.org/10.3390/ceramics8010003

consumption because it is generally polluted by anthropic activities. The consumption of this type of water by populations tends to expose them to many diseases such as typhoid, cholera, dysentery, etc. In order to produce drinking water, some people utilize decantation accompanied by chlorination and filtration using a filter, whose principal component is an organic or inorganic membrane.

Figure 1. Illustrating the use of water from a well, river and drill for drinking purposes.

Porous ceramics are mostly recommended because of their good chemical and thermal stability, good mechanical resistance, ease of cleaning and long lifespan [2–4]. In general, these membranes are made from metal oxides, such as alumina, silica, zirconia and titanium oxide. They present some disadvantages, which are related to their relatively high cost because of the expensive raw materials and the complexity of the fabrication process [5–7]. Today, porous ceramics are used for a wide variety of applications (Figure 2).

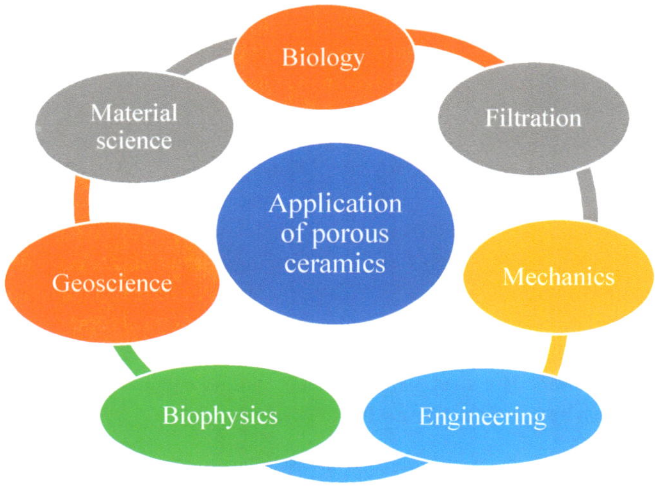

Figure 2. Applications of porous ceramics.

Porous ceramics are generally characterized by their pore size, surface, thickness and porosity [7]. Porous ceramics are divided into three main categories based on pore size according to the International Union of Pure and Applied Chemistry (IUPAC) nomenclature: microporous (pore diameter less than 0.002 μm), mesoporous (diameter ranging between 0.002 and 0.050 μm) and macroporous (diameter greater than 0.050 μm). Depending on the particular applications, different pore sizes may be desired. Liquid transport occurs only in mesoporous and macroporous ceramics, while the transport of gaseous molecules can take place in all three categories [8].

Almost 15,000 research articles have been published on porous ceramics since 2010, according to the Scopus database. Figure 3 shows the evolution of the number of research articles published each year obtained by searching the string "porous ceramics" since 2010. The processing techniques of porous ceramics have also experienced enormous evolution. Another area that is presently at the forefront of every technological sector globally is waste utilization through recycling. In this context, there is a circular use of porous ceramics, and it is essential to use the methods with filtration in mind. Thus, the aim of future studies will be to develop and produce conventional additive manufacturing processing-based porous silicate ceramics, with emphasis on the initial composition, pore-forming process and the final characteristics (connectivity) of the prepared porous ceramics.

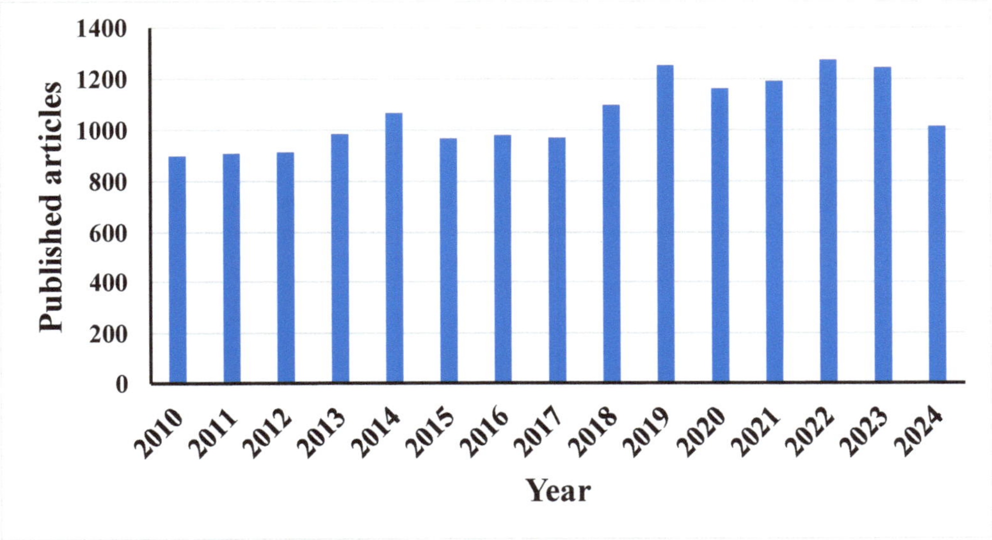

Figure 3. Evolution of the number of research articles published on "porous ceramics" according to the Scopus database since 2010 [3].

Previous work shows a variety of different methods for manufacturing porous ceramics. However, most of these methods seem to be adhere to the following path: choice of materials (clay and other additives), formulation, sintering and characterization of obtained products. Their properties are highly dependent on the powder or slurry (suspension) characteristics (plasticity, size and shape of particles, etc.), shaping method (uniaxial pressing, etc.) and sintering process (firing rate, temperature, etc.). Traditionally, porous ceramics are manufactured using techniques such as partial sintering, freeze-casting, tin plating, replica sacrificial template, direct foaming and gel casting [9–16]. Zhang et al. [17] prepared porous ceramics from lead zirconate titanate PZT (raw material), polyvinyl alcohol (organic binder) and acrylic acid (dispersant) by freeze-casting with water as a medium. As the solid content of PZT in the sample varies from 20 to 40%, porosity varies from 36 to 67%, and the compressive strength varies from ~1 to 73 MPa. They conclude that the freeze-casting method is favorable to improving the porous PZT ceramics properties. Zhu et al. [18] prepared porous ceramics with Tungsten ore tailings (raw material), CaO (binder), Kaolin (sintering additive) and polymethyl methacrylate (pore foaming agent) via pressing using the sintering method (1150–1210 °C, 5 °C/min and a dwell time of 1 h). A pore size of 0.18 μm, open porosity of 44.85% and compressive strength of 14.10 MPa were obtained at a sintering temperature of 1200 °C.

This paper gives a brief overview of the manufacture and properties of porous ceramics. In the last two decades, porous ceramics have been extensively studied because of their excellent performance characteristics and wide range of applications. In recent years, a novel approach for manufacturing porous ceramics has emerged, which is known as additive manufacturing. It is quite a challenging technique because it enables the manufacturing of ceramics with complex shapes and structures, needs less material and enables the recycling of material. This technique, which has been studied by several authors [19–21], consists of joining materials to make objects from three-dimensional model data, usually layer upon layer, as opposed to subtractive manufacturing [19].

Shaping is the key stage in ceramic production. To this end, supports can be shaped using a variety of techniques. These include pressing (uniaxial and isostatic), extrusion, casting (in porous molds, under pressure or in strips) and injection (Figure 4). Several consolidation processes are available for manufacturing ceramics, with the aim of controlling the pore volume. These processes belong to two main categories:

- Hydraulic setting, characterized by its ease of use. This generally involves mixing a powder with excess water during the mixing stage, resulting in residual porosity after the setting reaction, followed by a drying stage.
- Incomplete densification through partial sintering, used in the majority of industrial ceramics. By controlling the sintering cycle (temperature, time and/or pressure), materials with appropriate residual porosity are obtained.

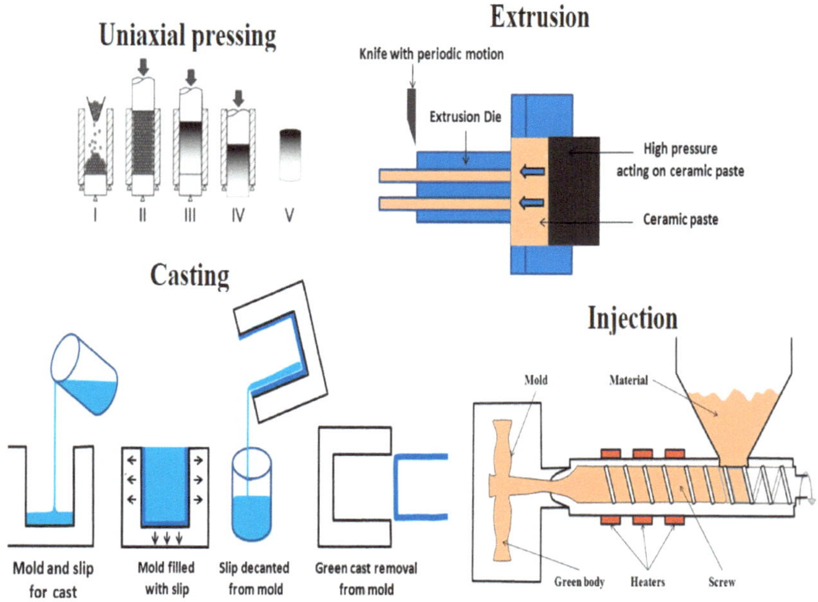

Figure 4. Some shaping techniques for ceramic production [22,23].

Furthermore, it is also possible to combine partial sintering with another process to obtain a material exhibiting hierarchical porosity.

Porous ceramic materials are produced by a variety of methods, the choice of which is mainly determined by the desired porous structure. The main production methods (Figure 5) are as follows: direct foaming, replication technique, partial sintering process or methods using pore-forming agents mixed with ceramic powders [6,7,24]. These processes enable a very wide range of porosities to be obtained, with controlled pore shape and size distribution.

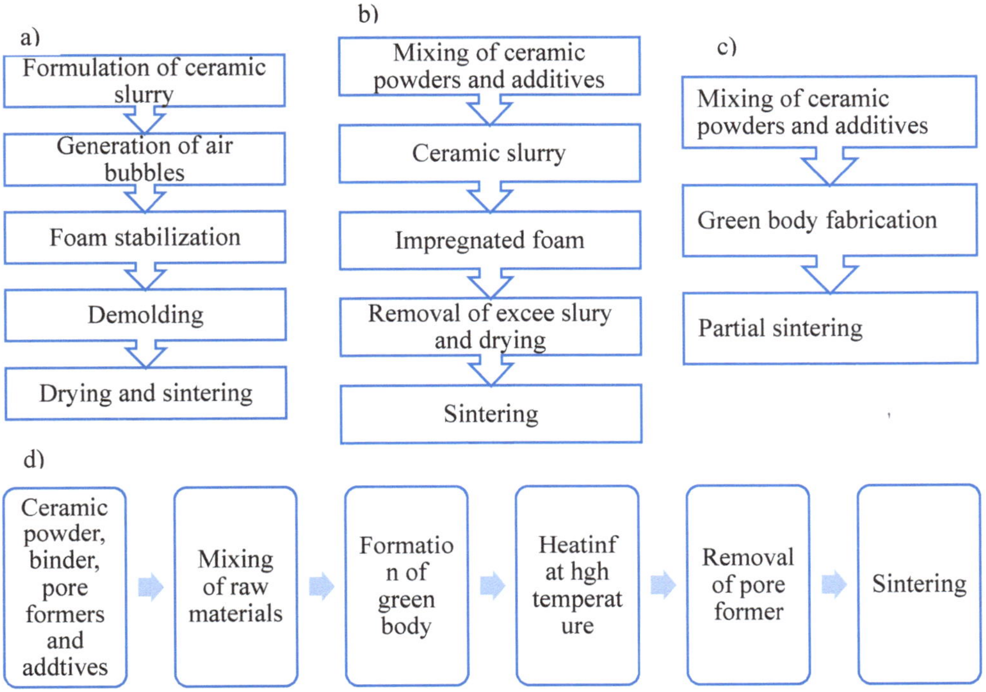

Figure 5. Main steps in the manufacturing of ceramics: direct foaming (**a**), replication technique (**b**), partial sintering process (**c**) and methods using pore-forming agents mixed with ceramic powders (**d**).

2. Experimental Procedure

2.1. Materials and Methods

To prepare porous ceramics, authors have used raw materials [9,25–27], synthesized materials [28–30] or raw materials and waste [24,31–33]. The formation of pores can be attributed to the embedment of air bubbles and the use of organic or inorganic waste, such as rice husks, banana peels, coffee husks, bovine bone ash, corn cob ash, cassava starch, eggshells, peanut shells, coal ash and cocoa cortex. The way in which ceramics are prepared and shaped is critical to the performance of the finished product.

Material preparation is an essential step, as the physical, chemical and technological properties of the materials have a major influence on the quality of the substrate. This step includes grinding, sieving and mixing.

In our work, two types clays were selected, namely kaolin and halloysite, noted K and H, respectively. Also, we choose to use a pore-forming agent mixed with clay powders for manufacturing our porous silicate ceramics. This porogenic agent is an agro-waste product, namely peanut shells (noted C). The clays were collected from the west region of Cameroun, while the waste (peanut shell) came from the southwest of Burkina Faso. Indeed, the properties of these groundnut shells are similar in most tropical regions. These materials were chosen because of their abundance in such developing countries, relative low cost, beneficial properties and for waste valorization under the scope of the circular economy. These materials were collected, then ball-milled for 1 h, sieved at 500 μm, air dried and oven dried at 105 °C for 48 h.

The powder mixtures (clay powder + peanut shell) were mixed using a rotating pelletizer at 32 rpm for 15 min. Then, the mixtures were humidified with progressive addition of water close to 15 mass% to obtain mixtures with good homogeneity. Subsequently, the pastes were stored for 12 h under high humidity to improve their rheological property, avoid premature drying and ensure complete diffusion of water within the pellets. The peanut shell content was 25% by mass. The samples were shaped using uniaxial pressing (this is advantageous because it produces dense samples with precise dimensions, very smooth surface quality and results in good homogeneity within the sample) and then sintered at 1100 °C, 5 °C/min with a soaking time of 1 h.

Evidently, there is a strong need to evaluate the circular economy of porous ceramics in the future in order to define a suitable tradeoff between their manufacturing and reducing the environmental impact [34,35]. As such, it is important to explore the circular economy by using biosourced additives for the preparation of porous ceramics.

2.2. Characterization

Here, we mention just a few of the characterization methods. A number of authors have studied these parameters in their works, with the aim of highlighting some correlations and improving the useful properties.

X-ray diffraction is a technique used in materials science to determine the atomic and molecular structure of a crystalline material [36]. This technique is usually complementary to the elemental analysis. It provides qualitative control over the elemental composition and crystalline structure of materials. Furthermore, more information can be obtained from a careful analysis of the diffraction patterns or by using specific XRD settings.

Fourier transform infrared spectroscopy (FTIR) is a technique used for identifying the functional groups of molecules on the basis of their vibrational modes and frequencies. The absorbed energy characterizes the functional group and close chemical environment. The absorbed energy is described by Beer-Lambert's Law.

Scanning electron microscopy (SEM) is an electron microscopy technique capable of producing high-resolution images of the surface of a sample using the principle of electron-matter interactions [37].

X-ray fluorescence spectrometry is a global non-destructive elemental analysis technique for identifying and determining the chemical elements that constitute a sample in a given order of concentration. It enables the elemental composition of a sample to be determined both qualitatively and quantitatively [38].

Specific surface area is the total surface area per unit mass of the membrane support accessible to atoms and molecules. The aim of this measurement is to assess the ability of the support to facilitate fluid flow through their structures. It is determined according to the theory of Brunauer, Emmett and Teller (BET) [39]. To achieve this, measurements are carried out on samples, and each measurement is preceded by degassing to remove the water molecules adsorbed in the porous structure of the sample.

Porosity is an important parameter for the characterization of the materials' microstructures. In general, it corresponds to the volume of the interstices that can contain fluid, related to the total volume of the material. Porosity is also influenced by the type of material used, the proportion of pore-forming agent and the sintering temperature. Some methods for porosity analyses are X-ray computed microtomography (MCT), mercury intrusion porosimetry (MIP), geometric calculation and tripled weighed method based on the principle of Archimedes.

- MCT is used to form three-dimensional images, thus observing the geometry of the pore space in samples, such as the shape, connectivity, size and distribution of pores. MCT imaging is performed using a micro-focused X-ray tube that obtains shadow images of the object transmitted by multiple X-rays from different angles [39].

- Mercury intrusion porosimetry method measures open porosity and is appropriate for pores with diameters ranging from 0.01 microns to 800 microns. Measuring the volume of mercury included in the sample at each pressure value gives the size distribution of its open pores. Considered a standard method of macropore analysis [40], it also enables the calculation of apparent density and total open porosity.
- Porosity is calculated from the absolute (or real) ρ_r and apparent bulk densities ρ_a. The real bulk density can be determined by pycnometry on powdered materials (obtained by a fine grinding of samples), and the apparent bulk density is calculated using both dimensions and mass of the dry material. Thus, total porosity *P* is hence determined by Equation (1) [41].

$$P = \left(1 - \frac{\rho_a}{\rho_r}\right) \times 100 \tag{1}$$

- The principle of Archimedes states that the force exerted on a body immersed in a fluid, whether fully or partially, is equal to the weight of the fluid that the body displaces. Thus, the tripled weighed method consists of weighing three different masses (M_1, M_2 and M_3) based on the Archimedes system, where M_1 is the dry mass of samples, M_2 is the mass of the sample completely immersed in liquid and M_3 is the wet mass of the samples. Hence, with clearly defined formulae, the open porosity P_o of the sample can be calculated using Equation (2) below [42]:

$$P_o = \frac{M_3 - M_1}{M_3 - M_2} \times 100 \tag{2}$$

Permeability is a parameter used to determine the membrane's ability to allow liquid to flow through its pores. It is mainly controlled by pore size, hydrophilic character and membrane thickness. However, this parameter is not an intrinsic characteristic of ceramics, as it also depends on the fluid's viscosity. It characterizes the surface area required for flow, the geometry of the internal pore network, the hydraulic conductivity of the pore network and the physicochemical interactions between the material and the fluid. Permeability L_p ($L/h/m^2$), according to Darcy's law, enables us to relate fluid flow to the pressure gradient applied by a parameter characteristic of the medium through which it passes, which is expressed by the relation (3) [43].

$$L_p = \frac{J}{\Delta P} \tag{3}$$

where J ($L/h/m^2/bar$) is the hydraulic flow and ΔP the transmembrane pressure (bar).

3. Results and Discussion

Some key properties reported in the previous studies are summarized in this section. These studies were devoted to the development of porous ceramics using either raw aluminosilicate materials, synthesized materials or waste combined with raw materials.

3.1. Porous Ceramics Obtained from Raw Materials

Table 1 summarizes the properties of some porous ceramic materials that were obtained by using raw materials. The data reveal that porous ceramic materials with satisfactory porosity (44–88%) and compressive strength can be obtained from a variety of raw materials when sintered in the range of 1200 to 1650 °C. Bai 2010, [26] prepared porous mullite ceramics from carbonaceous kaolinite clay, alumina powder and graphite as apore former using the sintering method. Porous ceramics sintered at 1500 °C led to 36.4% open porosity and 0.3–5 µm pore size (in a narrow range). This higher sintering temperature led to less porosity, and the mullitizaion was nearly completed. Li et al. [16] prepared porous mullite matrix ceramics from quartz powder, alumina powder, glass powder, ammonium polycrylated aqueous solution (dispersant) and propyl gallate (foaming agent), which were sintered from 1200 to 1650 °C for 1 h. When sintered at 1550 °C, the open porosity was 29.38% (porosity 81.37%), while the compressive strength and average pore size were 6.25 ± 0.91 MPa and 850 μm, respectively. Lui et al. [21] prepared porous ceramics from photocurable resin and zirconia slurry to perform multi-material printing experiments sintered at 1500 °C. They developed a ceramic multi-material 3D printing apparatus (CMM) with the aim of manufacturing porous ceramics with <100 μm pore size and designable distribution of pores directly. They obtained a pore size of 60 μm × 20 μm. This pore size is the smallest ceramic pore size known that can be directly printed, especially the pore size of 20 μm in the longitudinal direction, which cannot be achieved by other ceramic additive manufacturing technologies.

Table 1. Properties of porous ceramic materials obtained by using raw materials.

Composition	Final Product	Sintering Conditions	Apparent Porosity (%)	Pore Size (μm)	Compressive Strength (MPa)	Ref.
ZrO_2 powder + SiO_2 + boron carbide + carbon black	Porous ZrB_2-SiC ceramic	1500 °C/2 h	67–78		0.2–9.8	[15]
Quartz powder + alumina powder + glass powder + ammonium polycrylates solution + propyl gallate	Porous mullite matrix ceramic	1200 °C/1 h 1300 °C/1 h 1400 °C/1 h 1500 °C/1 h 1550 °C/1 h 1600 °C/1 h 1650 °C/1 h	~45 ~80 ~82 ~83 81.37 ~81.4 ~88	~750	88.3 ± 13.61 - - - 6.25 ± 0.91 - -	[16]
Tungsten ore tailing (75 wt%) + CaO + Kaolin + polymethyl methacrylate (18 wt%)	Porous ceramics	1200 °C/60 min	44.85	180	14.10	[18]
Alumina powder + milled carbon coal + stearin	Porous ceramic	1530 °C/2 h	-	60 × 20	-	[25]
Carbonaceous kaolinite clay + alumina powder	Porous ceramic	1400 °C/4 h 1450 °C/4 h 1500 °C/4 h 1550 °C/4 h	~43–46 ~44–48 ~36–38 ~17–32	- - - -	- - 42.1 75.9	[26]
Al_2O_3 + SiO_2 PHMs (0–40 wt%) + epoxy resin	Porous Al_2O_3 Ploy-hollow microspheres (PHM) ceramic	1550 °C/4 h	~77 ~76 ~71 65.0 ~66	51.95 45.51 38.25 38.25 30.14	~0.2 ~0.1 ~1 4 ~1.8	[27]

As a result, highly porous ceramics with good properties (high porosity, compressive strength, etc) can be obtained at sintering temperatures < 1200 °C, although it depends on the materials used. Also, additive manufacturing is a challenging technique because it has advantages such as high design freedom, a short manufacturing cycle and a low manufacturing cost.

Figure 6a,b shows the SEM images of halloysite- and kaolinite-based porous ceramics, respectively, obtained via uniaxial pressing and sintering at 1100 °C with a dwell time of 1 h. There is no defect on the support sections, as observed in these images. Ceramics (a) and (b) show a very compact texture and dense microstructure corresponding to the total porosity of 45 and 33%, respectively, and mechanical resistance to diametrical compression strength of 3.9 and 3.8 ± 0.5 MPa, respectively. The formation of intergranular contacts led to increased densification of the ceramic. This densification makes them more robust and does not prevent pore formation. Ceramic morphology shows granules with pores of various sizes.

Figure 6. SEM images of Halloysite (**a**) and Kaolin (**b**) based porous ceramics (shaped by uniaxial pressing and sintered at 1100 °C).

Barry et al. [9] prepared porous phyllosilicate-based ceramics by freeze-tape casting and sintering at 1200 °C. For that, they prepared four slurries labeled HCR (77% halloysite), KORS (29% kaolinite), KCR (98% kaolin) and KHCR (mixture of KCR and HCR). The SEM images (Figure 7) reveal a difference in the microstructure and texture of all four ceramic disks. According to this study, the pores in the KCR ceramic disks were more textured compared to others, and they noted the formation of the primary mullite in rigid skeleton shape. The SEM images of the KORS ceramic disks show the formation of secondary mullite needles with a relatively small quantity and a disordered structure in a vitreous phase. HCR led to the bulk formation of small mullite particles due to the relatively important vitreous phase in the porous material. X-ray diffraction patterns showed the presence of mullite in all four ceramics.

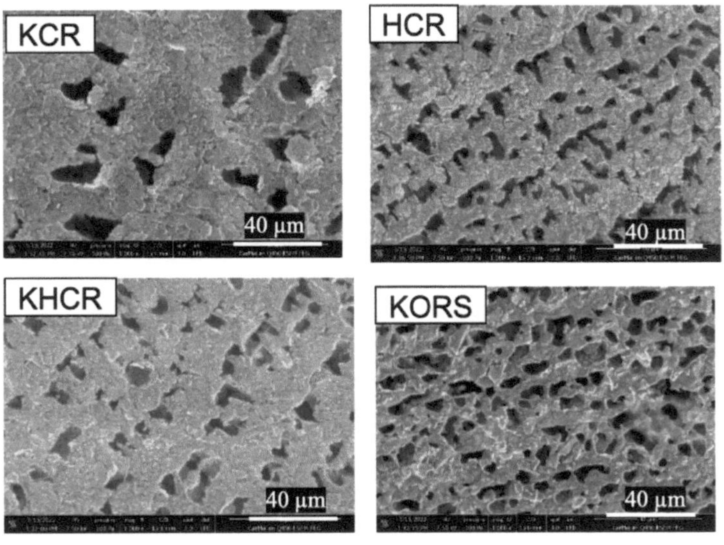

Figure 7. SEM images of porous kaolin- and halloysite-based ceramics (shaped by freeze-tape casting and sintered at 1200 °C).

3.2. Porous Ceramics Obtained from Synthesized Materials

Porous ceramics can be produced by using synthesized materials. Table 2 summarizes the properties of some porous ceramic materials that were obtained by using synthesized materials. Nowadays, the trend is to reduce the environmental impact by limiting the use of synthetized materials in the manufacture of porous ceramics and to optimize the utilization of naturally occurring raw materials (reuse and recycling), such as most clays. Xu et al. [28] prepared porous ceramics from a variety of materials and concluded that there is a possibility to control the pore volume, pore size and other properties of green bodies by adjusting the ratio of the starting materials used or changing the sintering cycle. Despite the high mechanical resistance, the consolidation of such mixtures requires very high temperatures (>1300 °C), thus leading to an increased carbon footprint.

Table 2. Properties of porous ceramic materials obtained by using synthesized materials.

Composition	Final Product	Sintering Conditions	Apparent Porosity (%)	Pore Size (μm)	Compressive Strength (MPa)	Ref.
$Al(NO_3)_3.9H_2O$ + $Mg(NO_3)_3.9H_2O$ + Urea + Soluble starch	Porous ceramics	1400 °C/2 h	35.2–51.6	~0.2–20	~36.1–454.7	[29]
Si_3N_4 + Y_2O_3 + $SrCO_3$ +Al_2O_3 + premix solution + polyacrylamide + methyletylenediamine + ammonium persulfate	Porous ceramics	1750 °C/1 h	~47–57	-	52.67–106.33	[28]

3.3. Porous Ceramics Obtained from Raw Materials and Waste

Waste management is a very important issue from both the public health perspective and the industrial point of view because an ever-increasing amount of hazardous materials needs to be disposed of in a safe and economical way. The development of new ceramic, glass and glass-ceramic materials made by recycling waste is acquiring particular importance [44]. A large number of articles have been published by researchers regarding the manufacture of porous ceramics using different types of waste. Table 3 summarizes the properties of some porous ceramic materials that were obtained by using waste combined with raw materials. The incorporation of rice husk ash promotes the formation of high porosity. According to Görhan et al. [45], an increase in rice husk leads to an increase in the apparent porosity and tends to decrease the unit weight. Dong et al. [46] manufactured porous ceramic membranes using waste fly ash, and Zhu et al. [47] manufactured porous ceramic membranes using coal fly. According to their studies, porous ceramic membranes with high porosities were obtained at sintering temperatures lower than 1200 °C. Thus, including some wastes in the formulation of green bodies can help in manufacturing porous ceramics at temperatures < 1200 °C with good properties (porosity, compressive strength, etc.).

Table 3. Properties of porous ceramic materials obtained by using clay materials and waste.

Composition	Final Product	Sintering Conditions	Apparent Porosity (%)	Compressive Strength (MPa)	Ref.
Ground rice husk (5–15 vol%) + brick raw material	Porous clay brick	800 °C/1 h 900 °C/1 h 1000 °C/1 h	~39–44.5 ~38–44 ~37–43.5	~9.5 ~5.75–9.25 ~6–9	[45]
Coffee waste (10–30 wt%) + red clay	Porous red ceramics	1150 °C	~30.2–63.8	~1.8–19.5	[48]

Table 3. *Cont.*

Composition	Final Product	Sintering Conditions	Apparent Porosity (%)	Compressive Strength (MPa)	Ref.
Wheat straw (4–8 wt%) + clay	Porous clay brick	950 °C	~40–50	-	[49]
Ceramic waste (80%) + flux + cement + clay	Ceramic foam	1000 °C/3 h	~40–83	~0–9.3	[50]
Polished stonewere residue (90%) + soda-lime glass	Highly porous glass ceramic	900 °C/1 h	75.1	2.5	[51]

According to Simonis et al. [52], porous ceramics consist of a high level of porosity, and that results in properties such as high specific area, high permeability and high tortuosity. These factors make ceramic filters effective by physically (non-chemical process) removing suspended solids, filamentous bacteria and protozoa having sizes of several micrometers to nanometers from water (thus used for water treatment). Moreover, the filtration of waste water rejected by many industries reduces emissions that are harmful to the environment [53]. Sobsey et al. [54], compared several household water treatment techniques based on several factors affecting the sustainability of the membrane. It was concluded that ceramic and bio-sand filters were potential options for a sustainable water treatment technique. Shirasaki et al. [55], suggested that due to the fact that the pore sizes of a ceramic filter are not sufficiently small to remove viruses efficiently, a coagulation process as a pre-treatment can be combined with a ceramic water filter for viral removal. They reported that such systems are used in Japan for the treatment of drinking water.

Figure 8 shows the SEM images of halloysite clay and peanut shells (25% mass percent)-based porous ceramic obtained via uniaxial pressing and sintering at 1100 °C, with a soaking time of 1 h. The image shows a very compact texture and dense microstructure related to the total porosity and diametrical compressive strength of 59.74% and 0.92 MPa, respectively. The creation of pores within materials is due to the incorporation of the pore-forming agent (peanut shells). The pores observed vary in size and geometry. The SEM image presents no defect on the analyzed support sections.

Figure 8. SEM image of halloysite clay and peanut shell-based porous ceramic (shaped by uniaxial pressing and sintered at 1100 °C).

Barry et al. [56] prepared porous ceramics from kaolinitic clay and peanut shells (25% mass percent) via unidirectional pressing and sintering at 900 and 1100 °C with a dwell time of 1 h. There is no defect on the support sections, as observed in these images (Figure 9). According to them, the porous ceramic sintered at 1100 °C (Figure 9b) had a porous microstructure that is fairly developed with interconnected pores. On the other hand, the sample fired at 900 °C (Figure 9a) had a poorly developed interconnected porosity and a relatively dense structure. The presence of cristobalite within the porous ceramic fired at 1100 °C (Figure 9b) had an effect on the microstructure by modifying the porous architecture.

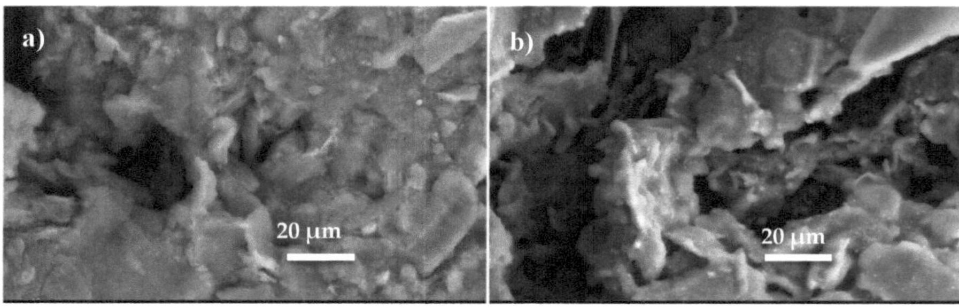

Figure 9. SEM images of kaolinitic clay and peanut shell-based porous ceramics sintered at 900 (**a**) and 1100 °C (**b**) [56].

Together with the mechanical properties, the efficiency of these filters for water treatment is of great importance. Water treatment is a process that enables wastewater to be returned to a quality defined by various articles of law and prefectural decrees. There are several water treatment methods, including physicochemical, biological and chemical methods [57–60].

Physicochemical methods, which include coagulation-flocculation, ion exchange, adsorption and ceramic filtration, are one of the most preponderant. Activated carbon was the first adsorbent used for water treatment. Today, other types of natural or synthetic adsorbents are available, such as laterites, clays, resins, etc. The adsorption technique is effective for low concentrations of adsorbate, but it can be relatively expensive depending on the adsorbent used. In general, the disadvantages of adsorption lie both in the competition between the molecules (adsorbate) of different sizes or similar properties and the relatively high price of the commercial adsorbent materials. In a water treatment process, filtration coupled with adsorption makes it possible to improve the retention properties of the filter material. This coupled treatment process is used in Bangladesh to remove arsenic from groundwater with significant efficiency [61]. Nd'e-Tchoup'e et al. [62] showed a removal rate of 90% in a column of multiple layers. Similary, Shafiquzzaman et al. [63] reported an arsenic removal rate of 86% on ceramic filters after double filtration at two different levels. The study by Barry et al. [56], reported an arsenic removal rate of 95% on the ceramic filters prepared from kaolinitic clay and peanut shells (sintered at 1100 °C). This was due to the fact that, the ceramic material of this filter had a more developed microstructure, and they formed crystalline mullite, which had a strong affinity with arsenic. The presence of mullite and the pores' architecture increases the retention capacity of this filter with respect to arsenic, thus highlighting the impact of mullite on the retention properties of the filters. Barry et al. [56] presented an example of a brief mechanism of arsenic absorption, as illustrated in Figure 10.

There are many types of filters, as presented in Table 4. Ceramic filters use the specific physical and chemical properties associated with porous ceramics, namely: filtration, ion exchange and adsorption on internal and external surfaces. Filtration is based on the gradient retention of particles according to their size and the size of the pores. At the same time, adsorption can be used since it enables the molecules to be attached to the surface of the solid by covalent bonds or specific interactions.

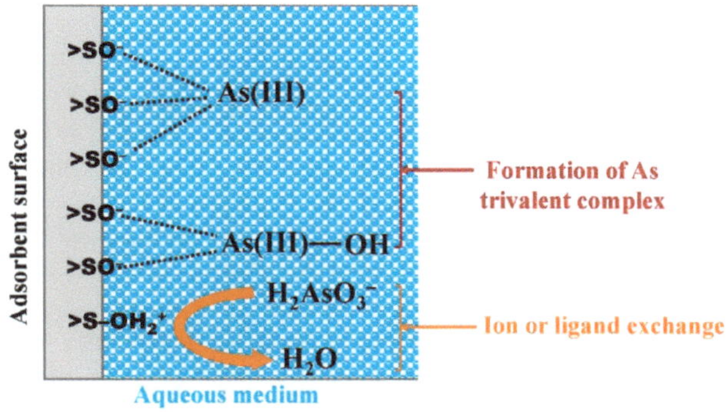

Figure 10. Example of arsenic (III) absorption mechanism onto a kaolinitic-based porous ceramic [56].

Table 4. Types of filters: costs, efficiency and pollutants that they eliminate.

Filters	Price (Euro)	Pollutant Eliminated	Yield Rate (%)
Reverse osmosis filter	60–400	Microorganisms, particles	85–95
UV filter	150–1000	Parasite, viruses	>86
Ion exchange filter		Heavy metals, limestone	82
Glass fiber filter		Suspended particles	95
Ceramic filter	150–600	Viruses, suspended particles, protozoaires, bacteria	98
Alkali water ionizer filter	400–2000	Suspended particles, bacteria, viruses	92

The permeability of a ceramic filter depends mainly on its physical and geometric characteristics, and it is essentially influenced by the morphological characteristics of the pore space: interconnectivity, pore size and many others. Table 5 shows some permeability values for porous ceramic materials. The difference in permeability values could be due to the presence or absence of particles that can easily foul the membrane pores and subsequently decrease the flow rate. Those with lower permeability values may be the result of lower connectivity between their pores and therefore the presence of a high or low proportion of dead-end pores.

Table 5. Permeability of some clay-based porous ceramics.

Composition	Permeability (L/h.m^2.Bar)	References
Clay + coconut husk + eggshell	14,013	[43]
Kaolinitic clay + peanut shell (sintered at 1100 °C)	53,802	[56]
Kaolinitic clay + peanut shell (sintered at 900 °C)	18,596	[56]
Metakaolin + corn cob ash waste	1359.93	[64]
Clays + cassava starch + bovine bone ash	2902	[65]

Fouling may occur on the surface during filtration, requiring specific cleaning, such as washing the top layer with hot water when the flow rate drops [66]. Fouling is a phenomenon that leads to a loss of performance through the deposition of suspended substances on its external surfaces, at the inlet or inside the pores [67]. As such, the efficiency of the ceramic filters depends on the porosity (the distribution, size and volume of the pores), the tortuosity of the porous network, and the density of the surface-active sites. Hence, the mastery of filter manufacturing technology is essential for controlling the filtration properties [68,69].

4. Conclusions

This paper presents a review on the manufacture and properties of porous ceramics for water filtration. The principle, applications and manufacturing strategies are discussed extensively. For the manufacture of porous ceramics, certain parameters such as compressive strength, porosity and pore size can be effectively increased by controlling the particle size of the material used, additives, sintering temperature and technique used. Concerning filters, it is necessary to improve their performance with antibiofilm substances. For instance, the impregnation of ceramic filters with bioactive compounds, such as silver, is often used to improve their performance, particularly in the removal of micro-organisms. Filters modified with an internal silver deposit are effective in removing at least 99% of protozoa and 90 to 99% of bacteria. The properties of filters can also be improved by moderate incorporation of organic adsorbent or by grafting organic molecules such as alkoxysilane.

Author Contributions: Conceptualization, G.L.L.-N., K.B., D.N., Y.E.H., C.P. and I.Q.M.N.; methodology, G.L.L.-N. and I.Q.M.N.; software, G.L.L.-N. and I.Q.M.N.; validation, G.L.L.-N. and I.Q.M.N.; formal analysis, G.L.L.-N., K.B. and I.Q.M.N.; investigation G.L.L.-N., K.B. and I.Q.M.N.; resources, G.L.L.-N., K.B., D.N., Y.E.H. and C.P.; data curation, G.L.L.-N., K.B. and I.Q.M.N.; writing-original draft preparation, G.L.L.-N. and I.Q.M.N.; writing-review and editing G.L.L.-N. and I.Q.M.N.; visualization, G.L.L.-N. and I.Q.M.N.; supervision, G.L.L.-N., K.B., D.N., Y.E.H. and C.P.; project administration, G.L.L.-N., K.B., D.N., Y.E.H. and C.P.; funding acquisition, G.L.L.-N., K.B., D.N., Y.E.H. and C.P. All authors have agreed to the published version of the manuscript.

Funding: This research received no external funding.

Acknowledgments: The authors of this work express their sincere gratitude to the staff of the Institute of Research on Ceramics (IRCER, UMR CNRS 7315) for their availability during this work. The CNRS-JRP GRAAL-EMC2 project is gratefully acknowledged.

Conflicts of Interest: The authors declare no conflict of interest.

References

1. Programme solidarité eau Conservation et Traitement de l'eau à Domicile. Available online: https://www.pseau.org/outils/ouvrages/ps_eau_conservation_et_traitement_de_l_eau_a_domicile_2018.pdf (accessed on 17 December 2024).
2. Misrar, W.; Loutou, M.; Saadi, L.; Mansori, M.; Waqif, M.; Favotto, C. Cordierite Containing Ceramic Membranes from Smectetic Clay Using Natural Organic Wastes as Pore-Forming Agents. *J. Asian Ceram. Soc.* **2017**, *5*, 199–208. [CrossRef]
3. Talidi, A.; Saffaj, N.; Kacemi, K.E.; Younssi, S.A.; Albizane, A.; Chakir, A. Processing and Characterization of Tubular Ceramic Support for Microfiltration Membrane Prepared from Pyrophyllite Clay. *Sci. Study Res. Chem. Chem. Eng. Biotechnol. Food Ind.* **2011**, *12*, 263–268.
4. Sarkar, S.; Bandyopadhyay, S.; Larbot, A.; Cerneaux, S. New Clay–Alumina Porous Capillary Supports for Filtration Application. *J. Membr. Sci.* **2012**, *392–393*, 130–136. [CrossRef]
5. Benito, J.M.; Sánchez, M.J.; Pena, P.; Rodríguez, M.A. Development of a New High Porosity Ceramic Membrane for the Treatment of Bilge Water. *Desalination* **2007**, *214*, 91–101. [CrossRef]
6. Achiou, B.; Elomari, H.; Ouammou, M.; Albizane, A.; Bennazha, J.; Alami Younssi, S.; El Amrani, I.E.; Aaddane, A. Elaboration and Characterization of Flat Ceramic Microfiltration Membrane Made from Natural Moroccan Pozzolan (Central Middle Atlas). *J. Mater. Environ. Sci.* **2016**, *7*, 196–204.
7. Amin, S.K.; Abdallah, H.A.M.; Roushdy, M.H.; El-Sherbiny, S.A. An Overview of Production and Development of Ceramic Membranes. *Int. J. Appl. Eng. Res.* **2016**, *11*, 7708–7721.
8. Li, N.N.; Fane, A.G.; Ho, W.S.W.; Matsuura, T. *Advanced Membrane Technology and Applications*; John Wiley & Sons: Hoboken, NJ, USA, 2008.
9. Barry, K.; Lecomte-nana, G.L.; Seynou, M.; Faucher, M.; Blanchart, P.; Peyratout, C. Comparative Properties of Porous Phyllosilicate-Based Ceramics Shaped by Freeze-Tape Casting. *Ceramics* **2022**, *5*, 75–96. [CrossRef]
10. Medri, V.; Mazzocchi, M.; Bellosi, A. ZrB2-Based Sponges and Lightweight Devices. *Int. J. Appl. Ceram. Technol.* **2011**, *8*, 815–823. [CrossRef]
11. Dai, P. Research Progress on Preparation of Porous Ceramics. *Interceram-Int. Ceram. Rev.* **2015**, *64*, 100–103.

12. Neeraj, V.S.; Wilson, P.; Vijayan, S.; Prabhakaran, K. Porous Ceramics with a Duplex Pore Structure by Compression Molding of Alumina-NaCl Paste in Molten Sucrose. *Ceram. Int.* **2017**, *43*, 14107–14113. [CrossRef]
13. Zhang, F.-Z.; Kato, T.; Fuji, M.; Takahashi, M. Gelcasting Fabrication of Porous Ceramics Using a Continuous Process. *J. Eur. Ceram. Soc.* **2006**, *26*, 667–671. [CrossRef]
14. Mohamed, M.; Dayirou, N.; Mohamed, H.; Andre, N.; Gisele Laure, L.N.; Daniel, N. Effect of Porogenic Agent Type and Firing Temperatures on Properties of Low-Cost Microfiltration Membranes from Kaolin. *Trans. Indian Ceram. Soc.* **2020**, *79*, 1692695. [CrossRef]
15. Wang, S.; Chen, H.; Li, Y.; Ma, H.; Liu, X.; Zhang, S.; Jia, Q. A Novel Strategy for Synthesizing Porous ZrB_2-SiC Ceramics via Boro/Carbothermal Reaction Process Templated Pore-Forming Approach. *J. Eur. Ceram. Soc.* **2023**, *43*, 3905–3916. [CrossRef]
16. Li, N.; Zhang, X.-Y.; Qu, Y.-N.; Xu, J.; Ma, N.; Gan, K.; Huo, W.-L.; Yang, J.-L. A Simple and Efficient Way to Prepare Porous Mullite Matrix Ceramics via Directly Sintering SiO_2-Al_2O_3 Microspheres. *J. Eur. Ceram. Soc.* **2016**, *36*, 2807–2812. [CrossRef]
17. Zhang, M.; Sun, H.; Liu, X.; Sui, H.; Xiao, S. Structural Design of PZT Porous Ceramics Obtained via Free-Casting by Ice-Templating and Performance Exploration. *Mater. Res. Bull.* **2020**, *127*, 110862. [CrossRef]
18. Zhu, Y.; Guo, B.; Zuo, W.; Jiang, K.; Chen, H.; Ku, J. Effect of Sintering Temperature on Structure and Properties of Porous Ceramics from Tungsten Ore Tailings. *Mater. Chem. Phys.* **2022**, *287*, 126315. [CrossRef]
19. Deckers, J.; Vleugels, J.; Kruth, J.-P. Additive Manufacturing of Ceramics: A Review. *J. Ceram. Sci. Technol.* **2014**, *5*, 245–260.
20. Zocca, A.; Colombo, P.; Gomes, C.M.; Gunster, J. Additive Manufacturing of Ceramics: Issues, Potentialities, and Opportunities. *J. Am. Ceram. Soc.* **2015**, 1983–2001. [CrossRef]
21. Liu, X.; Lv, C.; Liu, B.; Wang, G.; Duan, W.; Gu, Y. Digital Light Processing 3D Printing of Porous Ceramics Based on Multi-Materials Additive Manufacturing. *Chin. J. Struct. Chem.* **2023**, *42*, 100106. [CrossRef]
22. Melo, C.; Moraes, A.; Oliveira Rocco, F.; Montilha, F.S.; Canto, R. A Validation Procedure for Numerical Models of Ceramic Powder Pressing. *J. Eur. Ceram. Soc.* **2018**. [CrossRef]
23. Patro, P.; Lenka, R.; Mahata, T. Shape Forming and Sintering of Ceramics. In *Handbook on Synthesis Strategies for Advanced Materials: Volume-II: Processing and Functionalization of Materials*; Springer: Berlin/Heidelberg, Germany, 2022; pp. 1–54, ISBN 978-981-16-1802-4.
24. Zhou, L.; Li, Z.; Zhu, Y. Porous Silica/Mullite Ceramics Prepared by Foam-Gelcasting Using Silicon Kerf Waste as Raw Material. *Mater. Lett.* **2019**, *239*, 67–70. [CrossRef]
25. Maca, K.; Dobsak, P.; Boccaccini, A.R. Fabrication of Graded Porous Ceramics Using Alumina–Carbon Powder Mixtures. *Ceram. Int.* **2001**, *27*, 577–584. [CrossRef]
26. Bai, J. Fabrication and Properties of Porous Mulite Ceramics from Calcined Carbonaceous Kaolin and α-Al_2O_3. *Ceram. Int.* **2010**, 673–678. [CrossRef]
27. Wu, J.-M.; Li, M.; Liu, S.-S.; Shi, Y.-S.; Li, C.-H.; Wang, W. Selective Laser Sintering of Porous Al_2O_3-Based Ceramics Using Both Al_2O_3 and SiO_2 Poly-Hollow Microspheres as Raw Materials. *Ceram. Int.* **2021**, *47*, 15313–15318. [CrossRef]
28. Xu, X.; Fu, R.; Chen, K.; Ferreira, J.M.F. Cost-Effective Fabrication of Porous α-SiAlON Bonded β-SiAlON Ceramics. *Mater. Lett.* **2005**, *59*, 2601–2604. [CrossRef]
29. Bai, J.; Wei, C.; Meng, F.; Liu, J.; Wang, P.; Du, Q.; Tang, Z. Fabrication of Porous Al_2O_3–$MgAl_2O_4$ Ceramics Using Combustion-Synthesized Powders Containing in Situ Produced Pore-Forming Agents. *Mater. Lett.* **2011**, *65*, 1559–1561. [CrossRef]
30. Fan, F.; Xu, J.; Yang, R.; Lin, L.; Wang, H.; Shao, Y.; Zhang, P.; Gao, F. Enhanced Thermal Insulation and Mechanical Properties of Y-α-SiAlON Porous Ceramics with Sr Doping. *J. Mater. Res. Technol.* **2023**, *25*, 6074–6086. [CrossRef]
31. Xue, T.; Wang, Z.J.W. Preparation of Porous SiC Ceramics from Waste Cotton Linter by Reactive Liquid Si Infiltration Technique. *Mater. Sci. Eng. A* **2010**, *527*, 7294–7298. [CrossRef]
32. Huo, X.; Xia, B.; Hu, T.; Zhang, M.; Guo, M. Effect of MoO_3 Addition on Fly Ash Based Porous and High-Strength Mullite Ceramics: In Situ Whisker Growth and Self-Enhancement Mechanism. *Ceram. Int.* **2023**, *49*, 21069–21077. [CrossRef]
33. Ndzana, E.J.A.; Njuhou, S.; Pountouenchi, A.; Lecomte-Nana, G.; Njoya, D. Comparative Study of Alluvial Clay-Based Stoneware Ceramic's Properties Using Cocoa Cortex Ash and Syenite as Fluxing Agents. *Open Ceram.* **2024**, *17*, 100547. [CrossRef]
34. Winans, K.; Kendall, A.; Deng, H. The History and Current Applications of the Circular Economy Concept. *Renew. Sustain. Energy Rev.* **2017**, *68*, 825–833. [CrossRef]
35. Wietschel, L.; Halter, F.; Thorenz, A.; Schüppel, D.; Koch, D. Literature Review on the State of the Art of the Circular Economy of Ceramic Matrix Composites. *Open Ceram.* **2023**, *14*, 100357. [CrossRef]
36. Gonon, M. Case Studies in the X-Ray Diffraction of Ceramics. In *Encyclopedia of Materials: Technical Ceramics and Glasses*; Pomeroy, M., Ed.; Elsevier: Oxford, UK, 2021; pp. 560–577, ISBN 978-0-12-822233-1.
37. Carty, W.M.; Senapati, U. Porcelain - Raw Materials, Processing, Phase Evolution, and Mechanical Behavior. *J. Am. Ceram. Soc.* **1998**, *81*, 3–20. [CrossRef]
38. Truche, C. Caractérisation et Quantification des Minéraux Argileux dans les sols Expansifs par Spectroscopie Infrarouge aux Echelles du Terrain et du Laboratoire. Ph.D. Thesis, Université Paul Sabatier-Toulouse III, Toulouse, France, 2010.

39. da Silva, M.T.Q.S.; Perretto, F.; do Rocio Cardoso, M.; Mazer, W. Porosity: Some Characterization Techniques. *Mater. Today Proc.* **2023**. [CrossRef]
40. Rouquerol, J.; Baron, G.V.; Denoyel, R.; Giesche, H.; Groen, J.; Klobes, P.; Levitz, P.; Neimark, A.V.; Rigby, S.; Skudas, R.; et al. The Characterization of Macroporous Solids: An Overview of the Methodology. *Microporous Mesoporous Mater.* **2012**, *154*, 2–6. [CrossRef]
41. Doubi, H.G.; Kouamé, A.N.; Konan, L.K.; Tognonvi, M.; Oyetola, S. Thermal Conductivity of Compressed Earth Bricks Strengthening by Shea Butter Wastes with Cement. *Mater. Sci. Appl.* **2017**, *7*, 848–858. [CrossRef]
42. Wang, L.; Ma, B.; Ren, X.; Yu, C.; Tian, J.; Liu, C.; Deng, C.; Hu, C.; Liu, Z.; Yu, J.; et al. Phase-Engineering Strategy of ZrO_2 for Enhancing the Mechanical Properties of Porous Cordierite Ceramics. *Mater. Today Commun.* **2022**, *30*, 103032. [CrossRef]
43. Kamgang, S.P.; Njoya, D.; Kamseu, E.; Cyr, L.C. de S.; Marcano-Zerpa, A.; Balme, S.; Bechelany, M.; Soussan, L. Elaboration of a New Ceramic Membrane Support from Cameroonian Clays, Coconut Husks and Eggshells: Application for *Escherichia coli* Bacteria Retention. *Appl. Clay Sci.* **2020**, *198*, 105836. [CrossRef]
44. Scarinci, G.; Brusatin, G.; Barbieri, L.; Corradi, A.; Lancellotti, I.; Colombo, P.; Hreglich, S.; Dall'Igna, R. Vitrification of Industrial and Natural Wastes with Production of Glass Fibres. *J. Eur. Ceram. Soc.* **2000**, *20*, 2485–2490. [CrossRef]
45. Görhan, G.; Şimşek, O. Porous Clay Bricks Manufactured with Rice Husks. *Constr. Build. Mater.* **2013**, *40*, 390–396. [CrossRef]
46. Dong, Y.; Liu, X.; Ma, Q.; Meng, G. Preparation of Cordierite-Based Porous Ceramic Micro-Filtration Membranes Using Waste Fly Ash as the Main Raw Materials. *J. Membr. Sci.* **2006**, *285*, 173–181. [CrossRef]
47. Zhu, L.; Dong, Y.; Hampshire, S.; Cerneaux, S.; Winnubst, L. Waste-to-Resource Preparation of a Porous Ceramic Membrane Support Featuring Elongated Mullite Whiskers with Enhanced Porosity and Permeance. *J. Eur. Ceram. Soc.* **2015**, *35*, 711–721. [CrossRef]
48. Manni, A.; El Haddar, A.; El Amrani El Hassani, I.-E.; El Bouari, A.; Sadik, C. Valorization of Coffee Waste with Moroccan Clay to Produce a Porous Red Ceramics (Class BIII). *Bol. Soc. Esp. Ceram. Vidr.* **2019**, *58*, 211–220. [CrossRef]
49. Nigay, P.M.; Sani, R.; Cutard, T.; Nzihou, A. Modeling of the Thermal and Mechanical Properties of Clay Ceramics Incorporating Organic Additives. *Mater. Sci. Eng. A* **2017**, *708*, 375–382. [CrossRef]
50. Zhu, L.; Li, S.; Li, Y.; Xu, N. Novel Applications of Waste Ceramics on the Fabrication of Foamed Materials for Exterior Building Walls Insulation. *Constr. Build. Mater.* **2018**, *180*, 291–297. [CrossRef]
51. Monich, P.R.; Romero, A.R.; Rambaldi, E.; Bernardo, E. Case Studies of Up-Cycling of Partially Crystallized Ceramic Waste in Highly Porous Glass-Ceramics. *Constr. Build. Mater.* **2020**, *261*, 119971. [CrossRef]
52. Simonis, J.J.; Basson, A.K. Manufacturing a Low-Cost Ceramic Water Filter and Filter System for the Elimination of Common Pathogenic Bacteria. *Phys. Chem. Earth* **2012**, *50–52*, 269–276. [CrossRef]
53. Hammel, E.C.; Ighodaro, O.L.-R.; Okoli, O.I. Processing and Properties of Advanced Porous Ceramics: An Application Based Review. *Ceram. Int.* **2014**, *40*, 15351–15370. [CrossRef]
54. Sobsey, M.D.; Stauber, C.E.; Casanova, L.M.; Brown, J.M.; Elliott, M.A. Point of Use Household Drinking Water Filtration: A Practical, Effective Solution for Providing Sustained Access to Safe Drinking Water in the Developing World. *Environ. Sci. Technol.* **2008**, *42*, 4261–4267. [CrossRef] [PubMed]
55. Shirasaki, N.; Matsushita, T.; Matsui, Y.; Urasaki, T.; Oshiba, A.; Ohno, K. Evaluation of Norovirus Removal Performance in a Coagulation-Ceramic Microfiltration Process by Using Recombinant Norovirus Virus-like Particles. *Water Sci. Technol.* **2010**, *61*, 2027–2034. [CrossRef] [PubMed]
56. Barry, K.; Lecomte-Nana, G.L.; Sory, N.; Ouedraogo, M.; Sawadogo, L.; Sawadogo, M.; Sanou, I.; Seynou, M.; Zerbo, L.; Blanchart, P. Mullite Effect on the Ceramic Filters Effectiveness in the Removal of Arsenic from Borehole Water from Burkina Faso. *Open Ceram.* **2024**, *20*, 100679. [CrossRef]
57. Mansour, H.B.; Boughzala, U.; Dridi, D.; Barillier, D.; Chekir-Ghedira, L.; Mosrati, R. Textiles Dyes as a Source of Wastewater Contamination: Screening of the Toxicity and Treatment Methods; [Les Colorants Textiles Sources de Contamination de l'eau: CRI BLAGE de La Toxicité et Des Methods Detraitement]. *Rev. Sci. L'eau* **2011**, *24*, 209–238.
58. Pei Ling, Y.; Ooi, C.-H.; Matsumoto, A.; Yeoh, F.-Y. Properties Evaluation and Fabrication of Green Clay Reformulated from Water Sludge. *Ceram. Int.* **2018**, *44*, 1411–1419. [CrossRef]
59. Hartal, O.; Madinzi, A.; Khattabi Rifi, S.; Haddaji, C.; Agustiono Kurniawan, T.; Anouzla, A.; Souabi, S. Optimization of Coagulation-Flocculation Process for Wastewater Treatment from Vegetable Oil Refineries Using Chitosan as a Natural Flocculant. *Environ. Nanotechnol. Monit. Manag.* **2024**, *22*, 100957. [CrossRef]
60. Romphophak, P.; Chamnanmor, R.; Fagkaew, P.; Sairiam, S.; Painmanakul, P. Modelling the Predictive Analysis of Turbidity Removal Efficiency in the In-Line Coagulation and Flocculation Process. *Chem. Eng. Res. Des.* **2024**, *210*, 301–310. [CrossRef]
61. Halim, M.A.; Waliul Hoque, S.A.M.; Hossain, M.K.; Saadat, A.H.M.; Goni, M.A.; Saiful Islam, M. Arsenic Removal Properties of Laterite Soil by Adsorption Filtration Method. *J. Appl. Sci.* **2008**, *8*, 3757–3760. [CrossRef]
62. Ndé-Tchoupé, A.I.; Konadu-Amoah, B.; Gatcha-Bandjun, N.; Hu, R.; Gwenzi, W.; Noubactep, C. Kanchan Arsenic Filters for Household Water Treatment: Unsuitable or Unsustainable? *Water* **2022**, *14*, 2318. [CrossRef]

63. Shafiquzzaman, M.; Hasan, M.M.; Haider, H.; Bari, Q.H.; EL-Ghoul, Y.; Nakajima, J. Arsenic Removal by Household-Based Ceramic Filters: Evaluating Mode of Operations and Influence of Groundwater Compositions. *J. Water Process Eng.* **2022**, *46*, 102598. [CrossRef]
64. Kamarudin, N.H.; Harun, Z.; Othman, M.H.D.; Abdullahi, T.; Syamsul Bahri, S.; Kamarudin, N.H.; Yunos, M.Z.; Wan Salleh, W.N. Waste Environmental Sources of Metakaolin and Corn Cob Ash for Preparation and Characterisation of Green Ceramic Hollow Fibre Membrane (h-MCa) for Oil-Water Separation. *Ceram. Int.* **2020**, *46*, 1512–1525. [CrossRef]
65. Mouafon, M.; Lecomte-Nana, G.L.; Tessier-Doyen, N.; Njoya, A.; Njoya, D.; Njopwouo, D. Processing and Characterization of Low-Thermal Conductivity, Clay-Based Ceramic Membranes for Filtering Drinking Water. *Clays Clay Miner.* **2021**, *69*, 339–353. [CrossRef]
66. Shafiquzzaman, M.; Azam, M.S.; Nakajima, J.; Bari, Q.H. Arsenic Leaching Characteristics of the Sludges from Iron Based Removal Process. *Desalination* **2010**, *261*, 41–45. [CrossRef]
67. Ayral, A.; Hulea, V.; Jean-Pierre, J.; Julbe, A. Céramiques Pour l'environnement: Filtres, Membranes, Adsorbants et Catalyseurs. *Chim. Verte* **2014**. [CrossRef]
68. Rayner, J.; Skinner, B.; Lantagne, D. Current Practices in Manufacturing Locally-Made Ceramic Pot Filters for Water Treatment in Developing Countries. *J. Water Sanit. Hyg. Dev.* **2013**, *3*, 252–261. [CrossRef]
69. Salvinelli, C.; Elmore, A.C.; Reidmeyer, M.R.; Drake, K.D.; Ahmad, K.I. Characterization of the Relationship between Ceramic Pot Filter Water Production and Turbidity in Source Water. *Water Res.* **2016**, *104*, 28–33. [CrossRef]

ceramics

MDPI

Article

Analysis of the Structure and Durability of Refractory Castables Impregnated with Sodium Silicate Glass

Jurgita Malaiškienė *, Valentin Antonovič, Renata Boris, Andrius Kudžma and Rimvydas Stonys

Laboratory of Composite Materials, Institute of Building Materials, Vilnius Gediminas Technical University, 10223 Vilnius, Lithuania; valentin.antonovic@vilniustech.lt (V.A.); renata.boris@vilniustech.lt (R.B.); andrius.kudzma@vilniustech.lt (A.K.); rimvydas.stonys@vilniustech.lt (R.S.)

* Correspondence: jurgita.malaiskiene@vilniustech.lt; Tel.: +370-5-2512329

Abstract: This study examines the impact of the impregnation of fireclay-based conventional (CC) and medium-cement castables (MCCs) with liquid sodium silicate glass under vacuum conditions. The goal is to assess how this treatment affects the physical and mechanical properties and durability (alkali and thermal shock resistance) of these castables used in biomass combustion boilers, where they are exposed to temperatures up to 1100 °C. The research work employs standard test methods to evaluate the physical and mechanical properties. Additionally, advanced techniques such as scanning electron microscopy (SEM), energy-dispersive X-ray spectroscopy (EDS), X-ray diffraction (XRD), and specific tests for alkali resistance and thermal shock resistance are used. The research findings suggest that impregnation with sodium silicate glass under vacuum significantly enhances the alkali resistance of both CC and MCCs. This improvement is primarily due to the reduction in porosity and the increase in density. SEM images reveal that the impregnated samples are coated with a glassy layer and the pores are partially filled with sodium silicate. Tests for alkali resistance demonstrate the formation of a protective glassy layer (with a thickness of 0.9–1.5 mm) on the castable surfaces, thereby reducing the further penetration of alkali into deeper layers of the samples. However, it is important to mention that the impregnated refractory castables have reduced resistance to thermal shock cycles.

Keywords: refractory castable; impregnation; liquid sodium silicate glass; alkali resistance; physical and mechanical properties; thermal shock resistance

1. Introduction

In recent decades, there has been a widespread adoption of biomass combustion boilers used as a cost-effective option for energy production [1,2]. These boilers utilize renewable energy sources such as wood, straw, mixed municipal waste, and other materials to reduce the reliance on fossil fuels in order to align with ambitious targets set by the European Union aiming for at least 27% renewable energy by 2030 [3]. The lining of these biomass boilers is constructed using various refractory materials, including castables, which are carefully selected for their ability to withstand high temperatures and to maintain structural integrity over extended periods [4,5]. For instance, in the demanding conditions of some species of biomass combustion where temperatures can range up to 1200 °C, it is advised to opt for chemically resistant refractory castables containing SiC, or alumina-chrome, engineered to withstand extreme operating conditions [4].

However, wood biofuel boilers predominantly operate at temperatures below 1100 °C, making them prime candidates for castables enriched with aluminosilicate aggregates like fireclay or mullite [5]. Regrettably, these castables exhibit limited resistance to alkali corrosion, which often leads to premature degradation attributed to the corrosive influence of alkalis present in the refractory materials and fuel [6]. The genesis of alkali corrosion lies in the formation of potassium/sodium oxides and their corresponding salts, a consequence of biomass combustion. The alkaline content of the resultant ash is influenced by a myriad of factors, including wood type, grade, and quality [7]. Furthermore, the use of alkaline salts, despite preventing wood fuel freezing in winter conditions, significantly amplifies the concentration of alkaline compounds (sodium and potassium), thereby lowering the melting point of the ash and potentially causing it to fall below the boiler's operational temperature [8].

The intricate reactions responsible for reducing the ash/slag melting point are rooted in the interaction between potassium and carbon in various forms, culminating in the formation of alkaline compounds [9]. Corrosion, in this context, ensues from the dissolution of the refractory material and its intricate interplay with the alkaline molten phase. Additionally, it can manifest as reactions with alkaline vapors, the presence of alkaline compounds in liquid or solid-state, and the infiltration of alkaline vapors or the molten phase into the material's porous structure, thereby giving rise to altered zones within the material matrix [10]. These alkali reaction byproducts, including leucite ($KAlSi_2O_6$) and kalsilite ($KAlSiO_4$), demonstrate a peculiar characteristic by occupying volumes greater than that of the original material. This phenomenon is commonly referred to as "alkali bursting" [11–13]. Importantly, the extent of alkali reaction damage is linked to the porosity of the material [14], with materials of lower porosity and superior gas impermeability displaying greater resistance to such forms of damage [15,16].

A prevalent choice for binders in refractory castables is calcium aluminate cement [17]. In comparison to conventional castables, low-cement and ultra-low-cement castables have surged in popularity, thanks to their lower porosity, higher mechanical strength within critical temperature ranges, impressive erosion resistance, and elevated refractoriness [18–20]. These castables incorporate various ultra-dispersive additives, such as silica fume and calcined alumina, alongside deflocculants that effectively diminish cement content to a mere fraction compared to conventional refractory castables [21]. However, it is important to note that these low and ultra-low-cement castables often prove sensitive to factors such as ambient conditions, water content, mixing parameters, and more, making the production process challenging [22–24].

Meanwhile, medium-cement castables have gained recognition for their capacity to maintain consistent quality even under varying manufacturing conditions, setting them apart as a favorable choice for biofuel boiler installations [22,23]. While the lower porosity of castables can undoubtedly enhance their durability within alkaline environments, it is important to acknowledge that over time, alkali may still infiltrate the castable structure and engage with aluminosilicate aggregates [21]. One promising avenue for mitigating alkali-induced damage involves the addition of ground quartz sand to fireclay refractory castables, a practice known to stimulate the formation of a protective glassy layer on the castable surface when subjected to alkali salts at elevated temperatures [24,25]. However, it is imperative to exercise caution when determining the quantity of the quartz sand added due to the polymorphic transformations of SiO_2 that transpire at specific temperature thresholds [26].

Impregnation (immersion of the samples in the special fluid) of the material could also be an effective method to substantially decrease the porosity of refractory castables and increase their alkali resistance. In the field of impregnation technology, suspensions of various materials are employed [22,23], and the specific materials used for impregnation can vary in their roles. The authors in [27] reported a significant enhancement in corrosion resistance of a chromium–magnesium material after impregnation with suspensions containing nano-Cr_2O_3 and nano-Fe_2O_3 particles. The nanoparticles within the material pores dissolved when exposed to the slag melt, resulting in an increase in its viscosity and a reduction in the ability of the slag melt to penetrate. Research on the resistance of refractory concrete to alkali attack at 1100 °C temperature revealed that incorporating 2.5% milled quartz sand and impregnating the samples with a SiO_2 sol can lead to the formation of a protective glassy barrier. This barrier effectively blocks the further penetration of potassium and prevents additional cracking and disintegration of refractory concrete [21].

A cheaper material, such as liquid sodium silicate glass, could also be used as an impregnant for fireclay refractory castables [28]. Liquid sodium silicate glass, when employed as a binder in refractory castables, serves as a flux when subjected to high temperatures and reduces the required operating temperature for aluminosilicate refractory castables by approximately 200–300 °C. Research findings, however, indicate that refractory castables containing a sodium silicate liquid glass binder can achieve operating temperatures of 1100–1300 °C [28]. This means that the permitted use temperatures of such materials fully satisfy the conditions in which wood biofuel boilers are operated.

The purpose of this study was to assess how the properties (density, cold crushing strength, porosity, microstructure, as well as alkali and thermal shock resistance) of fireclay refractory castables change when they are impregnated with liquid sodium silicate glass under vacuum.

2. Materials and Methods

The characteristics and chemical composition of the used raw materials are presented in Table 1.

Table 1. Characteristics and chemical composition of raw materials.

Mark	Standard	Calcium Aluminate Cement		Reactive and Calcined Alumina		Microsilica RW-Fuller	Fireclay		Milled Quartz Sand
		Górkal 70	Istra-40	CTC 20	CT 19		Bos 135	Bos 145	
Producer	-	Górka Cement	Calucem GmbH	Almatis		RW Silicon GmbH	Tabex-Ozmo		AB Anykščių kvarcas
Country	-	Poland		Germany			Poland		Lithuania
Abbreviation	-	G70	ISTR	RA	CA	MS	FA35	FA45	MQS
Specific surface area, m^2/kg	ASTM C204-18	450	295	2100	400	-	-	-	490
Particle size, μm	ISO 13320 EN 12620	<63	<90	<20	<63	0.15	<4000	<4000	<50
The refractoriness, °C	ISO 1893	1630	1250	-	-	-	1710	1750	-
Oxides	**Chemical analysis, %**								
Al_2O_3		71.0	40.8	99.7	99.8	0.2	36.6	44.3	0.6
CaO		28.5	38.5	0.02	0.02	0.25	3.1	0.6	-
SiO_2		0.3	4.5	0.03	0.05	96.06	53.7	49.5	99.2
Fe_2O_3		0.2	15.0	0.03	0.04	0.05	3.3	2.1	0.05
MgO		-	1.2	0.01	-	0.4	0.7	0.9	-
K_2O		-	-	-	-	1.2	1.0	0.7	-
TiO_2		-	-	-	-	-	1.3	1.4	0.1

Additionally, the same fireclay after 1 h of milling and sieving through a sieve with a mesh size of 0.14 mm was used as fine aggregate. Milled quartz sand (MQS) was used to improve the fireclay refractories resistance to alkali attack [29]. The following deflocculants were used: Castament FS20 and FS30 from BASF Construction Solutions GmbH (Trostberg, Germany) and technical sodium tripolyphosphate NT.

Liquid sodium silicate glass with a SiO_2/Na_2O molar ratio of 3.3 and a density of 1.33 g/cm^3 was used for the impregnation of the samples. Drinking water was used to prepare the mixtures. The chemical reagent K_2CO_3 (99.0%, Sigma Aldrich, St. Louis, MI, USA) was used for the testing of alkali resistance.

Samples sized 70 × 70 × 70 mm and 160 × 40 × 40 mm (for thermal shock resistance) were prepared from the raw materials listed above. The compositions of the samples are presented in Table 2. For conventional refractory castables CC-1 and CC-2, a higher amount of ISTR cement was used and there was a reduced content of dispersive additives compared to the medium-cement refractory castables, in which a smaller amount of cement, additional RA and CA, and a higher amount of MS was used. The difference in marking 1 and 2 is the class of fireclay used, either FA35 or FA45.

Table 2. The compositions of refractory castables.

Composition, %	G70	ISTR	MS	FA35 <0.14 mm	FA45 <0.14 mm	FA35 <4 mm	FA45 <4 mm	RA	CA	MQS	Water
CC-1	-	25	2.5	10	-	60	-	-	-	2.5	11.3
CC-2	-	25	2.5	-	10	-	60	-	-	2.5	7.4
MCC-1	12	-	5	8	-	60.5	-	5	7	2.5	8.9
MCC-2	12	-	5	-	8	-	60.5	5	7	2.5	7.1

For CC deflocculant FS30 0.1% was used; for MCC, FS20 0.1% and NT 0.1% were used. Deflocculants and water were added up to 100%.

The dry components presented in Table 2 were mixed in a Hobart planetary mixer for 5 min and then mixed with water for 3 min. The samples were kept for 72 h at 20 ± 1 °C, relative humidity of about 50%, and were then dried at 110 °C for 48 h and fired at 1100 °C for 5 h. The preparation and treatment of the refractory concrete samples (curing, drying, and thermal treatment) were carried out according to the requirements of LST EN ISO 1927-5. Some of the fired samples were subsequently impregnated with liquid sodium silicate glass under vacuum (1 atm), dried, and fired in a similar manner.

Physical and mechanical characteristics (density, cold crushing strength) were determined according to LST EN ISO 1927-6:2013; three samples 70 × 70 × 70 mm from each composition were used for the testing. Density was calculated according to mass and volume (established according to the dimensions) ratio. The cold crushing strength was measured via hydraulic press ALPHA3-3000S (Riedlingen, Germany). The open porosity of the castable was determined according to LST EN ISO 10545–3. Porosity, expressed as a percentage, is the relationship of the volume of the open pores of the test specimen to its exterior volume. Thermal shock resistance was determined by evaluating the changes in the ultrasonic pulse velocity [28] when the samples are cyclically heated at 1100 °C and cooled between 2 metal plates filled with cold water. Before testing (*Uo*), the ultrasonic pulse velocity was measured after 3 (*U3c*) and 7 (*U7c*) cycles, and the relative thermal shock resistance R was calculated according to the formula:

$$R = \frac{\sqrt{U3c \times U7c}}{Uo} \times 100\% \tag{1}$$

The propagation time of ultrasound waves was determined using the Pundit 7 instrument (converter frequency = 54 kHz, Schleibinger Geräte Teubert u. Greim GmbH, Buchbach, Germany) to calculate the ultrasonic pulse velocity (*UPV*, m/s):

$$UPV = \frac{l}{\tau} \tag{2}$$

where l is the length of the specimen, m; τ is the signal propagation time, s.

The modulus of elasticity E was calculated from Equation (3):

$$E = U^2 \cdot \rho \frac{(1+\mu)(1-2\mu)}{1-\mu} \tag{3}$$

where U is the UPV (m/s), ρ is the density of concrete (kg/m^3), and μ is Poisson's coefficient, which is 0.17 for all types of concrete [30,31].

The crucible method (ASTM C 454–83) was used to evaluate the resistance of refractory castables to alkali corrosion. Castable cubes of 70 mm × 70 mm × 70 mm with a hole of diameter 20 mm and height 30 mm were produced (3 specimens per composition). The formed samples were cured in natural conditions for 72 h and then dried at a temperature of 110 ± 5 °C for 48 h and fired at 1100 ± 5 °C for 5 h. A total of 9 g of K_2CO_3 was added to the hole, which was covered with a castable plate of the same composition, and the specimens were fired at a temperature of 1100 ± 5 °C for 5 h. After the tests were repeated (9 g of K_2CO_3 was added for each test), the specimens were visually examined for microcracks. Some specimens were cut into two pieces along the cylindrical axis after 6 and 20 cycles, and the affected area of the castable was examined.

The chemical composition of the materials was determined using the X-ray fluorescence spectrometer ZSX Primus IV (Rigaku, Tokyo, Japan) equipped with a Rh tube with an anode voltage of 4 kV. Tablet-shaped samples with a diameter of 40 mm were prepared and compressed at 200 kN for analysis.

X-ray diffraction (XRD) analysis was performed using a DRON-7 diffractometer (Bourevestnik, Saint Petersburg, Russia) with Cu Kα (λ = 0.1541837 nm) radiation. The following test parameter values were set: 30 kV voltage, 12 mA current, and 2θ diffraction angle ranging from 4° to 60° with increments of 0.02° measured every 2 s. The existing phases were identified by comparing the XRD diffractograms with standard diffraction patterns provided by the International Centre for Diffraction Data (ICDD).

Microstructural analysis was performed using a JSM-7600F scanning electron microscope (JEOL, Tokyo, Japan). The analysis was performed at an accelerating voltage of 10 kV in the secondary electron mode for image formation. Initially, the surface was covered with a layer of electrically conducting material using a QUORUM Q150R ES device (Quorum Technologies Ltd., Lewes, UK). Some polished samples were used to evaluate the thickness and chemical composition of the potassium-treated surface layer. X-ray microanalysis was performed with the Energy Dispersion Spectrometer (EDS) Inca Energy 350 (Oxford Instruments, Oxford, UK) using Silicon Drift type detector X-Max20 (Oxford Instruments, Oxford, UK). The INCA software package (Part number 51-1720-001, Version 4, Oxford Instruments, Oxford, UK) was used. Three points per plane were tested, starting from the surface exposed to corrosion attack to going down every 100 μm, with 3 more points (1 μm^3) selected in the next plane. In this way the elemental composition of the test sample was determined down to 3000 μm. An area of corrosion-free material at 10 mm distance from the exposed surface was also tested (Figure 1).

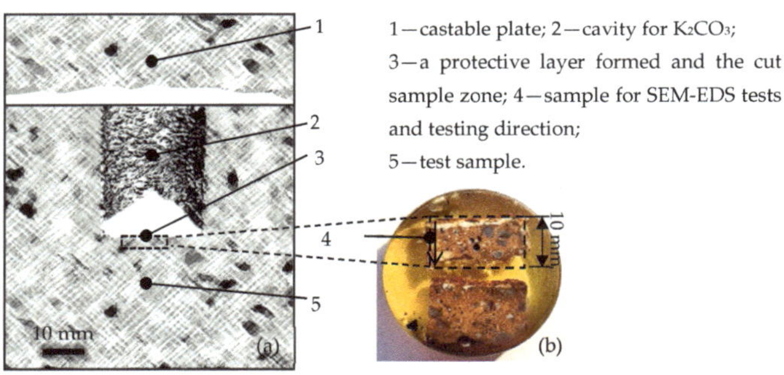

Figure 1. The scheme of the "crucible method" of a test sample (**a**) and a polished samples for SEM–EDS tests (**b**).

3. Results

3.1. The Impact of Impregnation on Refractory Castable Chemical and Mineral Composition and Microstructure after Firing at 1100 °C

The chemical composition of the samples (Table 3) showed that, irrespective of the type of concrete, there is a five times higher sodium oxide content on the surface of the samples impregnated with sodium silicate solution. A higher silica content up to 13–69%, depending on the composition of the concrete, was also observed.

It was determined that after firing at 1100 °C, sodium silicate influences the mineralogical composition of the refractory castable. XRD analysis (Figure 2) showed that the same compounds were identified in the control and impregnated samples (anorthite, gehlenite, mullite, tridymite, quartz, corundum, grossite), but the intensities of anorthite were much higher in the impregnated samples. According to the literature, anorthite has good physical properties, such as a low thermal expansion coefficient and good thermal shock resistance at high temperatures; thus, it could improve the properties of refractory castables [32].

Table 3. Chemical composition of control and impregnated castables.

Composition of Castable	Chemical Composition, Mass %									
	Al_2O_3	SiO_2	CaO	Fe_2O_3	K_2O	Na_2O	MgO	ZrO_2	TiO_2	P_2O_5
Control samples										
CC-1	29.3	35.4	22.6	9.20	0.27	0.20	0.51	0.09	1.59	0.43
CC-2	29.2	32.1	25.5	10.3	0.30	0.12	0.55	0.04	1.24	0.14
MCC-1	51.1	36.2	8.36	1.27	1.14	0.55	0.30	0.03	0.40	0.24
MCC-2	52.8	34.6	8.15	1.10	0.64	0.46	0.40	0.02	0.77	0.73
Impregnated samples										
CC-1	15.3	40.0	33.1	7.42	0.22	0.91	1.34	0.11	1.00	0.28
CC-2	11.5	41.3	35.2	8.41	0.19	0.63	1.20	0.06	0.93	0.36
MCC-1	30.2	51.7	12.2	1.15	1.10	2.60	0.37	0.05	0.00	0.14
MCC-2	22.0	58.6	12.6	1.42	2.11	1.96	0.53	0.02	0.23	0.38

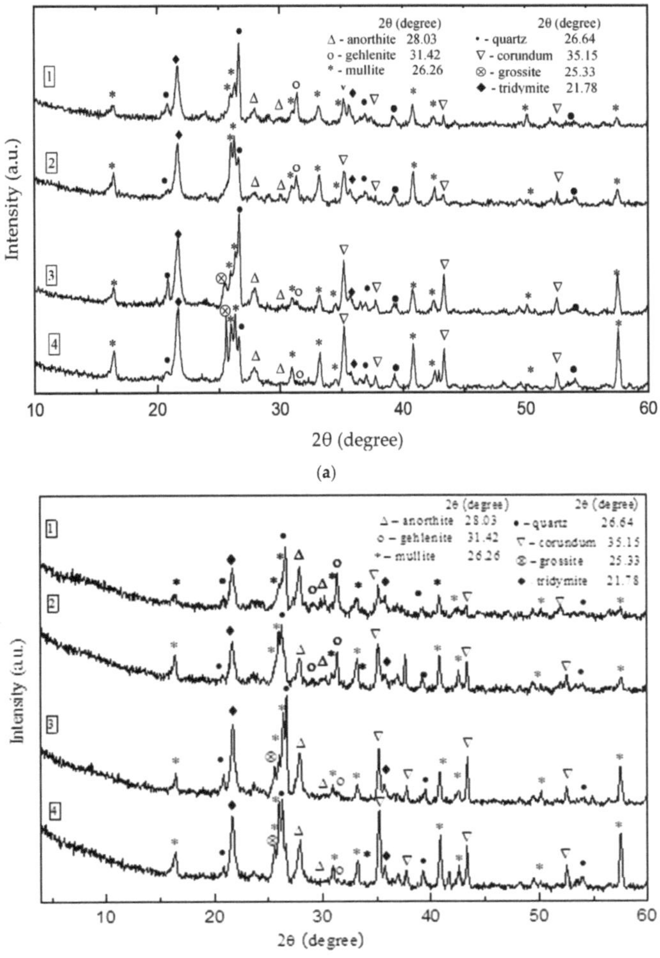

Figure 2. Xray analysis of control (**a**) and impregnated (**b**) samples. (1—CC-1, 2—CC-2, 3—MCC-1, 4—MCC-2).

SEM images show that the microstructure of the impregnated and fired at 1100 °C samples, regardless of composition, is denser (Figure 3a–d) compared with the control samples (Figure 3e–h). The porosity and diameter of many pores reduced when the refractory castable was exposed to liquid glass under vacuum as the resulting sodium silicate partially filled the pores and capillaries. After firing, water from liquid glass evaporated, forming hardened sodium silicate films that prevented the vaporization of residual moisture and could cause the bloating of sodium silicate mass. Sodium silicate at 300–500 °C includes amorphous silica, the crystallization of which starts at 600 °C; the crystalline phase vanishes completely at 900 °C, and then porosity drops with vitrification [33]. Additionally, at 900 °C and above, sodium silicate starts reacting with other refractory raw materials and forms new compounds, or a higher quantity of them, for example, anorthite (Figure 2).

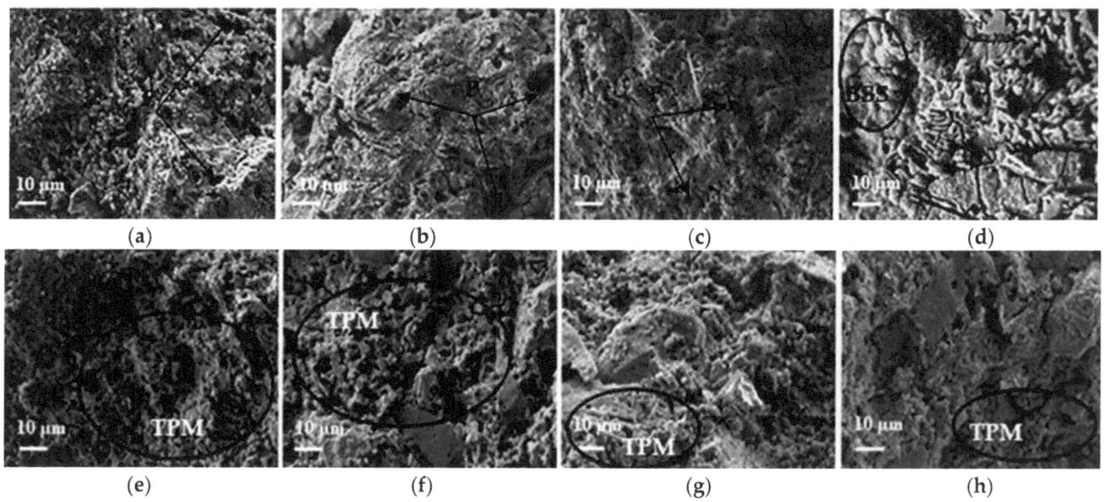

Figure 3. The images of control and impregnated samples (P—pore, BSS—blown sodium silicate, TPM—typical porous matrix): (**a**) CC-1- impregnated, (**b**) CC-2- impregnated, (**c**) MCC-1- impregnated, (**d**) MCC-2- impregnated, (**e**) CC-1 control, (**f**) CC-2 control, (**g**) MCC-1 control, (**h**) MCC-2 control.

The most visible changes in the microstructure were observed in the impregnated samples MCC-2, where the blown sodium silicate zones in the pores are visible (Figure 3d).

3.2. The Impact of Impregnation on Physical and Mechanical Properties of Refractory Castables after Firing at 1100 °C

The density and *UPV* values are shown in Figure 4. It is seen that both of the properties of the castables improved after impregnation. Density did not increase significantly, only 0.5–1.9%, but *UPV* increased from 2.6% to 6.5% due to the formation of a denser structure and the filling of some pores with liquid silicate glass. A more significant improvement in these properties was observed in samples CC-1 and MCC-1, where a lower-grade fireclay (FA35) was used, and this improvement can be related to the higher initial porosity of these samples compared to CC-2 and MCC-2 (Figure 5).

The cold crushing strength (Figure 6) of all types of refractory castables impregnated with liquid sodium silicate glass increased from approximately 10% to 18%. The increase in the cold crushing strength of the impregnated samples resulted from the formation of a denser structure and a decrease in porosity (Figure 3, Figure 5).

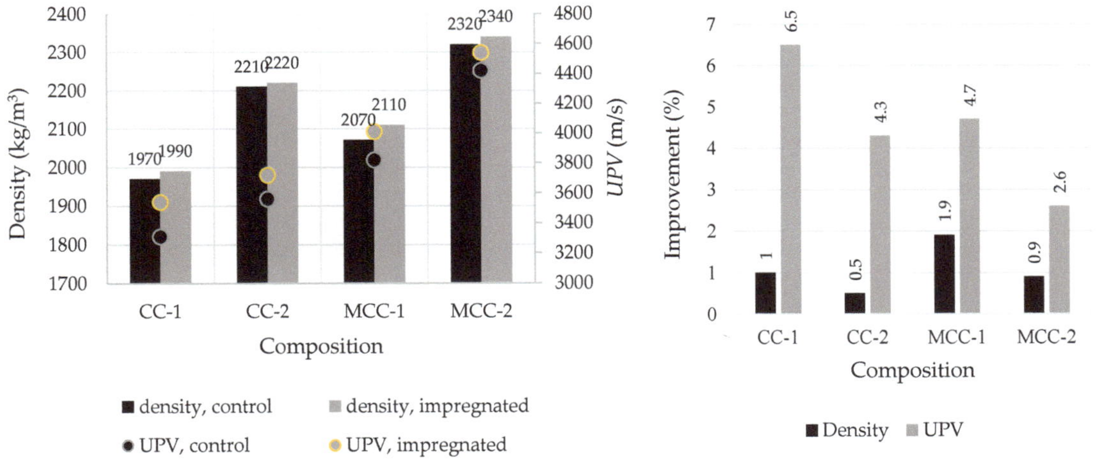

Figure 4. Results of density and *UPV* of control and impregnated samples.

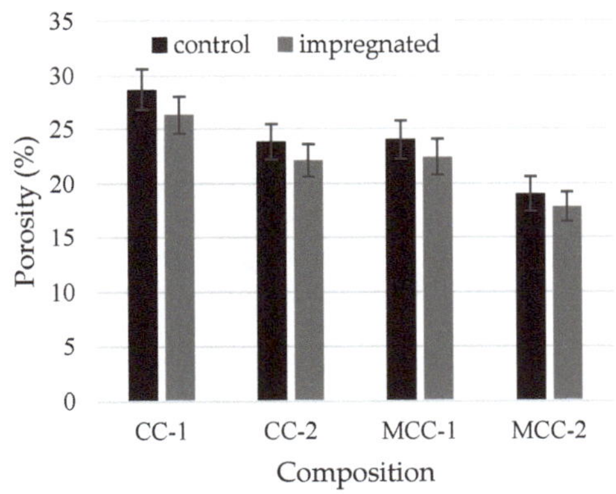

Figure 5. Porosity of control and impregnated samples.

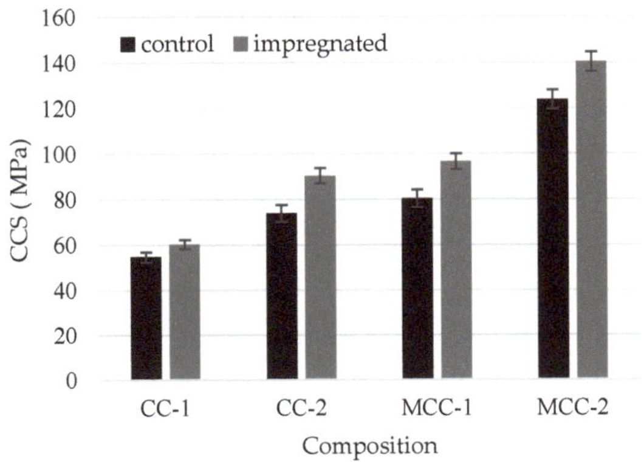

Figure 6. CCS of control and impregnated samples.

After impregnation, it was found that the porosity (Figure 5) of all types of refractory castables decreased by approximately 10% because the pores were filled with sodium silicate glass. Similar results are reported in article [34], but the authors used silica sol for the impregnation of fireclay refractory castable. In that case, the compressive strength increased approximately 40% because of SiO_2 sol filling the cracks and the presence of open pores in the samples, as well as the increased amount of mullite at high temperatures through the reaction of ultrafine Al_2O_3 with nano-sized SiO_2 particles. In this work, it was determined that cold crushing strength increases till 18%, but porosity decreases about 10%, and the improvement in strength could have resulted from chemical reactions and the formation of higher quantities of some thermally stable compounds, such as anorthite (Figure 2).

The diagram above (Figure 5) illustrates that the porosity of refractory castables varies significantly among different compositions, ranging from approximately 16% to around 29%. This difference in the compositions of the refractory castables was chosen to evaluate the effectiveness of impregnation for different refractory castables.

3.3. The Impact of Impregnation on Alkali Resistance at 1100 °C

According to conclusions presented in the literature [17–19], the densified microstructure of the surface should increase the alkali resistance of refractory castables with fireclay fillers at high temperatures. Figure 6 presents the images of the cut control and impregnated refractory castable samples after 20 cycles of alkali exposure. A clearly visible white glassy barrier with a width of about 1–2 mm was formed on the surface of impregnated sample holes, into which the potassium carbonate was added. At high temperatures, sodium from impregnant and potassium from reagent reacted with refractory compounds and formed sodium alumina silicate and potassium alumina silicate layers. Then, the viscosity of the formed glassy layer on the surface of the sample increased and alkali reagent (potassium) could not penetrate deeper into the material. Thus, the glassy layer acts as a protective barrier, making it difficult for the alkali to penetrate the material. According to the $Na_2O-Al_2O_3-SiO_2$ phase diagram [35], liquid sodium silicate can further react to form albite ($NaAlSi_3O_8$) and nepheline ($NaAlSiO_4$). Additionally, it was found [36] that Al_2O_3 can influence viscous glass melting temperature and its viscosity; therefore, the melting temperature and viscosity values of formed glass on the surface of the sample tend to increase [37].

Milled quartz additive participated in the building of the protective barrier in the control samples as well, but after 20 alkali exposure cycles, this barrier became weaker and the samples expanded, leading to expansion caused deformations at the top of the samples (Figure 7a–d).

A comparison of the depth of potassium penetration between the control and impregnated samples after six alkali corrosion cycles showed (Figure 8) that the impregnated samples had a significantly lower potassium content of about 40% in the surface layer and 30–80% at 10 mm depth from the surface. Hence, sample impregnation significantly reduces potassium penetration into refractory castables and increases the resistance to alkali corrosion due to the formation of a more stable protective barrier.

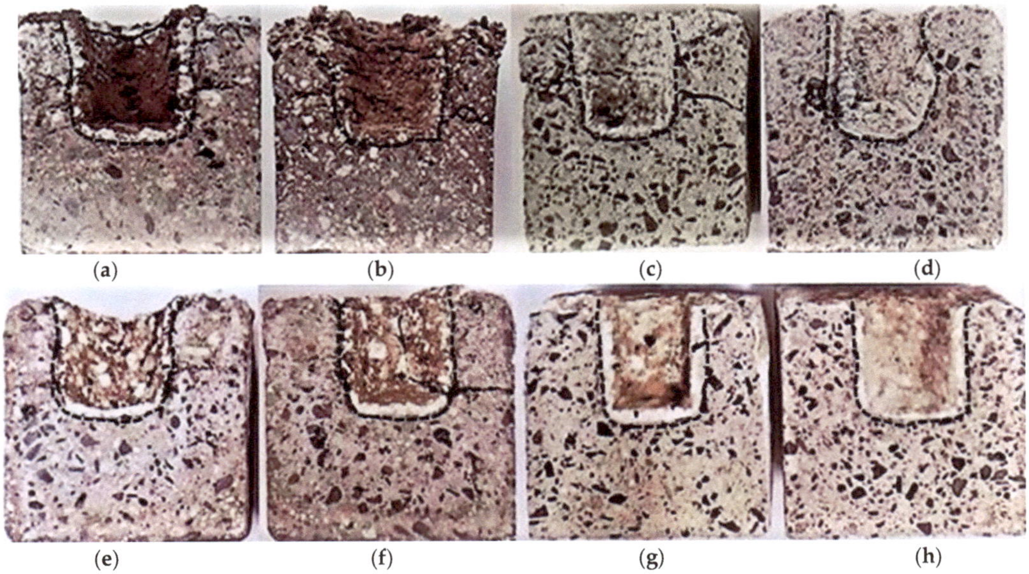

Figure 7. Images of impregnated refractory castable samples after 20 cycles of alkaline exposure (dashed line—zone of protestive glassy barrier formation): (**a**) CC-1 control, (**b**) CC-2 control, (**c**) MCC-1 control, (**d**) MCC-2 control, (**e**) CC-1- impregnated, (**f**) CC-2- impregnated, (**g**) MCC-1- impregnated, (**h**) MCC-2- impregnated.

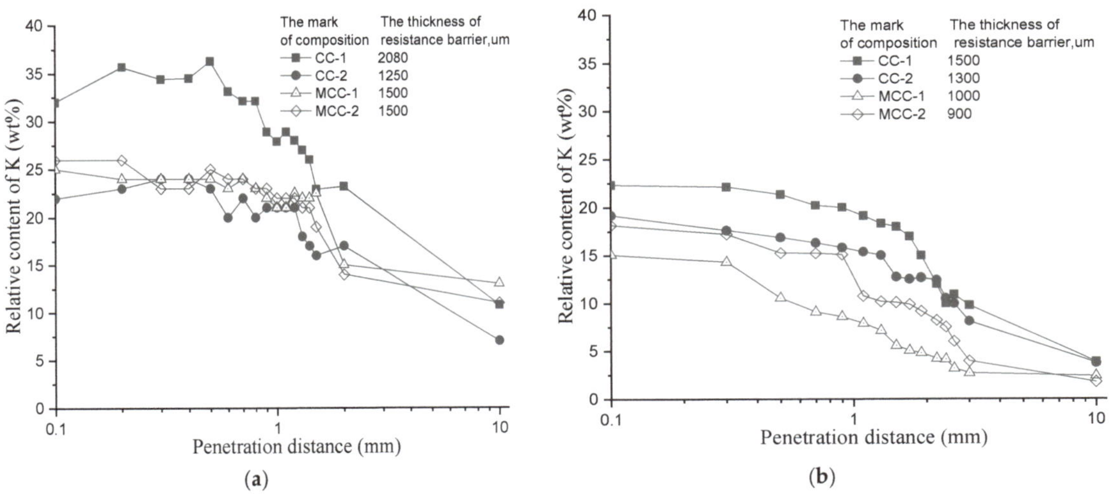

Figure 8. Relative content of potassium at different penetration distances: (**a**) control samples; (**b**) impregnated samples.

3.4. The Impact of Impregnation on Elasticity Modulus and Thermal Shock Resistance at 1100 °C

After the impregnation, the modulus of elasticity of the refractory castable (Figure 9) samples CC-1, CC-2, MCC-1, and MCC-2 increased by 16%, 10%, 13%, and 7%, respectively. Following the principles of thermoelastic theory concerning crack nucleation, engineers and researchers often employ the stress-to-elastic modulus ratio (σ/E) as a measure to evaluate the capacity of a material to withstand the initiation of cracks [33,38]. However, a material possessing a higher modulus of elasticity tends to exhibit diminished thermal shock resistance [21,34,39], as well as the result of the work illustrated in Figure 10.

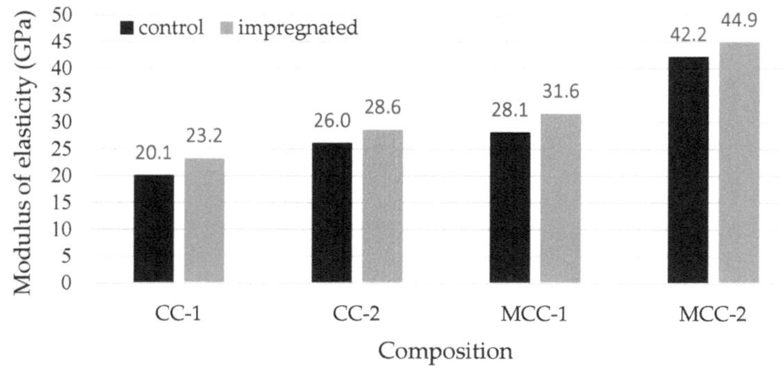

Figure 9. Modulus of elasticity of control and impregnated samples.

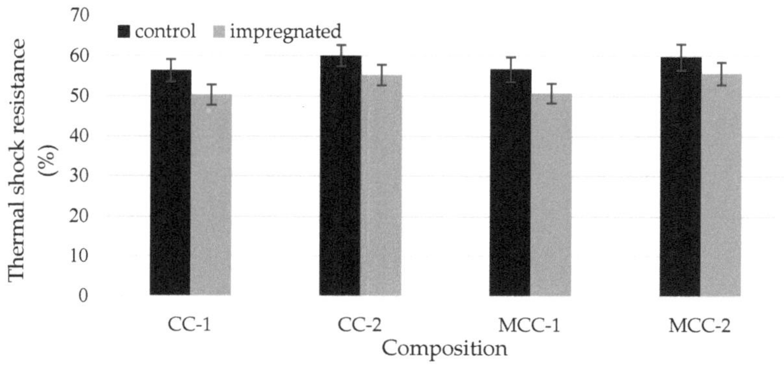

Figure 10. The comparison of thermal shock resistance of control and impregnated samples.

The thermal shock resistance (Figure 10) of the impregnated samples compared to the control samples decreased from approximately 8% (CC-2, MCC-2) to 12% (CC-1, MCC-1).

In this study, impregnation technology was used for a more accurate laboratory testing, but in practice, liquid sodium silicate glass can be applied on the refractories by spraying or coating with a brush. Such application methods can be used easily in combustion plants boilers.

4. Conclusions

Regardless of the refractory castable type (conventional (CC) or medium-cement castable (MCC)), impregnation with liquid sodium silicate glass under vacuum increased the density, ultrasonic pulse velocity, cold crushing strength, modulus of elasticity and reduced porosity, compacted the microstructure, and increased resistance to alkali as a result of a former barrier (0.9–1.5 mm) that prevented the further penetration of alkali and stopped the further disintegration of refractory castable.

The main properties of conventional and medium-cement fireclay refractory castable impregnated with liquid sodium silicate glass improved because the impregnation densifies the microstructure and decreases the porosity of refractory castables by partially filling the pores and connecting capillaries, healing the microcracks and accelerating anorthite formation, which has good thermal stability properties.

The thermal shock resistance of the impregnated samples compared to the control samples can decrease about 10% because of the improved modulus of elasticity (7–16%).

Author Contributions: Conceptualization, J.M. and V.A.; methodology, J.M., R.B. and A.K.; validation, J.M., V.A. and R.S.; formal analysis, J.M., R.B. and A.K.; investigation, J.M., R.B. and A.K.; resources, R.S.; data curation, J.M. and R.B.; writing—original draft preparation, J.M.; writing—review and editing, J.M., V.A. and R.S.; visualization, J.M. and R.B.; supervision, V.A.; project administration, R.S.; funding acquisition, R.S. All authors have read and agreed to the published version of the manuscript.

Funding: This research received no external funding.

Institutional Review Board Statement: Not applicable.

Informed Consent Statement: Not applicable.

Data Availability Statement: The data presented in this study are available in article.

Acknowledgments: The equipment and infrastructure of Civil Engineering Scientific Research Center of Vilnius Gediminas Technical University was employed for investigations.

Conflicts of Interest: The authors declare no conflict of interest.

References

1. Vares, V.; Kask, U.; Muiste, P.; Pihu, T.; Soosaar, S. *Biofuel User Manual*; TutPress: Singapore, 2005; p. 178.
2. Rosendahl, L. *Biomass Combustion Science, Technology and Engineering*; Woodhead Publishing: Cham, The Netherland, 2013; Volume 40, p. 301.
3. European Council—Conclusion, Brussels, Belgium, 2014. Available online: https://www.consilium.europa.eu/media/24561/145397.pdf (accessed on 15 September 2023).
4. Klinger, W.; Weber, W.; Zimmermann, H. Application concepts of shaped and unshaped refractories for combustion plants. *Proc. UNITECR* **2007**, *2007*, 440–443.
5. Antonovič, V. Refractory materials for biofuel boilers: Chapter 22. In *Frontiers in Bioenergy and Biofuels*; Antonovič, V., Szczerba, J., Kerienė, J., Stonys, R., Boris, R., Eds.; InTech: Rijeka, Croatia, 2017; pp. 443–464.
6. Mahapatra, M.K. Review of corrosion of refractory in gaseous environment. *Int. J. Appl. Ceram. Technol.* **2020**, *17*, 606–615. [CrossRef]
7. Alibasic, E.; Oldin, J.; Kannabiran, S. Design of castables and their relevance to alkali resistance applications. In Proceedings of the 57th International Colloquium on Refractories EUROGRESS, Aachen, Germany, 24–25 September 2014; pp. 67–69.
8. Taylor, J.R.; Bull, A.C. *Ceramics Glaze Technology*; A. Wheaton & Co. Ltd.: Exeter, UK, 1986; 276p.
9. Parr, C.; Simonin, F.; Touzo, B.; Wohrmeyer, C.; Valdelievre, B.; Namba, A. The impact of calcium aluminate cement hydration upon the properties of refractory castables. *Kerneos Aluminate Technol.* **2004**, *64*, 30–48.
10. Scudeller, L.A.M.; Longo, E.; Varela, J.A. Potassium Vapor Attack in Refractories of the Alumina-Silica System. *J. Am. Ceram. Soc.* **1990**, *73*, 1413–1416. [CrossRef]
11. Kerienė, J.; Boris, R.; Antonovič, V.; Stonys, R.; Škamat, J. Action of the products of biofuel combustion on the phase composition and structure of refractory material. *Glass Ceram.* **2016**, *72*, 345–350. [CrossRef]
12. Carlborg, M.; Bostrom, D.; Ohman, M.; Backman, R. Reactions between ash and ceramic lining in entrained flow gasification of wood—Exposure studies and thermodynamic considerations. In Proceedings of the 21st European Biomass Conference and Exhibition, Copenhagen, Denmark, 3–7 June 2013; pp. 446–449.
13. Brunk, F. Silica refractories. *CN Refract. Special. Issues* **2001**, *5*, 27–30. Available online: https://www.pd-refractories.com/files/Dokumente/Publikationen/Silica-Refractories_2001.pdf (accessed on 10 June 2021).
14. Goswami, G.; Sanu, P.; Panigrahy, P.K. Estimation of thermal expansion of silica refractory based on its mineralogy. *Interceram Refract. Man.* **2015**, *64*, 174–176. [CrossRef]
15. Baspinar, M.; Serhat, K.F. Optimization of the corrosion behaviour of mullite refractories against alkali vapour via $ZrSiO_4$ addition to the binder phase. *Ceram. Silik.* **2009**, *53*, 242–249. Available online: https://www.ceramics-silikaty.cz/2009/pdf/2009_04_242.pdf (accessed on 11 December 2022).
16. Vargas, F.; Restrepo, E.; Rodríguez, J.E.; Vargas, F.; Arbeláez, L.; Caballero, P.; Arias, J.; López, E.; Latorre, G.; Duarte, G. Solid-state synthesis of mullite from spent catalysts for manufacturing refractory brick coatings. *Ceram. Int.* **2018**, *44*, 3556–3562. [CrossRef]
17. Ren, B.; Shaobai, S.; Yawei, L.; Shengli, J. Correlation of pore structure and alkali vapor attack resistance of bauxite-SiC composite refractories. *Ceram. Int.* **2015**, *41*, 14674–14683. [CrossRef]
18. Ren, B.; Shaobai, S.; Yawei, L.; Yibiao, X. Effects of oxidation of SiC aggregates on the microstructure and properties of bauxite-SiC composite refractories. *Ceram. Int.* **2015**, *41*, 2892–2899. [CrossRef]
19. Butzbach, K. Hasle have a long tradition for producing highly alkali resistant refractory castables. *Hasle Refract.* **2019**, *125*, 20.
20. Zawraah, M.F.M.; Khalil, N.M. Effect of mullite formation on properties of refractory castables. *Ceram. Int.* **2001**, *27*, 689–694. [CrossRef]
21. Malaiškienė, J.; Antonovič, V.; Boris, R.; Stonys, R. Improving the physical and mechanical properties and alkali resistance of fireclay-based castables by modifying their structure with SiO_2 sol. *Ceram. Int.* **2022**, *48*, 22575–22585. [CrossRef]

22. Kobayashi, W.T.; Paulo, S.; Da Silva, E.L.; Carlos, S.; Paskocimas, C.A. Method for Producing Corrosion Resistant Refractories. U.S. Patent US 6,667,074 B2, 23 December 2003.
23. Khlystov, A.; Konnov, M.; Shirokov, V. Resource and energy saving technologies of refractory linings of thermal units. *MATEC Web Conf.* **2017**, *106*, 06011. [CrossRef]
24. Lianying, Z.; Bo, P.; Jiandong, W.; Xiaofeng, Y.; Lianwei, F. Waterproof Treatment Method for Magnesium Refractory Bricks. Patent No. CN103172405, 4 November 2013.
25. Ben, D.; David, N. Refractory Brick. Patent No. US3489580, 13 January 1970.
26. Moiseevich, L.A.; Yurevich, M.T.; Yarushina, T.; Aleksandrovich, O.M. Refractory Product and Method of Its Production. Patent No. RU2638065, 11 December 2017.
27. Fard, F.G.; Talimian, A. Improving Corrosion Behaviour of Magnesia-Chrome Refractories by Addition of Nanoparticles. *Refract. World Forum* **2014**, *6*, 93–98.
28. Malaiškienė, J.; Antonovič, V.; Boris, R.; Stonys, R. Effect of the impregnation with liquid glass on the properties of refractory castable. In *The Carbon Challenge. Steps & Leaps to Master the Future, Proceedings of the Unified International Technical Conference on Refractories (UNITECR) 18th Biennial Worldwide Congress on Refractories, Frankfurt am Main, Germany, 26–29 September 2023*; European Centre for Refractories gGmbH (ECREF): Höhr-Grenzhausen, Germany, 2023; pp. 834–838. ISBN 9783981581393.
29. Antonovič, V.; Stonys, R.; Zdanevičius, P.; Mačiulaitis, R.; Boris, R.; Malaiškienė, J. Analysis of the formed protective layer inhibiting alkali corrosion in aluminosilicate refractory castables. *Ceram. Spec. Issue Adv. Ceram.* **2022**, *5*, 1051–1065. [CrossRef]
30. Antonovich, V.; Shyukshta, M.; Pundene, I.; Stonis, R. Procedural elements in estimation of the thermal shock resistance of different types of refractory concrete based on chamotte filler. *Refract. Industr. Ceram.* **2011**, *52*, 70–74. [CrossRef]
31. Niyogi, S.K.; Das, A.C. Prediction of thermal shock behaviour of castable refractories by sonic measurements. *Refractories* **1994**, *43*, 453–457.
32. Csaki, S.; Lukac, F.; Húlan, T.; Veverka, J.; Knapek, M. Preparation of anorthite ceramics using SPS. *J. Eur. Ceram. Soc.* **2021**, *41*, 4618–4624. [CrossRef]
33. Nishikawa, A. *Technology of Monolithic Refractories*; Japan by Toppan Printing Company: Tokyo, Japan, 1984; 598p.
34. Malaiškienė, J.; Antonovič, V.; Boris, R.; Stonys, R. The analysis of the change in the structure and properties of refractory concrete impregnated with SiO_2 sol. In *The UNITECR 2022 Proceedings, Proceedings of the 17th Unified International Technical Conference on Refractories (UNITECR 2022), Chicago, IL, USA, 15–18 March 2022*; The American Ceramic Society: Westerville, OH, USA, 2022; pp. 1–6.
35. Magliano, M.V.M.; Prestes, E.; Medeiros, J.; Veiga, J.L.B.C.; Pandolfelli, V.C. Colloidal silica selection for nanobonded refractory castables. *Refract. Appl. News* **2010**, *15*, 14–17.
36. Li, N.; Vainio, E.; Hupa, L.; Hupa, M.; Zabetta, E.C. Interaction of high Al_2O_3 refractories with alkaline salts containing potassium and sodium in biomass and waste combustion. *Energy Fuels* **2018**, *32*, 12971–12980. [CrossRef]
37. Niu, Y.; Wang, Z.; Zhu, Y.; Zhang, X.; Tan, H.; Hui, S. Experimental evaluation of additives an K_2O-SiO_2-Al_2O_3 diagrams on high temperature silicate melt-induced slagging during biomass combustion. *Fuel* **2016**, *179*, 52–59. [CrossRef]
38. Ulbricht, J.; Dudczig, S.; Tomšu, F.; Palco, S. Technological measures to improve the thermal shock resistance of refractory materials. *Interceram Refract. Man* **2012**, *2*, 103–106.
39. Szczerba, J.; Pedzicha, Z.; Nikiel, M.; Kapuscinska, D. Influence of raw materials morphology on properties of magnesia-spinel refractories. *J. Eur. Ceram. Soc.* **2007**, *27*, 1683–1689. [CrossRef]

ceramics MDPI

Article

Fabrication of Dicarboxylic-Acid- and Silica-Substituted Octacalcium Phosphate Blocks with Stronger Mechanical Strength

Yuki Sugiura [1,2,*], Yasuko Saito [3], Etsuko Yamada [2] and Masanori Horie [2]

1 Health and Medical Research Institute, National Institute of Advanced Industrial Science and Technology (AIST), 1-1-1, Higashi, Tsukuba 305-0045, Ibaraki, Japan
2 Health and Medical Research Institute, National Institute of Advanced Industrial Science and Technology (AIST), 2217-14, Hayashi, Takamatsu 761-0395, Kagawa, Japan; potatostarch2@gmail.com (E.Y.); masa-horie@aist.go.jp (M.H.)
3 Research Institute for Sustainable Chemistry, National Institute of Advanced Industrial Science and Technology (AIST), 3-11-32 Kagamiyama, Higashi-Hiroshima 739-0046, Hiroshima, Japan; saitou.y4@aist.go.jp
* Correspondence: yuki-sugiura@aist.go.jp

Abstract: Octacalcium phosphate (OCP) is an attractive base material to combine into components developed for medical purposes, especially those used in bone replacement procedures, not only because of its excellent biocompatibility but also because of its ability to intercalate with multiple types of molecular layers such as silica, dicarboxylic acid, and various cations. On the other hand, there are no examples of simultaneous substituting for several different compounds on OCPs. Therefore, in this study, the physical and mechanical strength (DTS: diametral tensile strength) of OCPs substituted with both silica and dicarboxylic acids (thiomalate: SH-malate) were evaluated. By optimizing the amount of SH-malate, we were able to prepare a block consisting of OCPs with both silica and SH-malate supported in the interlayer. The composition of the OCP-based compound comprising this block was $Ca_8Na_{1.07}H_{6.33}(PO_4)_{4.44}(SiO_4)_{1.32}(\text{SH-malate})_{2.40}\cdot nH_2O$. Interestingly, the low mechanical strength, a drawback of silica-substituted OCP blocks, could be improved by dicarboxylic acid substituting. The dicarboxylic acid addition increased the mechanical strength of silica-substituted OCP blocks, and the acid successfully incorporated into the interlayer, even with the presence of silica. These results are expected to advance the creation of better silica-substituted OCPs and improved bone replacement materials.

Keywords: octacalcium phosphate (OCP); bioceramics; silica; dicarboxylate; co-substitute

1. Introduction

A prevalent health issue among the elderly is hard tissue syndrome [1,2]. This is because as individuals age, they develop hard tissues that have relatively low healing abilities. As typified by bone fracture and periodontal diseases, hard tissue syndrome leads to a greatly decreased quality of life. In particular, alveolar bone recovery as well as the creation of effective bone regeneration materials are required to restore oral functions. Therefore, medical infrastructures that help with the rapid healing from diseases related to hard tissue syndrome are essential for the maintenance of a vibrant society, especially among elderly populations.

Octacalcium phosphate [OCP: $Ca_8(PO_4)_4(HPO_4)_2\cdot 5H_2O$], the main inorganic component of immature bone, exhibits excellent biocompatibility and can be medically replaced in the bone of a patient through the bone remodeling process [3–8]. In addition, the OCP crystal structure can be incorporated within various ions and molecules such as transition metal ions and dicarboxylic acids [5,9–14]. Thus, OCP has been an attractive base material to combine with components for the development of medical products, especially those utilized in bone regeneration. Furthermore, OCP is a precursor of other calcium phosphates that aid in bone healing and regeneration such as hydroxyapatite [HAp: $Ca_{10}(PO_4)_6(OH)_2$] and carbonate apatite [CO_3Ap: $Ca_{10-a}(PO_4)_{6-b}(CO_3)_c(OH)_{2-d}$] [15–18]. Depending on the molecule, the OCP unit lattice-substituted materials interlayer might be maintained or released through the phase conversion process [19–22].

Silica enhanced the viabilities of osteoblasts and increases their bone production [23–25]. In fact, there have been attempts to demonstrate high bone regeneration ability by combining silica with vaterite, a polymorph of calcium carbonate, in bone replacement materials [26–29]. In addition, bioceramics containing silica nanoparticles were also introduced for bone regeneration materials [30–33]. We have investigated the usefulness of this material as a bone replacement material by substituting silica onto OCP (OCP-silica) without using organic silica such as tetraethyl silicate (TEOS), which possess the harmful problem of residual organic matter material feasibility tests for application in bone regeneration because silica enhances osteoblastic activities [34–36]. It was found that OCP-silica exhibited several times higher bone regeneration ability compared to CO_3Ap, which is presently used in commercial bone substitutes for dental implant [35]. On the other hand, the low formability and mechanical strength of OCP-silica blocks are serious drawbacks for bone substitute applications and further equipment. Since silica acts as an inhibitor of OCP crystal growth, the interlocking process results in low mechanical strength [34,35]. Moreover, when OCP blocks were converted into apatite blocks through a solution medicated phase conversion process, the mechanical strength of treated blocks further decreased [36].

In a previous related study, the results showed that dicarboxylic-acid-substituted OCP blocks fabricated by the OCP cementing process displayed a mechanical strength that tended to be higher than that of normal OCP blocks [37,38]. This knowledge can be applied in the preparation of OCP-silica blocks that exhibit high strength and serves as a precursor for other silica-substituted calcium phosphates. To further investigate the capabilities of molecular-substituted OCP blocks for practical application, this study prepared silica-dicarboxylic acid molecule-supported OCP blocks using thiomalate acid [SH-malate: $C_4H_6O_4S$] as the dicarboxylic acid molecule.

2. Materials and Methods

2.1. Preparation of SH-Malate and Silica-Substituted Blocks through the Cement-Setting Reaction

The reagents used for this study were obtained as follows: all reagents (except SH-malate) were purchased as reagent grade from FUJI Film Wako Pure Chemical Inc., Co., Osaka, Japan; SH-malate was purchased from Tokyo Chemical Industrial Co., Tokyo, Japan; H_3PO_4 was diluted by distilled water and then prepared in 4.0 mol/L H_3PO_4. In addition, the SH-malate and silica-substituted blocks were prepared through the cement-setting reaction [35,37,38].

Next, the OCP component fabrication process is detailed. First, 0.0–0.6 g SH-malate was dissolved into 2.0 mL 4.0 mol/L H_3PO_4 at an agate motor. Then, 1.2 g of $CaCO_3$ powder was gradually mixed into the solution with mixing by agate pestle. Afterwards, 1.08 mL, 38 percentage by weight (wt%) Na_2SiO_3 solution was subsequently added by stirring with a pestle for several ten-second intervals until the bubbling process was stabilized. Using a stainless spatula, the mixture was placed in a silicon rubber mold ($\varphi 6 \times 3$ mm) that was tightly sealed by a 0.3 mm thick polypropylene sheet to avoid evaporation and induce the interlocking of formed crystals during incubation. The packed molds with samples were placed at 60 °C for 1 d. Finally, the treated materials were removed from the mold; then, they were thoroughly washed with distilled water three times and finally placed in a 40 °C drying oven for 1 d.

2.2. Evaluation Process

To evaluate the experimental results, crystallographic information from the samples was obtained through X-ray diffraction (XRD; MiniFlex600, Rigaku Co., Tokyo, Japan) at an acceleration voltage and amplitude of 40 kV and 15 mA, respectively. For characterization, the diffraction angle was continuously scanned over 2θ values ranging from 3° to 70° at a scanning rate of 5°/min. Meanwhile, for the crystallographic parameter analysis, it was scanned from 2° to 12° at a rate of 1°/min.

Next, the chemical bonding of the samples was analyzed with Fourier transform infrared spectroscopy (FT-IR) (IRTracer-100, Shimadzu Co., Kyoto, Japan) and a triglycine sulfate detector (30 scans, resolution of 2 cm^{-1}) that has an attenuated total reflection diamond prism. Also, the atmosphere provided the measurement background.

After Os sputtering, the fine structures of the samples were examined by field emission scanning electron microscopy (FE-SEM; JSM-6700F, JEOL Co., Tokyo, Japan) at 3 kV acceleration voltage.

The ^{31}P chemical shifts of the samples were determined by using solid state nuclear magnetic resonance (NMR) spectroscopy (Varian FT-NMR, 400 MHz, Agilent Technologies Co., Santa Clara, CA, USA) at a 161.8 MHz resonance frequency. Next, cross-polarization magic angle spinning (CP-MAS) ^{31}P NMR spectroscopy was performed at 10 kHz. In addition, an Agilent 4 mm T3 CP-MAS HXY solid probe and zirconia rotors were used. Each sample weighed ≈0.03 g. Meanwhile, the contact time for the ^{31}P CP-MAS measurements was 3 ms, with an acquisition time of 50 ms and relaxation delay of 40 s for each measurement interval. Each measurement was repeated 400 times. The ^{31}P chemical shift of $(NH_4)H_2PO_4$ was used as the external reference (1.0 ppm).

The Si and S content of the samples were measured using energy-dispersive X-ray fluorescence spectroscopy (EDX-8100, Shimadzu Co., Japan) at 15 kV acceleration voltage under a vacuum. Next, the Si/Ca ratios of the samples were calibrated using inner standard samples, which consisted of a hydroxyapatite–SiO_2 mixture in disk-shaped sintered blocks (Ca/Si = 100:0, 90:10, 75:25, 50:50, and 0:100) measured in the same manner as the experimental samples. The S/Ca ratio was also calibrated using a disk-shaped mixture of $CaCO_3$–$CaSO_4{\cdot}2H2O$ set blocks (Ca/S = 100:0, 90:10, 75:25 and 50:50).

Afterwards, the mechanical strengths of five cylindrical specimens were evaluated. Their respective diameter and thickness were measured using a micrometer (MDC-25MU, Mitsutoyo Co., Ltd., Kawasaki, Japan). These specimens were then submitted under compression testing by crushing at a constant crosshead speed of 1 mm/min using a universal testing machine (AGS-J, Shimadzu, Kyoto, Japan). Next, the mean DTS values of the specimens were estimated and reported as the mean standard deviation. The results were then assessed by *t*-test analysis (with a significance level of $p < 0.05$).

3. Results and Discussion

Figure 1 shows photographs of the set samples with various SH-malate amounts. Note that in the study of dicarboxylic acid molecule substituting, the amount of $CaCO_3$ is increased compared to the conventional OCP-silica preparation. This is because dicarboxylic acid is acidic, and it is necessary to increase the amount of $CaCO_3$ to neutralize it. The mud mixture foamed more strongly because of this. Through the cement-setting reaction, all the samples revealed block shapes. It should be noted that in low SH-malate amounts (SH-malate: 0.00–0.15 g), the bubbling process was continued during setting reactions.

Next, the XRD evaluation of the obtained blocks was performed. Figure 2 shows the XRD patterns of the set blocks with different SH-malate amounts. As the SH-malate concentration increased, the block composition became from apatite to OCP, and monophasic OCP peaks were observed around 0.45 g and 0.60 g of SH-malate content. The peak corresponding to *d*100 of OCP shifted slightly to the low angle, indicating that SH-malate intercalated with the OCP interlayer. Therefore, the sample with 0.60 g of SH-malate was found to be OCP monophase and thereby can be referred to as an OCP-silica:SH-malate block.

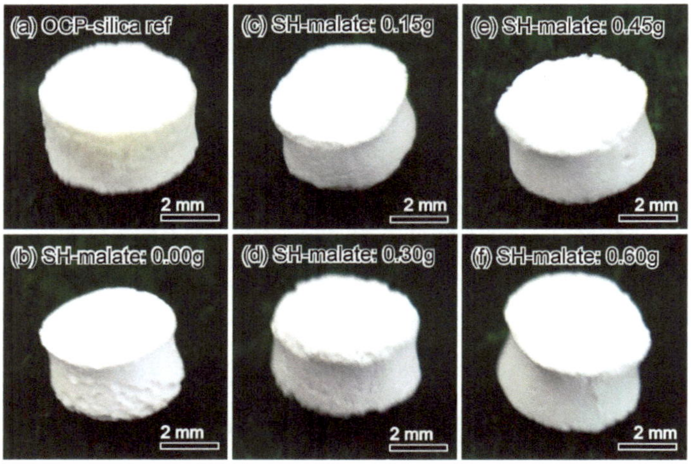

Figure 1. Photographs of the set blocks with different SH-malate amounts. (**a**) OCP-silica block as reference. (**b**) SH-malate: 0.00 g. (**c**) SH-malate: 0.15 g. (**d**) SH-malate: 0.30 g. (**e**) SH-malate: 0.45 g. (**f**) SH-malate: 0.60 g.

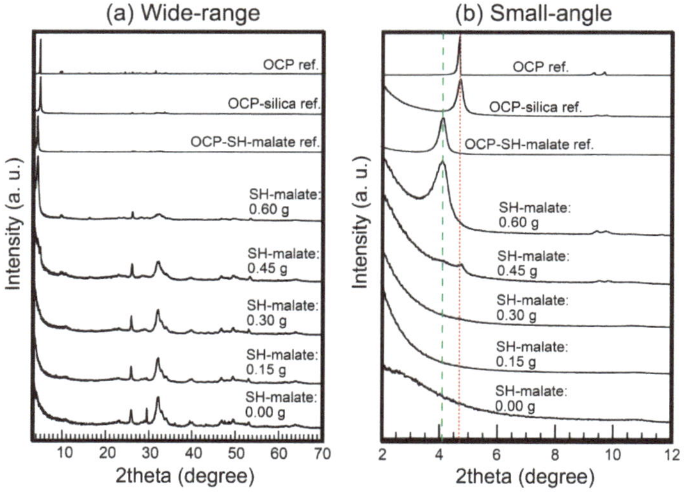

Figure 2. XRD patterns of the set blocks. (**a**) Wide range. (**b**) Small angle. Green broken line: OCP-SH-malate *d*(100)′ peak. Red dotted line: OCP *d*(100) peak.

The conditions for substituted silica and SH-malate in OCP were further evaluated. Figures 3 and 4 show the OCP-silica:SH-malate block FT-IR and NMR spectra, respectively. In FT-IR spectra, the OCP-silica:SH-malate block clearly exhibited -COOH band ranging from 1400 to 1600 cm^{-1}, which is the same as in OCP-SH-malate. In addition, a weak band ranging from 1100 to 1200 cm^{-1}, which corresponds to the silanol band, was observed. Then, the OCP-silica:SH-malate block revealed both OCP-silica and OCP-SH-malate. In NMR spectra, the $P2/P4$ peaks were clearly attenuated, while $P5/P6$ corresponded to the hydrous layer and shifted slightly toward the higher wavenumber. Meanwhile, bands corresponding to P_A/P_B were observed at lower wavenumbers. These results were attributed to the inclusion of dicarboxylic acid molecules [35,36], as well as indicating that both silica and SH-malate are contained in the OCP crystals.

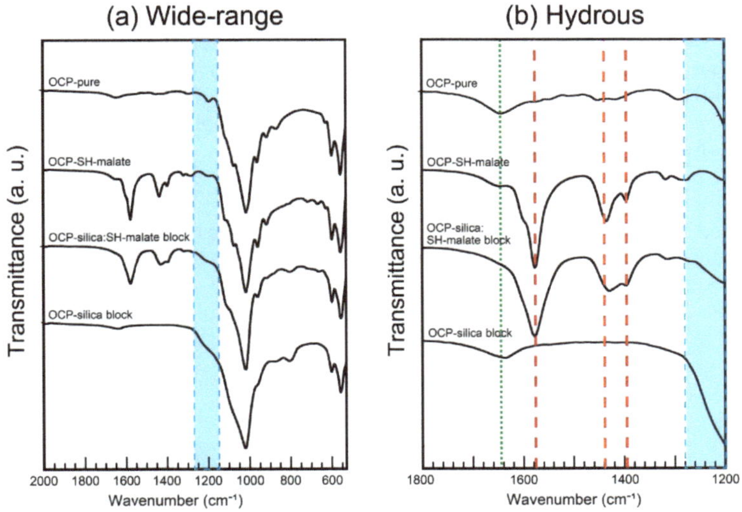

Figure 3. FT-IR spectra of OCP-silica:SH-malate block with OCP–silica and OCP-SH-malate reference for facilitate comparison. (**a**) Wide range. (**b**) Hydrous. Blue band: silanol vibration. Red broken lines: -COOH vibration. Green dotted line: HPO_4 vibration [39].

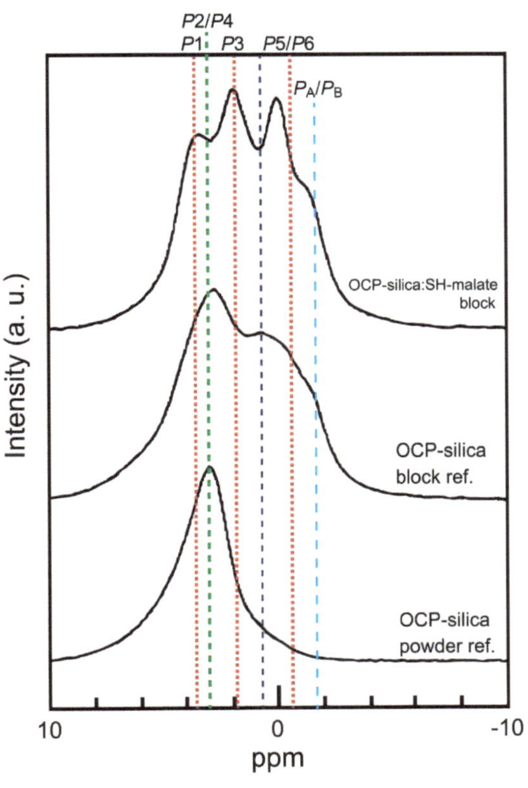

Figure 4. ^{31}P solid-state NMR spectra of OCP-silica:SH-malate block with conventional OCP, OCP-silica block, and OCP-silica powder for facilitate comparison. Broken lines corresponded each PO_4 state. Dark blue broken line: PO_4 of OCP-silica block state [35,40].

The OCP-silica:SH-malate block (SH-malate: 0.60 g) contains of Si (8.35 ± 0.68 at%) and S (15.20 ± 3.38 at%). However, the OCP-silica:SH-malate block Si content was significantly lower than that of OCP-silica powder and block. On the other hand, the OCP-silica:SH-malate block S content was significantly higher than that of OCP-SH-malate powder. Then, the possibly chemical composition of OCP-silica:SH-malate block was described as $Ca_8Na_{1.07}H_{6.33}(PO_4)_{4.44}(SiO_4)_{1.32}(SH\text{-malate})_{2.40}\cdot nH_2O$.

Next, the mechanical strength of the samples was evaluated. Figure 5 shows the DTS values of the set samples. As SH-malate was increased, the block mechanical strength was attenuated, but increased again. For the OCP-silica:SH-malate block, the mechanical strength of samples was significantly higher than that of blocks without SH-malate and OCP-silica blocks.

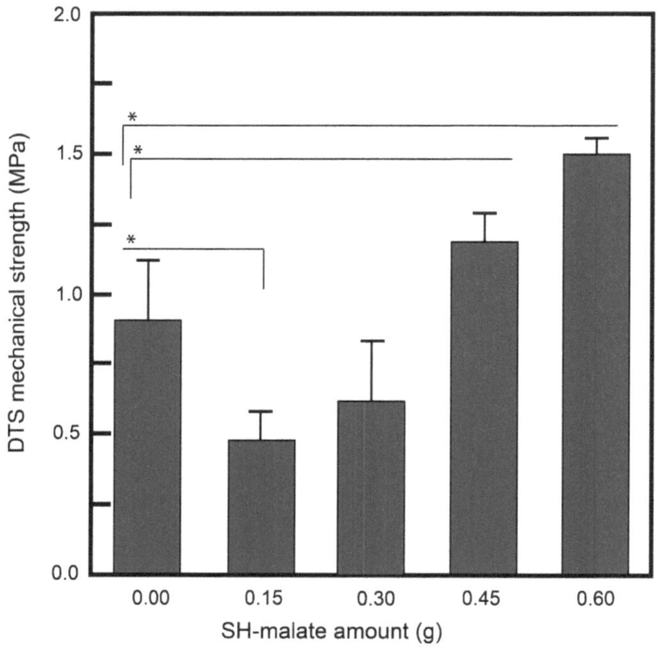

Figure 5. DTS values of the set blocks with different SH-malate amounts. *: $p < 0.05$.

Meanwhile, the fine structure of the samples indicated that they possessed higher mechanical strength. Figure 6 shows SEM micrographs of the OCP-silica:SH-malate block and two types of OCP-silica block ($CaCO_3$-H_3PO_4-Na_2SiO_3 and DCPD-MCPM-Na_2SiO_3 systems) for facilitate comparison. In the OCP-silica:SH-malate block, much denser and well-developed crystals were observed compared to both OCP-silica blocks. This suggests that a high degree of entanglement results in higher strength. When dicarboxylic acid molecules and other molecules are substituted on OCP, developing in the *a*-axis direction is suppressed [40–43]. As a result, the aspect ratio of OCP crystals increases. This is assumed to increase the entanglement and form a denser and stronger structure. Furthermore, since the dicarboxylic acid molecules have carboxy groups, an adhesive effect of free dicarboxylic acid molecules binding to Ca ions on the OCP surface is also expected [44,45]. Since the same phenomenon is observed even when silica is not supported, it is considered that dicarboxylic acid contributes to both aspect ratio and adhesive strength [34,35].

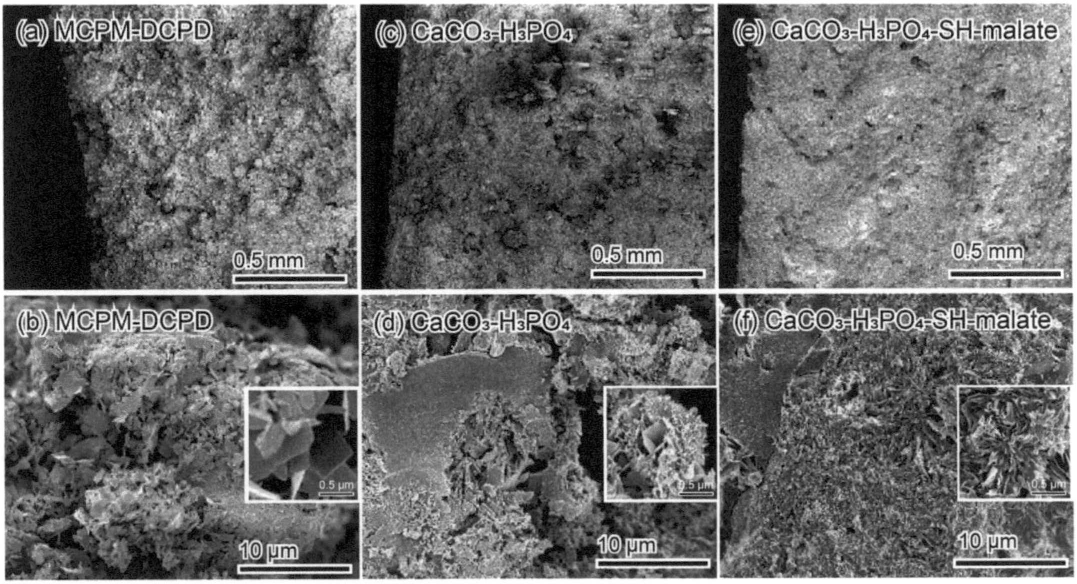

Figure 6. SEM micrographs of OCP-silica:SH-malate block and reference materials in 0.5 mm and 10 μm, respectively. (**a**,**b**) OCP-silica fabricated from MCPM-DCPD-Na_2SiO_3 system. (**c**,**d**) OCP-silica fabricated from $CaCO_3$-H_3PO_4-Na_2SiO_3. (**e**,**f**) OCP-silica:SH-malate block.

The results clearly show that both silica and SH-malate can be simultaneously substituted in the OCP crystal structure. Furthermore, we think that the prepared OCP-silica:SH-malate block is a material in which both silica and SH-malate are substituted between the layers of OCP. This is shown by FT-IR and NMR spectral measurements. In addition, compared to conventional OCP–silica, the strength of the prepared blocks is improved. One of the possible reasons for this is the morphological change of OCP crystals due to the incorporation of SH-malate into the OCP interlayer: OCP crystals generally assume a ribbon-like crystal form, while dicarboxylic acid molecules, including SH-malate, are distributed along the *a*-axis direction, that is, the orientation in which the crystal is thinnest [46,47]. Therefore, the *a*-axis direction growth is inhibited, while growth in the *b*- and *c*-axis directions is relatively less inhibited, resulting in the formation of crystals with a large aspect ratio, i.e., crystals that are easily entangled.

For further investigation, it was also tested as to whether OCP is useful for dicarboxylic acid molecules other than SH-malate. As shown in Figure 7, succinic acid ($C_4H_6O_4$) substituting was considered. It was found that any dicarboxylic acid molecule, such as succinic acid, can be incorporated into OCP crystals and blocked. Therefore, these would have the same effect as that observed in SH-malate. Note that the mechanical strength of succinic-acid-substituted OCP-silica blocks was found to be significantly higher than that of SH-malate. This is likely due to the higher acidity of succinic acid and how it might compare to SH-malate, as well as the stronger adhesion of the carboxy groups of the dicarboxylic acid molecules, resulting in a stronger block.

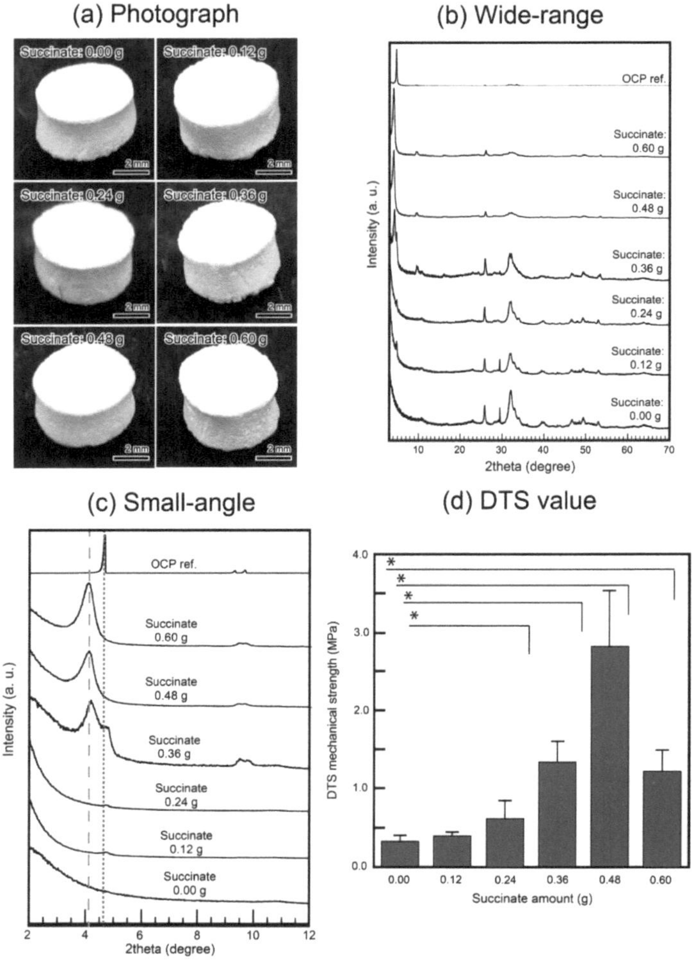

Figure 7. Evaluation results of OCP-silica:succinate block fabrication. (**a**) Photographs of the set blocks with different succinate amounts. (**b**) Wide-range XRD patterns of the set blocks with different succinate amounts. (**c**) Small-angle XRD patterns of the set blocks with different succinate amounts. Green broken line: OCP-SH-malate *d*(100)′ peak. Red dotted line: OCP *d*(100) peak. (**d**) DTS values of the set blocks with different succinate amounts. *: $p < 0.05$.

4. Conclusions

Because of OCP's ability to intercalate with multiple types of molecular layers such as silica, dicarboxylic acid, and various cations, OCP is an attractive base material to combine into components developed for medical purposes. On the other hand, there are no examples of simultaneous substituting for several different compounds on OCPs. Of particular interest is whether low mechanical strength, a drawback of OCP-silica blocks, could be improved by dicarboxylic acid substituting. Therefore, the preparation of silica-substituted OCP blocks with the addition of SH-malate, a dicarboxylic acid molecule, was investigated. If the formulation ratio is appropriate, it is possible to prepare monophasic OCP blocks substituted with dicarboxylic acid molecules and silica. The mechanical strength of the resulting blocks tends to be higher than that of the system without dicarboxylic acid molecules substituted. Based on these results, the findings of this study have promising applications for the development of improved medical components used in bone replacement that aid in more efficient as well as effective bone recovery and regeneration.

Author Contributions: Conceptualization, Y.S. (Yuki Sugiura) and M.H.; methodology, Y.S. (Yuki Sugiura) and E.Y.; NMR measurement Y.S. (Yasuko Saito). All authors have read and agreed to the published version of the manuscript.

Funding: This study was financially supported by the AMED Seeds H program (Keio University; grant number H424TS), AMED Seeds A program (Keio University; grant number A424TR), Kazuchika Okura Memorial Foundation 2022FY Grant-in-Aid, and JST-CREST (grant number JPMJCR22L5).

Institutional Review Board Statement: Not applicable.

Informed Consent Statement: Not applicable.

Data Availability Statement: Data are contained within the article.

Acknowledgments: We thank T. Nakanishi, RIST, Kagawa, Japan, for performing the FT-IR measurements.

Conflicts of Interest: The authors declare no conflicts of interest.

References

1. Safiri, S.; Kolahi, A.-A.; Smith, E.; Hill, C.; Bettampadi, D.; Mansournia, M.A.; Hoy, D.; Ashrafi-Asgarabad, A.; Sepidarkish, M.; Almasi-Hashiani, A.; et al. Global, regional and national burden of osteoarthritis 1990–2017: A systematic analysis of the Global Burden of Disease Study 2017. *Ann. Rheum. Dis.* **2020**, *79*, 819–828. [CrossRef] [PubMed]
2. World Health Organization. WHO's Annual World Health Statistics Reports, 2020. In *WHO's Division for Data, Analytics and Delivery*; WHO Press: Geneva, Switzerland, 2020.
3. Simon, P.; Grüner, D.; Worch, H.; Pompe, W.; Lichte, H.; El-Khassawna, T.; Heiss, C.; Wenisch, S.; Kniep, R. First evidence of octacalcium phosphate@osteocalcin nanocomplex as skeletal bone component directing collagen triple–helix nanofibril mineral-ization. *Sci. Rep.* **2018**, *8*, 13696. [CrossRef] [PubMed]
4. Suzuki, O.; Kamakura, S.; Katagiri, T. Surface Chemistry and Biological Responses to Synthetic Octacalcium Phosphate. *J. Biomed. Mater. Res. Part B Appl. Biomater.* **2006**, *77B*, 201–212. [CrossRef] [PubMed]
5. Shi, H.; He, F.; Ye, J. Synthesis and structure of iron- and strontium-substituted octacalcium phosphate: Effects of ionic charge and radius. *J. Mater. Chem. B* **2016**, *4*, 1712–1719. [CrossRef] [PubMed]
6. Sai, Y.; Shiwaku, Y.; Anada, T.; Tsuchiya, K.; Takahashi, T.; Suzuki, O. Capacity of octacalcium phosphate to promote osteoblastic differentiation toward osteocytes in vitro. *Acta Biomater.* **2018**, *69*, 362–371. [CrossRef] [PubMed]
7. Kamakura, S.; Sasano, Y.; Homma, H.; Suzuki, O.; Kagayama, M.; Motegi, K. Implantation of Octacalcium Phosphate (OCP) in Rat Skull Defects Enhances Bone Repair. *J. Dent. Res.* **1999**, *78*, 1682–1687. [CrossRef]
8. Sugiura, Y.; Munar, M.L.; Ishikawa, K. Fabrication of octacalcium phosphate foam through phase conversion and its histological evaluation. *Mater. Lett.* **2018**, *212*, 28–31. [CrossRef]
9. Markovic, M.; Fowler, B.O.; Brown, W.E. Octacalcium phosphate carboxylates. 1. Preparation and identification. *Chem. Mater.* **1993**, *5*, 1401–1405. [CrossRef]
10. Monma, H.; Goto, M. Thermal alteration of succinate-complexed octacalcium phosphate. *J. Mater. Sci. Lett.* **1980**, *4*, 147–150. [CrossRef]
11. Yokoi, T.; Kamitakahara, M.; Ohtsuki, C. Continuous expansion of the interplanar spacing of octacalcium phosphate by incor-poration of dicarboxylate ions with a side chain. *Dalton Trans.* **2015**, *44*, 7943–7950. [CrossRef]
12. Davies, E.; Müller, K.H.; Wong, W.C.; Pickard, C.J.; Reid, D.G.; Skepper, J.N.; Duer, M.J. Citrate bridges between mineral platelets in bone. *Proc. Natl. Acad. Sci. USA* **2014**, *111*, E1354–E1363. [CrossRef] [PubMed]
13. Boanini, E.; Gazzano, M.; Rubini, K.; Bigi, A. Collapsed Octacalcium Phosphate Stabilized by Ionic Substitutions. *Cryst. Growth Des.* **2010**, *10*, 3612–3617. [CrossRef]
14. Yamada, I.; Noda, D.; Shinozaki, K.; Galindo, T.G.P.; Tagaya, M. Synthesis of Luminescent Eu(III)-Doped Octacalcium Phosphate Particles Hybridized with Succinate Ions and Their Reactive Behavior in Simulated Body Fluid. *Cryst. Growth Des.* **2021**, *21*, 2005–2018. [CrossRef]
15. Chickerur, N.S.; Tung, M.S.; Brown, W.E. A Mechanism for Incorporation of Carbonate into Apatite. *Calcif. Tissue Int.* **1980**, *32*, 55–62. [CrossRef]
16. Sugiura, Y.; Ishikawa, K. Fabrication of carbonate apatite blocks from octacalcium phosphate blocks through different phase conversion mode depending on carbonate concentration. *J. Solid State Chem.* **2018**, *267*, 85–91. [CrossRef]
17. Horváthová, R.; Müller, L.; Helebrant, A.; Greil, P.; Müller, F.A. In vitro transformation of OCP into carbonated HA under physiological conditions. *Mater. Sci. Eng. C* **2008**, *28*, 1414–1419. [CrossRef]
18. Ito, N.; Kamitakahara, M.; Yoshimura, M.; Ioku, K. Importance of nucleation in transformation of octacalcium phosphate to hydroxyapatite. *Mater. Sci. Eng. C* **2014**, *40*, 121–126. [CrossRef]

19. Yokoi, T.; Goto, T.; Kitaoka, S. Formation of Hydroxyapatite Crystals from Octacalcium Phosphate with Incorporated Succinate Ion under Hydrothermal Conditions. *Chem. Lett.* **2019**, *48*, 855–858. [CrossRef]
20. Sugiura, Y.; Makita, Y. Intercalated molecule releasing process of thiomalate substituted octacalcium phosphate crystals during phase conversion. *J. Cryst. Growth* **2022**, *583*, 126545. [CrossRef]
21. Yokoi, T.; Goto, T.; Kato, T.; Takahashi, S.; Nakamura, J.; Sekino, T.; Ohtsuki, C.; Kawashita, M. Hydroxyapatite Formation from Octacalcium Phosphate and Its Related Compounds: A Discussion of the Transformation Mechanism. *Bull. Chem. Soc. Jpn.* **2020**, *93*, 701–707. [CrossRef]
22. Sugiura, Y.; Horie, M. Fabrication of silver-doped apatite powders from silver-substituted octacalcium phosphate powders via solid–solid phase-conversion process. *Ceram. Int.* **2021**, *47*, 25614–25621. [CrossRef]
23. Carlisle, E.M. Silicon: An Essential Element for the Chick. *Science* **1972**, *178*, 619–621. [CrossRef] [PubMed]
24. Xynos, I.D.; Edgar, A.J.; Buttery, L.D.; Hench, L.L.; Polak, J.M. Ionic Products of Bioactive Glass Dissolution Increase Proliferation of Human Osteoblasts and Induce Insulin-like Growth Factor II mRNA Expression and Protein Synthesis. *Biochem. Biophys. Res. Commun.* **2000**, *276*, 461–465. [CrossRef] [PubMed]
25. Reffitt, D.M.; Ogston, N.; Jugdaohsingh, R.; Cheung, H.F.J.; Evans, B.A.J.; Thompson, R.P.H.; Powell, J.J.; Hampson, G.N. Orthosilicic acid stimulates collagen type 1 synthesis and osteoblastic differentiation in human osteoblast-like cells in vitro. *Bone* **2003**, *32*, 127–135. [CrossRef] [PubMed]
26. Obata, A.; Tokuda, S.; Kasuga, T. Enhanced in vitro cell activity on silicon-doped vaterite/poly(lactic acid) composites. *Acta Biomater.* **2009**, *5*, 57–62. [CrossRef]
27. Kasuga, T.; Maeda, H.; Kato, K.; Nogami, M.; Hata, K.-I.; Ueda, M. Preparation of poly(lactic acid) composites containing calcium carbonate (vaterite). *Biomaterials* **2003**, *24*, 3247–3253. [CrossRef] [PubMed]
28. Nakamura, J.; Kasuga, T. Enhancement of crystalline plane orientation in silsesquioxane-containing vaterite particles towards tuning of calcium ion release. *J. Mater. Chem. B* **2014**, *2*, 1250–1254. [CrossRef] [PubMed]
29. Obata, A.; Ozasa, H.; Kasuga, T.; Jones, J.R. Cotton wool-like poly(lactic acid)/vaterite composite scaffolds releasing soluble silica for bone tissue engineering. *J. Mater. Sci. Mater. Med.* **2013**, *24*, 1649–1658. [CrossRef]
30. Manchón, A.; Alkhraisat, M.; Rueda-Rodriguez, C.; Torres, J.; Prados-Frutos, J.C.; Ewald, A.; Gbureck, U.; Cabrejos-Azama, J.; Rodriguez-González, A.; López-Cabarcos, E. Silicon calcium phosphate ceramic as novel biomaterial to simulate the bone regenerative properties of autologous bone. *J. Biomed. Mater. Res. Part A* **2014**, *103*, 479–488. [CrossRef]
31. Dong, G.; He, L.; Wu, G.; Deng, C. The effect of silicon doping on the transformation of amorphous calcium phos-phate to silicon-substituted α-tricalcium phosphate by heat treatment. *Ceram. Int.* **2016**, *42*, 883–890. [CrossRef]
32. Huang, S.-H.; Chen, Y.-J.; Kao, C.-T.; Lin, C.-C.; Huang, T.-H.; Shie, M.-Y. Physicochemical properties and biocompatibility of silica doped b-tricalcium phosphate for bone cement. *J. Dent. Sci.* **2015**, *10*, 282–290. [CrossRef]
33. Ha, S.-W.; Viggeswarapu, M.; Habib, M.M.; Beck, G.R., Jr. Bioactive effects of silica nanoparticles on bone cells are size, surface, and composition dependent. *Acta Biomater.* **2018**, *82*, 184–196. [CrossRef] [PubMed]
34. Sugiura, Y.; Niitsu, K.; Saito, Y.; Endo, T.; Horie, M. Inorganic process for wet silica-doping of calcium phosphate. *RSC Adv.* **2021**, *11*, 12330–12335. [CrossRef] [PubMed]
35. Sugiura, Y.; Ono, F.; Nohara, M.; Takechi, A.; Kutara, K.; Kanda, T.; Saito, Y.; Yamada, E.; Oowada, K.; Endo, T.; et al. Inorganic silica hybrid octacalcium phosphate bone substitute: Harmonics to acceleration in biological metabolism and its curing process. *Materialia* **2023**, *28*, 101771. [CrossRef]
36. Sugiura, Y.; Saito, Y.; Yamada, E.; Endo, T.; Horie, M.; Makita, Y. Fabrication of silica-substituted carbonate apatite blocks through phase transformation process from silica-substituted octacalcium phosphate blocks. *J. Ceram. Soc. Jpn.* **2023**, *131*, 458–465. [CrossRef]
37. Sugiura, Y.; Yamada, E.; Horie, M. Fabrication of Octacalcium Phosphate Block through the Reaction between $CaCO_3$ Powder and Phosphate Acid. *Chem. Lett.* **2022**, *51*, 851–853. [CrossRef]
38. Sugiura, Y.; Yamada, E.; Horie, M. Fabrication of dicarboxylic-acid-substituted octacalcium phosphate blocks via cementing. *Ceram. Int.* **2023**, *49*, 9719–9724. [CrossRef]
39. Berry, E.; Baddiel, C. Some assignments in the infra-red spectrum of octacalcium phosphate. *Spectrochim. Acta Part A Mol. Spectrosc.* **1967**, *23*, 1781–1792. [CrossRef]
40. Tsai, T.W.T.; Chan, J.C.C. Recent Progress in the Solid-State NMR Studies of Biomineralization. *Ann. Rep. NMR Spectros.* **2011**, *73*, 1–61.
41. Sugiura, Y.; Makita, Y. Tris(hydroxymethyl)aminomethane Substitution into Octacalcium Phosphate. *Chem. Lett.* **2019**, *48*, 1304–1307. [CrossRef]
42. Bigi, A.; Boanini, E.; Borghi, M.; Cojazzi, G.; Panzavolta, S.; Roveri, N. Synthesis and hydrolysis of octacalcium phosphate: Effect of sodium polyacrylate. *J. Inorgn. Biochem.* **1999**, *75*, 145–151. [CrossRef]
43. Sugiura, Y.; Yamada, E.; Horie, M. Interlayer expansion of octacalcium phosphate via forced oxidation of the intercalated molecules within its interlayers. *Phys. Chem. Chem. Phys.* **2023**, *25*, 26640. [CrossRef] [PubMed]
44. Ince, D.E.; Johnston, C.T.; Moudgil, B.M. Fourier transform infrared spectroscopic study of adsorption of oleic acid/oleate on surfaces of apatite and dolomite. *Langmuir* **1991**, *7*, 1453–1457. [CrossRef]
45. Dong, L.; Wei, Q.; Qin, W.; Jiao, F. Selective adsorption of sodium polyacrylate on calcite surface: Implications for flotation separation of apatite from calcite. *Sep. Purif. Technol.* **2020**, *241*, 116415. [CrossRef]

46. Tsai, T.W.T.; Chou, F.-C.; Tseng, Y.-H.; Chan, J.C.C. Solid-state P-31 NMR study of octacalcium phosphate incorporated with succinate. *Phys. Chem. Chem. Phys.* **2010**, *12*, 6692–6697. [CrossRef]
47. Li, Y.; Reid, D.G.; Duer, M.J.; Chan, J.C.C. Solid state NMR—An indispensable tool in organic-inorganic biocomposite characterization; refining the structure of octacalcium phosphate composites with the linear metabolic di-acids succinate and adipate. *Solid State Nucl. Magn. Reson.* **2018**, *95*, 1–5. [CrossRef]

ceramics

MDPI

Article

Modification of Sepiolite for Its Catalytic Upgrading in the Hydrogenation of Furfural

Antonio Guerrero-Torres [1], Rocío Maderuelo-Solera [1], Cristina García-Sancho [1,2], José J. Quirante-Sánchez [3], Ramón Moreno-Tost [1,2], Pedro J. Maireles-Torres [1,2] and Juan A. Cecilia [1,2,*]

1 Departamento de Química Inorgánica, Cristalografía y Mineralogía, Facultad de Ciencias, Universidad de Málaga, Campus de Teatinos, 29071 Málaga, Spain; guerrerotorresantonio@gmail.com (A.G.-T.); rocioms@uma.es (R.M.-S.); cristinags@uma.es (C.G.-S.); rmtost@uma.es (R.M.-T.); maireles@uma.es (P.J.M.-T.)

2 Instituto de Investigación en Biorrefinerías "I3B", Facultad de Ciencias, Universidad de Málaga, Campus de Teatinos s/n, 29071 Málaga, Spain

3 Departamento de Química Física, Facultad de Ciencias, Universidad de Málaga, Campus de Teatinos, 29071 Málaga, Spain; quirante@uma.es

* Correspondence: jacecilia@uma.es

Abstract: In this work, a new methodology for dispersing metal particles supported in clay has been described. For this purpose, a sepiolite has been modified by a microwave-assisted treatment to increase the surface area and pore volume due to the progressive leaching of the Mg^{2+}-species located in the octahedral sheet. These materials have been used as support to incorporate Cu species on the surface by a precipitation-deposition process from the thermal decomposition of urea at 95 °C. Once calcined and reduced, the Cu-based catalysts showed a Cu^0-particle size lower than 5 nm in the case of catalysts whose support is a sepiolite subjected to acid treatment. On the other hand, when raw sepiolite is used as a support, the Cu^0-crystal size is much larger (15–20 nm). This difference in Cu^0-crystal size showed a variable catalytic behavior for the hydrogenation reaction of furfural in gas-phase. Catalysts with larger particle size promote the hydrogenation reaction, obtaining a yield towards furfuryl alcohol close to 60% after 5 h at 190 °C. In contrast, catalysts with smaller particle size promote the hydrogenation reaction and subsequently the hydrogenolysis reaction, obtaining methylfuran as a product with a yield of 58% after 1 h of reaction at 190 °C; however, the sites where hydrogenolysis are involved are more prone to be deactivated.

Keywords: Cu-based catalysts; supported sepiolite; acid treatment; furfural hydrogenation; furfuryl alcohol; methylfuran

1. Introduction

Phyllosilicates are a subclass of silicates with layered structure, which have been used in a wide range of research fields [1]. Among them, one of the most widely used phyllosilicates is sepiolite [2]. Around half of the world reserves of this clay mineral is in Spain, although there are notable deposits in Turkey, Morrocco, Greece and the United States [3]. The greatest significance of this clay mineral lies in its fibrous morphology due to the periodical inversion of its tetrahedral sheet leading to nanocavities with dimensions of 1.06 nm × 0.37 nm [4,5]. These nanocavities can host small molecules or cations to counterbalance the charge deficiency [4,6]. Sepiolite can be used, under appropriate pretreatments, as adsorbent, deodorant, filter air, catalytic support, decolorizing agent, additive in paints, anticaking agent, drilling fluids, cosmetics, detergents, asphalt coating, animal nutrition, agriculture, and for phytosanitary and pharmaceutical uses [2,5,7–11].

https://doi.org/10.3390/ceramics8010021

A strategy to modify the structure of sepiolite is through acid treatment [12–15]. Using this treatment, it is possible to significantly improve its textural properties due to a partial leaching of Mg^{2+} species located in the octahedral layer [12]. The acid treatment was performed with concentrated solutions for many hours, even days [13,14]. However, it has recently been shown that microwave-assisted treatment can reduce this time to a few minutes, using solutions that are an order of magnitude more dilute [12]. The characterization of sepiolite modified with acid treatment has been widely studied [12–15]. However, its application has been primarily limited to CO_2 capture processes [6] photocatalysis [16] and cementitious materials [17].

The interest in this acid treatment is related to a partial leaching of the octahedral layer, which produces an increase in the specific area and pore volume but also modifies its chemical properties due to the loss of Mg^{2+}-species of its octahedral sheet causing vacancies in the fibrous structure [12]. Considering these premises, modified sepiolite could be an interesting material to be studied as a catalytic support with high available surface area to disperse a certain active phase.

The model reaction carried out for this study has been the furfural (FUR) hydrogenation in gas phase. This reaction has been selected since FUR molecule is considered a building block molecule since it is possible to obtain a high range of valuable products from several reactions [18,19]. FUR is obtained from lignocellulosic biomass, specifically from the hemicellulose fraction, which is a polymer consisting of pentosans, mainly xylose [20]. In turn, xylose is dehydrated to obtain FUR [21]. FUR is considered the second most produced product in the sugar platform, after bioethanol. From FUR, advanced products used in the polymer field, fuel additive, fine chemical, fragrances, paints among others have been obtained [21,22]. Most FUR is preferably treated by reduction processes using transition metals as active phase [22], although other reactions have also been carried out to obtain high added value products [18,19].

Industrially, FUR reduction has been carried out with copper chromite as a catalyst [23–26]. However, restrictive environmental regulations have led to the scientific community developing new Cr-free catalysts. In this sense, Cu-based catalysts are the most inexpensive active phase to be implemented on a large scale [22]. This active phase only promotes the reaction with the carbonyl group, obtaining furfuryl alcohol (FOL) and 2-methylfuran (MF) as products [27,28]. Both compounds are considered high value-added products, as FOL is used in the polymer industry for its excellent thermochemical properties, while MF can be used as a fuel additive [29]. Previous studies have reported that the interaction between FUR and Cu-based catalysts plays an important role in the activity and the catalytic behavior [28]. Thus, several authors have proposed the use of a neutral or basic support to obtain higher conversion and stability, minimizing the formation of carbonaceous deposits [30–33]. In the same way, the presence of small Cu^0-crystals as active phase also promotes the formation of MF through a hydrogenolysis reaction [30,34–36]. Considering these assumptions, this work aims to analyze the influence of the modification of the support, through a microwave-assisted acid treatment in a sepiolite, on the active phase-support interaction and in turn on the size of the copper crystals. Likewise, this work also studies how these modifications of the support influence the catalytic behavior in the hydrogenation reaction of FUR in the gas phase.

2. Materials and Methods

2.1. Synthesis of Cu-Based Catalysts

Sepiolite, selected as support, was supplied by Tolsa S.A and marketed as Pangel S9. This fibrous phyllosilicate was collected in Vallecas deposits (Madrid, Spain) and was used without a prior treatment.

For each microwave-assisted acid treatment, 5 g of sepiolite was suspended in a solution of 50 mL HNO_3 (0.2 M). Then, each sample was subjected to microwave radiation in a commercial microwave system EMS20100OX (Electrolux, Stockholm, Sweden) using an open glass cylinder reactor, operating at 800 W and 2.45 GHz for between 1 and 16 min. According to previous studies, the microwave radiation was discontinuously applied to avoid a temperature above 100 °C. Thus, once the sample was irradiated for 1 min, then, the suspension was cooled for 5 min [12,37,38]. This procedure was repeated until the desired radiation time was reached. Once the microwave-assisted treatment was performed, the sample was centrifuged and washed several times to remove NO_3^- species used in the acid treatment. Following this methodology, previous studies have described that the octahedral layer is progressively dissolved in magnesian phyllosilicates [12,37,38].

The Cu species were deposited on the surface of the sepiolite using copper nitrate trihydrate $Cu(NO_3)_2 \cdot 3H_2O$ (Sigma-Aldrich, Darmstadt, Germany, 99%) as copper source and urea as precipitant agent. Thus, for each synthesis, 2 g of the raw sepiolite or sepiolite subjected to the microwave-assisted acid treatment was suspended in 6 g of urea and the appropriate amount of $Cu(NO_3)_2 \cdot 3H_2O$ to obtain catalysts with a Cu loading between 5 and 20 wt.%. Later, the suspension was heated at 95 °C for 2 h to promote thermal decomposition of urea into CO_2 and NH_3, which promotes the precipitation and deposition of the Cu species on the surface of the sepiolite. In the next step, the solid was filtered, discarding the presence of Cu species in the solution. Then, the samples were washed with distilled water two times and dried at 90 °C overnight. Finally, the solid was calcined at 400 °C for 2 h and reduced to the desired temperature considering their respective H_2-TPR profiles.

The uncalcined samples were labeled as Prec-XCuO-Sep-Y, the calcined samples XCuO-Sep-Y and the reduced samples were labeled has XCu-Sep-Y where X indicates the Cu content in wt.%, while Y indicates the time of microwave-assisted acid treatment at which was subjected the sepiolite.

2.2. *Characterization Techniques*

Crystallinity of the samples was determined by X-Ray diffraction (XRD). The chemical composition of the samples was carried out by X-Ray fluorescence (XRF), attenuated total reflectance (ATR) and diffuse reflectance spectroscopy. The morphology of the catalysts was studied by Transmission Electron Microscopy (TEM). The analysis of the textural properties was evaluated from the analysis of their N_2 adsorption–desorption isotherms at −196 °C. The superficial charge of the modified sepiolites was performed by the analysis of their zeta-potential. The analysis of the chemical composition of the surface of the catalyst was performed by X-ray photoelectron spectroscopy (XPS). The reducibility of the Cu-based precursors was carried out by the analysis of their H_2-thermoprogrammed reduction (H_2-TPR) profiles. The detailed description of each characterization technique is provided in the Supplementary Material.

2.3. *Catalytic Studies*

The hydrogenation of FUR has been performed in a vapor-phase atmospheric-pressure fixed-bed continuous flow quartz catalytic reactor with a diameter of 6.35 nm. Prior to each study, the samples were pelletized with a range of 215–400 μm. The sample was placed in the middle of the quartz reactor, being retained with two layers of quartz wool. The temperature was controlled with a thermocouple, which was inserted in the quartz reaction, being in direct contact with the fixed bed. Prior to each study, the precursor was reduced under H_2-flow, following the H_2-TPR profiles studied previously. This temperature was maintained for 1 h to ensure the total reduction in the Cu^{2+}-species to Cu^0. After the reduction, the catalyst was cooled until the selected temperature.

FUR was pumped with a Gilson 307SC piston pump (model 10SC) with a flow of 3.87 mL/h of a FUR solution in CPME (5 vol.%) to reach a weight hourly space velocity (WHSV) for FUR of 1.5 h^{-1} with a flow of 10 mL/min.

Both reagents and products were quantified in a Shimadzu GC-14B gas chromatograph equipped with a flame ionization detector (FID) and a CP-Wax 52CB capillary column. The furfural conversion and product selectivity were calculated as follows:

$$Conversion\ (\%) = \frac{mol\ of\ furfural\ converted}{mol\ of\ furfural\ fed} \times 100 \quad (1)$$

$$Selectivity\ (\%) = \frac{mol\ of\ product}{mol\ of\ furfural\ fed} \times 100 \quad (2)$$

3. Results and Discussion

Characterization of the Catalysts

Prior to the synthesis of the catalysts, the raw sepiolite was subjected to a microwave-assisted acid treatment with a range of time between 1 and 16 min. Figure 1 compiles some representative diffractograms of the sepiolite after this treatment. The diffractogram of the raw sepiolite reveals that the starting material displays high purity (PDF N: 98-015-6199), discarding the presence of impurities [12,39]. Microwave-assisted acid treatment causes a clear change in the diffraction profile, which becomes more pronounced as the treatment time increases. Thus, (110) reflection decreases clearly while the set of reflections located between 2θ (°) of 15–40 also decrease noticeably, forming a very broad and poorly defined signal, which is attributed to the formation of amorphous silica [12,40].

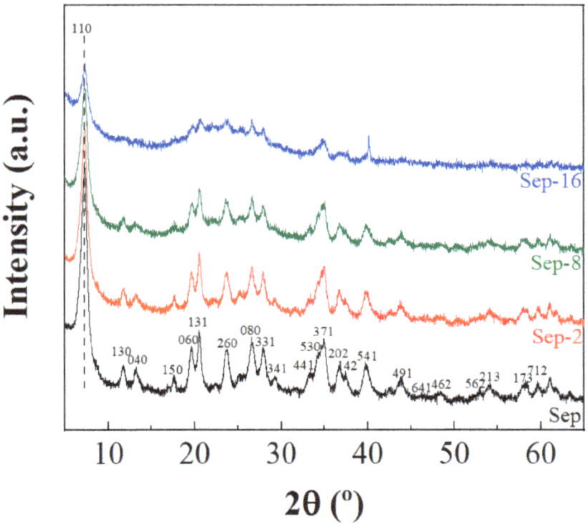

Figure 1. X-ray diffractograms of the raw sepiolite and sepiolite subjected at different times of microwave-assisted acid treatment.

The chemical composition of the raw clay and those samples subjected to microwave-assisted acid treatment was determined by XRF (Figure 2 and Supplementary Material, Table S1). These data reveal how the raw clay is mainly composed of Si and Mg, which suggests the presence of a trioctahedral clay such as sepiolite. After the microwave-assisted acid treatment, the Mg content notably decreases while the Si-content increases progressively. These data confirm that Mg species located in the octahedral sheet of sepiolite is partially leached, leading to the formation of a silicious material with poorer crystallinity, as was suggested by XRD (Figure 1).

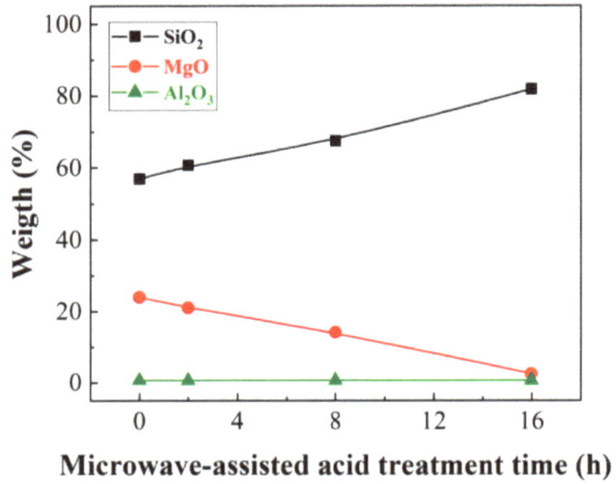

Figure 2. Evaluation of the SiO_2, MgO and Al_2O_3 with the microwave-assisted acid treatment. (Data obtained from XRF).

Infrared spectroscopy has also been used to study how the acid treatment affected the sepiolite. As can be seen in Figure 3, this treatment causes changes in the spectrum of the starting sepiolite. In the region between 4000 and 3000 cm^{-1}, the sepiolite shows different bands associated with the stretching vibrations of the -OH group, such as the bands at 3687 cm^{-1} associated with the stretching of the OH bound to octahedral Mg^{2+} (Mg_3OH) [41], the band at 3618 cm^{-1} associated with the Si-OH groups [41], the band at 3566 cm^{-1} associated with structural water [41] and the band at 3369 cm^{-1} associated to zeolitic water [41]. The intensity of these bands is modified, and some of them even disappear, with the acid treatment [12,37]. Thus, the bands associated with the vibrations of the -OH groups associated with Mg^{2+} disappear, at 3687 and 3566 cm^{-1}, which would indicate the solubilization of the octahedral layer with the treatment time, while the bands due to the Si-OH groups and adsorbed water remain [12,37]. These changes are corroborated when analyzing the bands due to the sepiolite structure in the region between 1400 and 400 cm^{-1} [12,41]. In this region, the bands corresponding to the vibrations of the Si-O bonds remain, such as the bands at 1213, 1078 and 976 cm^{-1}, corresponding to the Si-O bond in the tetrahedral sheet [41], the Si-O-Si vibration at 1004 cm^{-1} [41] and the band at 791 cm^{-1} corresponding to the symmetrical strain of the Si-O-Si group [41], while the bands associated with octahedral Mg^{2+} disappear at 689 and 642 cm^{-1}, corresponding to the bending of the OH group in Mg_3OH and attached to octahedral Mg^{2+} [41,42], respectively. As the acid treatment time increases, the bands due to the Si-O bonds become wider and less defined, indicating an increase in the amorphization of the sepiolite [4,5].

The determination of the textural properties of the raw sepiolite and sepiolite subjected to microwave-assisted acid treatment was carried out through their N_2 adsorption–desorption isotherms at -196 °C (Table 1 and Supplementary Material, Figure S1). According to the IUPAC classification, all isotherms can be adjusted to type II [43], which are typical of microporous materials; although, the high N_2-adsorption at low relative pressure suggests that these materials also display microporosity [43]. These data agree with those reported in the literature previously where the periodic inversion of its tetrahedral sheet leads to the formation of nanocavities of 1.06 nm $\times$ 0.37 nm [4,5], which is related to the microporosity of this fibrous phyllosilicate [6]. On other hand, all samples hardly display hysteresis loops, which confirms a pore diameter lower than 4 nm [43]. The analysis of the N_2 adsorption–desorption isotherms reveals that the microwave-assisted acid treatment promotes a progressive increase in the N_2-adsorbed at low and intermediate relative pressures because of the Mg-leaching of the octahedral sheet causes an increase in the nanochannels.

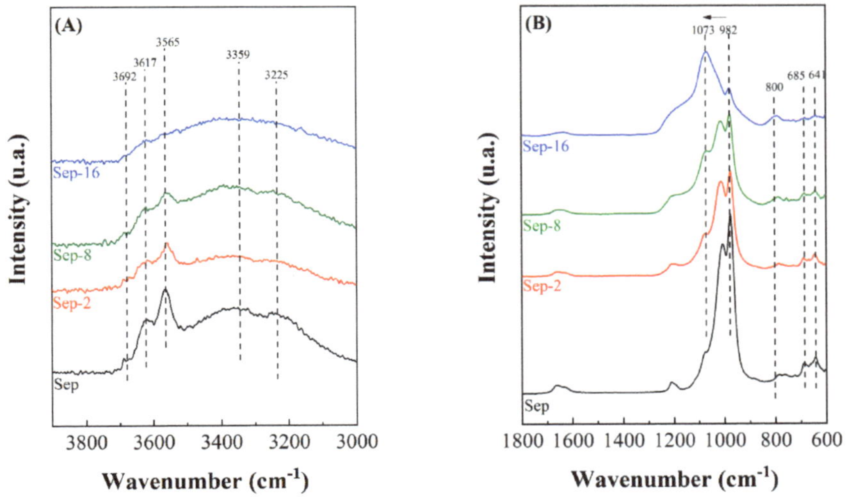

Figure 3. ATR of the raw sepiolite and sepiolite subjected at different times of microwave-assisted acid treatment. Wavenumber region 3900–3000 cm^{-1} (**A**) and wavenumber region 1800–600 cm^{-1} (**B**).

Table 1. Textural properties obtained from the N_2 adsorption–desorption isotherms at −196 °C for the raw sepiolite and sepiolite subjected to different times of microwave-assisted acid treatment.

Sample	S_{BET} (m^2/g)	t-Plot (m^2/g)	S_{ext} (m^2/g)	V_p (cm^3/g) [1]	V_{mp} (cm^3/g) [2]
Sep	283	149	134	0.3678	0.1382
Sep-1	333	160	173	0.4345	0.1609
Sep-2	347	155	192	0.4513	0.1736
Sep-4	370	175	195	0.4904	0.1813
Sep-8	452	246	206	0.5312	0.2164
Sep-12	468	244	224	0.5512	0.2246
Sep-16	509	266	243	0.5959	0.2421

[1] V_p: Pore volume. [2] V_{mp}: Micropore volume.

The quantification of the surface area determined by the BET equation [44] (Table 1) shows how the microwave-assisted acid treatment causes an increase in the surface area from 283 m^2/g to 509 m^2/g after 16 min of microwave-assisted acid treatment. In the same way, the microporosity and the external surface improve after this time. On the other hand, the analysis of the pore volume follows a similar trend since the porosity rises notably after the microwave-assisted acid treatment. The increase in the surface area and pore volume is attributed to the progressive leaching of the Mg species as well as the amorphization of the remaining silica [12,37]. The analysis of the pore size distribution (Supplementary Material, Figure S2) reveals that the microwave-assisted acid treatment mainly affects the micro- and mesoporosity, as indicated by the shift in the pore size distribution at higher values, while the macroporosity is hardly modified. This suggests that acid treatment on sepiolite may take place either across the fiber edges in the octahedral sheet or in the nanocavities formed by periodical inversion of the tetrahedral layer [45], causing an increase in the micro- and mesoporosity.

As the textural properties and chemical composition of sepiolite are modified along the microwave-assisted acid treatment, the next was the determination of the point of zero charge for the raw sepiolite and those samples modified by acid treatment. The data reported in the Supplementary Material, in Figure S3, show that the microwave-assisted acid treatment causes a shift in the point of zero charge at lower pH values when Mg species are progressively extracted from the octahedral layer of sepiolite, confirming the chemical modifications of this fibrous clay after the treatment. The study of the samples at higher pH values indicates that the samples where the Mg^{2+}-species have been partially extracted (Figure 2 and Supplementary Material, Table S1) are those samples with a higher content of negative charges in their structure.

After characterizing the raw sepiolite and the sepiolite subjected to microwave-assisted acid treatment at different times, the incorporation of Cu species on the supports with different composition was carried out following the methodology of thermal decomposition of urea, where the presence of NH_3 favors the precipitation of the Cu species.

This methodology was chosen because the decomposition of urea favors the formation of more homogeneous metallic particles because of a more homogeneous increase in pH compared to other methodologies where basic solutions were added [46]. The study of the Prec-15CuO-Sep-Y samples by XRD (Figure 4) reveals that this treatment hardly modifies the crystallinity of the supports. It is striking that Cu species are only observed in the Prec-15CuO-Sep, appearing to reflections located at 2θ (°) of 35.2 and 38.4, which are assigned to the presence of CuO (PDF N. 01-089-2531). However, those samples where sepiolite modified by acid treatment is used as support do not show diffraction peaks assigned to Cu species.

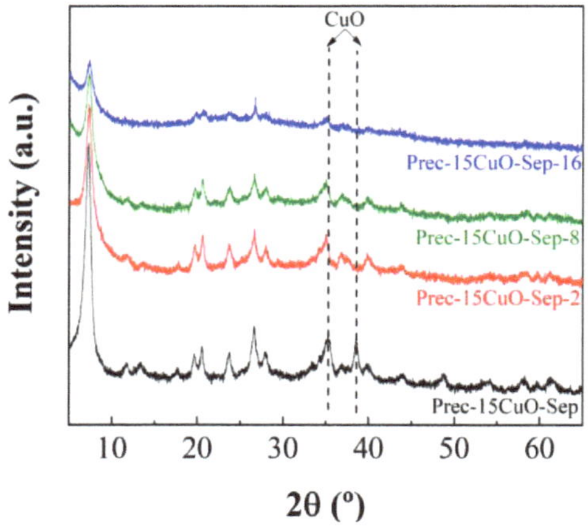

Figure 4. X-ray diffractograms of the raw sepiolite and sepiolite subjected at different times of microwave-assisted acid treatment with Cu species deposited on its surface.

During synthesis, urea hydrolysis provides the necessary hydroxide ions, derived from the acid-base equilibrium of released ammonia, for $Cu(OH)_2$ formation [47]. In the presence of nitrate ions, this leads to Cu(II) hydroxide nitrate formation ($Cu_2(OH)_3NO_3$), as nitrate ions have a strong tendency to incorporate into the $Cu(OH)_2$ crystal structure [48]. This compound has been confirmed through XRD (Supplementary Material, Figure S4) by precipitating it following the experimental procedure described above, leading to $Cu_2(OH)_3NO_3$ as the only crystalline phase (PDF N: 00-045-0594). This $Cu_2(OH)_3NO_3$ consists of positively charged layers with $Cu_2(OH)_3{}^+$ stoichiometry, neutralized by nitrate ions in the interlayer space [47]. Due to the basic synthesis medium, the sepiolite surface becomes activated through the reaction of its surface hydroxyl groups with hydroxide anions, allowing the surface to become negatively charged and thus interact with $Cu_2(OH)_3NO_3$. On the other hand, upon filtering, a dark solid was obtained, indicating that $Cu_2(OH)_3NO_3$ likely did not exist on the sepiolite surface, as confirmed by XRD analysis. Additionally, XPS analysis (Supplementary Material, Table S1) confirmed the absence of $Cu_2(OH)_3NO_3$, as the surface N content in sepiolite was negligible (Supplementary Material, Figure S5). The appearance of CuO instead of $Cu_2(OH)_3NO_3$ on the surface may be due to its decomposition under reaction conditions. The synthesis temperature (95 °C) likely promotes this decomposition to CuO, which is the phase detected on untreated sepiolite. Van der Grift et al. [49] indicated that when solution pH exceeded 6.8 at 90 °C, $Cu_2(OH)_3NO_3$ became unstable and decomposed, depositing as hydrated octahedral Cu(II) on SiO_2, rather than as CuO as in our case. Lee et al. [50] showed that $Cu_2(OH)_3NO_3$ or CuO formation depended on both $[OH^-]/[Cu(NO_3)_2]$ ratio and temperature.

Regarding acid-treated sepiolite, for any treatment duration, all showed a blue-green color after Cu(II) deposition, indicating that CuO formation was not favored, but rather Cu(II) likely deposited as isolated Cu(II) ions or forming small aggregates on the support (Supplementary Material, Figure S6). In this case, XRD only showed peaks ascribed to treated sepiolite with no other phases detected. XPS analysis showed no detectable N quantity, as with untreated sepiolite, ruling out Cu(II) species having N-containing compounds like ammonia or nitrate in their coordination sphere (Supplementary Material, Figure S5). The acid treatment removed $Mg(OH)_2$ from sepiolites, exposing the tetrahedral SiO_2 layers of sepiolite. Cu catalyst synthesis on SiO_2 was studied by Toupance et al. [51–53], evaluating the Cu species present. Their method, termed "cation exchange" or "selective adsorption", involved the synthesis of the octhaedral $[Cu(NH_3)_4(H_2O)_2]^{2+}$ complex and suspending SiO_2 in a complex solution maintaining pH above 9 to avoid $Cu_2(OH)_3NO_3$ precipitation. They observed two distinct species on SiO_2: isolated Cu species, $(\equiv SiO)_2Cu(NH_3)_2(H_2O)_2$, and Cu phyllosilicate formation. They proposed Cu phyllosilicate formation occurred in solution, as at basic pH, silica dissolved as $Si(OH)_4$ and condensed with neutral $[Cu(OH)_2(H_2O)_4]^0$ species. These species were more abundant at pH 9, while Cu(II) tetraamine complex was more abundant at higher pH. The authors demonstrated that bridging -OH ligands were necessary for phyllosilicate formation, while ligands like ammonia or ethylenediamine prevented it. Additionally, neutral Cu species were necessary for phyllosilicate formation as they could condense to form a solid phase. Considering this background, we can postulate that acid-treated sepiolite, in the catalyst synthesis medium, may have partially dissolved silica as silicic acid, which would condense with $Cu_2(OH)_3NO_3$ species forming Cu phyllosilicate on the sepiolite surface.

In order to analyze the chemical composition of the obtained materials on their surface, XPS experiments were carried out (Figure 5 and Supplementary Material, Table S2). The analysis of the different core level spectra confirms the typical bands of their respective oxides. The acid treatment hardly causes a shift in the binding energy values in each region. However, the analysis of the Mg 1s core level spectra confirms that the Mg content progressively decreases from 10.12% until 0.74% after 16 min of treatment (Supplementary Material, Table S2). This decrease was accompanied by an increase in the Si-content, confirming the formation of a SiO_2-rich material with prolonged acid treatment. The analysis of the Cu 2p core level spectra shows a main contribution with a maximum about 933.2 eV for Prec-15CuO-Sep while the Cu-based materials subjected to a microwave-assisted acid treatment the contribution appears about 935.0 eV. In both cases, this contribution is ascribed to the presence of Cu(II)-species. In the same way, the presence of a broad band located about 942 eV, attributed to the shake-up satellite, confirms the presence of Cu as a divalent species. The shift in the binding energy for those samples subjected to the acid treatment must be related to a modification of the interaction of the Cu(II)-species with the support by the formation of a Cu phyllosilicate on the surface of the sepiolite subjected to the acid treatment. The analysis of the atomic concentration on the surface of the catalysts reveals that Prec-15CuO-Sep sample displays lower amount of Cu species on the surface (Supplementary Material, Table S2). According to the XRD data (Figure 4), the lowest level of Cu species in this sample must be ascribed to Cu(II) species with higher agglomeration. The analysis of the N-content on the surface of the precursors is similar in all cases 0.31–0.39%, discarding the deposition of nitrate species and/or the amine-Cu complex. These values are below those obtained for Cu on the surface by XPS. Thus, it is expected that the ammonia formed in the decomposition of urea produces a rise in pH. However, the diffraction and XPS data rule out the formation of Cu complexes on the surface, at least for the most part.

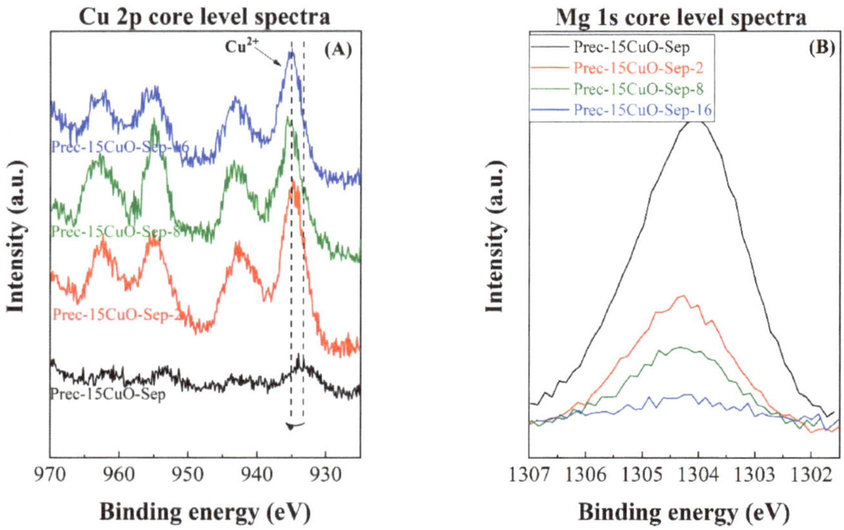

Figure 5. Cu 2p (**A**) and Mg 1s (**B**) core level spectra of the raw sepiolite and sepiolite subjected at different times of microwave-assisted acid treatment with Cu species deposited on its surface (estimated by XPS).

Once the samples were calcined, the precursors show similar colors. Thus, the 15CuO-Sep sample maintains a black color (Supplementary Material, Figure S7). However, the color evolves to green, for the same Cu content, as the microwave-assisted treatment time increases so the sintering during the calcination step must be ruled out.

Figure 6A shows the UV-vis spectra of Cu-based catalysts supported on sepiolite after acid treatment and calcination at 450 °C. These spectra show unequivocally that the Cu species present in the acid-treated sepiolite are completely different from those present in the untreated sepiolite. The latter shows bands 204, 206 and 299 cm^{-1} of charge transfer from the O^{2-} ligand to isolated Cu^{2+} ions [54], others at longer wavelengths (322 and 359 cm^{-1}) which are associated with oligomeric Cu species with different degrees of oligomerisation, indicating the presence of Cu-O clusters [55] and a broad band extending beyond 800 nm, as was observed by previous authors [55], due to d-d transitions of the Cu^{2+} ion in an octahedral environment, characteristic of bulk CuO particles [56]. As for Cu catalysts supported on sepiolite treated in acid medium, an intense broad band extending from 300 nm to approximately 700 nm can be distinguished, in which different contributions associated with Cu(II) species forming different cluster sizes are observed.

The magnification of this band may indicate a large tetragonal distortion of Cu(II) in these clusters increasing the intensity as they are less forbidden transitions by Laporte's rule [57]. For these catalysts, the absence of the band associated with the transitions due to CuO is remarkable, which would indicate a large dispersion of the CuO particles. The band gap increases significantly in the case of the catalysts based on sepiolite treated in acid medium, approximately 2 eV (Supplementary Material, Figure S8) with respect to the untreated one, which shows a modification of the copper species present in the catalysts and a lower semiconductor characteristic.

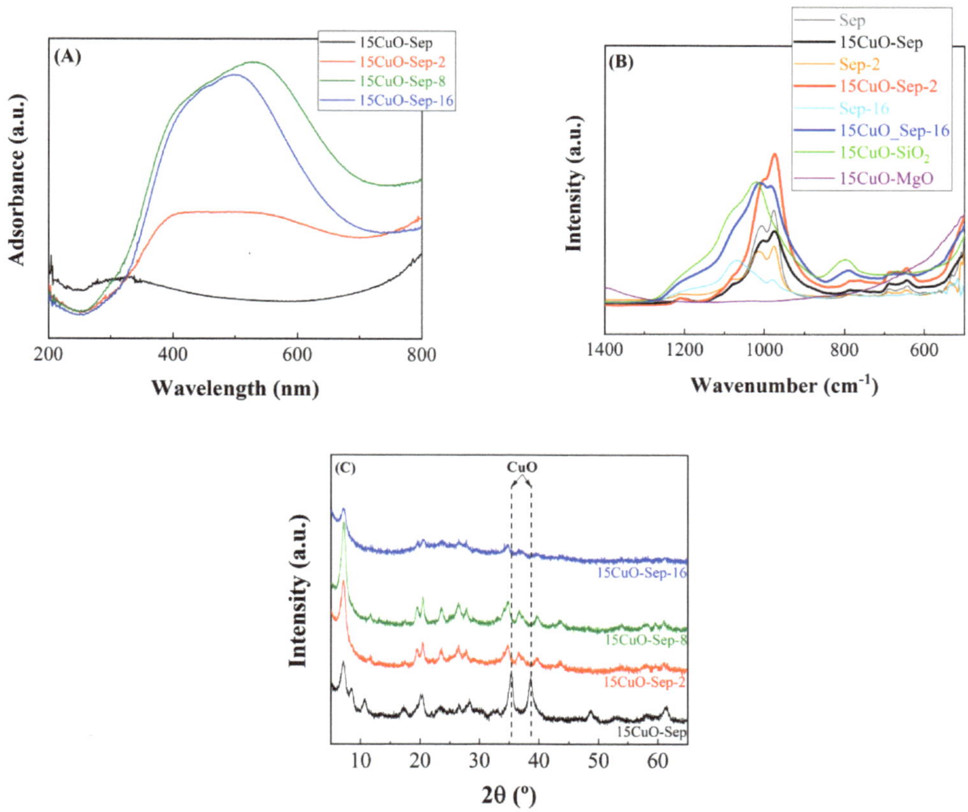

Figure 6. Diffuse reflectance spectra (**A**), ATR spectra (**B**) and XRD (**C**) of the raw sepiolite and sepiolite subjected at different times of microwave-assisted acid treatment with Cu species deposited on its surface and calcined at 450 °C.

When Cu^{2+} is incorporated into the untreated or acid-treated sepiolite, no remarkable changes are observed by ATR after Cu deposition (Figure 6B), both in the region of the hydroxyl group tension and in the region of the bands associated with the sepiolite structure. However, some bands have increased in intensity, such as the band at 974 cm^{-1}, which would indicate the formation of the Si-O-Cu bond [58]. It should also be noted that the bands associated with the Mg-OH bonds of the untreated sepiolite are recovered again in the case of the sepiolite treated for 16 min, which could indicate that Cu(II) occupies the sites left by the Mg(II). The presence of Cu phyllosilicate cannot be assured since the characteristic bands of this compound, above 1020–1030 cm^{-1} ($\nu_{Si\text{-}O}$) and 675 cm^{-1} (δ_{OH}) are obscured by those due to Si-O or Mg-OH bonds [41,59]. The presence of $Cu(OH)_2$ could be ruled out by the absence of the bands at 938 and 694 cm^{-1} corresponding to δ_{OH}, as well as the presence of Cu(II) oxynitrate, since the bands of the nitrate ion at 1423 and 1347 cm^{-1} are not observed [41,60]. Finally, the bands corresponding to CuO that should appear between 575 and 460 cm^{-1} are not observed, ruling out the presence of segregated CuO, as the XRD and UV-vis spectra demonstrated. As a conclusion, it could be indicated that Cu^{2+} is highly dispersed on the treated sepiolite forming Si-O-Cu.

The analysis of the calcined samples by XRD (Figure 6C) shows a decrease in the intensity of the reflections attributed to sepiolite, probably due to the thermal treatment causes a partial amorphization of the sepiolite. In the same way, the diffractograms also reveal that the reflections ascribed to Cu species in the form of CuO-crystals are only detected for the sample 15CuO-Sep, which suggests the formation of bigger crystals in this precursor. On the other hand, it is striking that the other precursors, which were subjected to microwave-assisted acid treatment, do not present reflections attributed to Cu species, confirming the high dispersion as well as good resistance to sintering after the thermal treatment at 450 °C for 2 h.

Once the supports and the precursors were characterized, the next step was the study of the reducibility of the Cu-based precursors by H_2-TPR (Figure 7). From the reduction profiles, it is not possible to discern the reduction steps $Cu^{2+} \rightarrow Cu^{+} \rightarrow Cu^{0}$. In this sense, previous authors have indicated that the reducibility is directly related to the interaction precursor-support and the crystal size of the precursor [61,62]. The data reported in Figure 7 shows how the reduction of 15CuO-Sep sample takes place between 150 and 350 °C while those catalysts where the support was subjected to a microwave-assisted acid treatment display a reduction in a narrower temperature range. These data suggest that 15CuO-Sep sample presents CuO-particles with variable particle size, which should affect its reducibility, while the particle size of the precursors, where the support was modified by an acid treatment, seems to be more homogeneous. On the other hand, the supports modified with acid treatment also display a maximum value of reducibility 20–30 °C lower than that observed for 15CuO-Sep sample. This shift supposes a poorer interaction between CuO and sepiolite which could be to the modification of the support because of the progressive loss of Mg in the sepiolite along the acid treatment (Figure 2 and Supplementary Material, Table S2. The H_2-TPR profiles also suggest that the presence of smaller CuO crystals promotes easier reducibility than that observed in 15CuO-Sep precursor where larger CuO clusters were obtained, as was observed by XRD (Figure 6C). Considering the H_2-TPR profiles, all Cu-based catalysts were reduced at 350 °C for 1 h to ensure a total reduction in the Cu species.

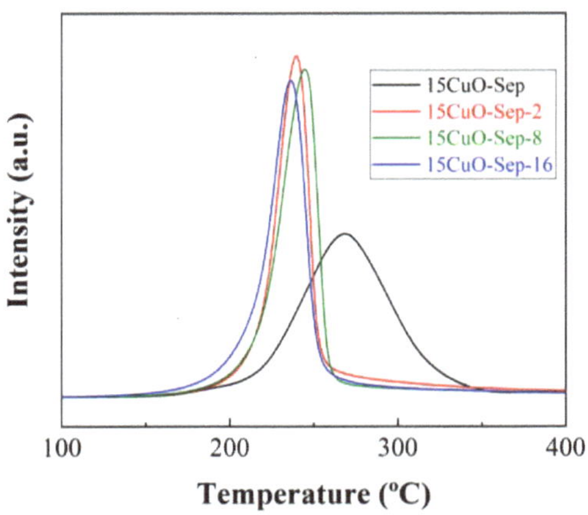

Figure 7. H_2-TPR profiles of the Cu-based precursors supported on raw sepiolite and sepiolite modified with microwave-assisted acid treatment at different times.

The analysis of the crystallinity of the Cu-based catalysts after its reduction (Figure 8) shows the typical reflection lines ascribed to sepiolite, which were described above. In addition, two new reflection lines located at 2θ (°) of 43.2 and 50.4° for 15Cu-Sep catalyst, which are ascribed to the formation of metallic copper (Cu^0) (PDF N°: 00-002-1225). However, the diffractograms of the other Cu-based catalysts, where sepiolite was treated with acid treatment, do not show diffraction peaks assigned to copper species, confirming a higher dispersion of Cu^0 for these catalysts after its reduction.

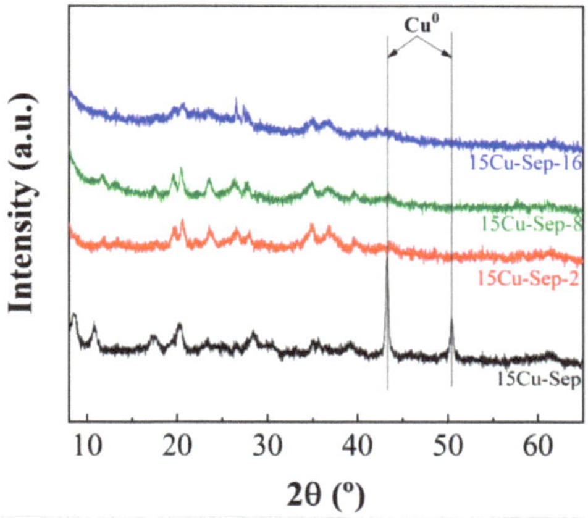

Figure 8. X-ray diffractograms of the Cu-based catalysts supported on raw sepiolite and sepiolite modified with microwave-assisted acid treatment at different times.

The morphology for the Cu-based catalysts was analyzed by TEM (Figure 9 and Supplementary Material, Figure S9). These micrographs reveal how the fibrous morphology of the sepiolite is progressively damaged due to the acid treatment causes a collapse of their structure by a progressive leaching of the Mg species located in the octahedral sheet. Regarding the Cu^0-particles, it is striking how all catalysts present metallic particles with spherical morphology, which are homogeneously dispersed on the surface of the sepiolite. However, the size of the metallic particles differs among them. Thus, the 15Cu-Sep catalyst displays a Cu^0 particle size between 15 and 25 nm (Figure 9A). However, those catalysts where the sepiolite was subjected to acid treatment (Figure 9B–D) show a lower Cu^0 particle size, being lower than 5 nm in all cases, confirming the data observed by XRD (Figure 8). The analysis of crystal size average shows how the modification of the sepiolite by microwave-assisted acid treatment is directly related to the dimensions of the Cu particles since the use of longer acid treatments promote smaller crystals after the calcination and reduction in the Cu species (Supplementary Material, Figure S10).

The analysis of the catalysts after their reduction by XPS shows how the spectra of each region are maintained, except in the Cu 2p core level, where the contributions attributed to the formation of CuO have disappeared. Thus, the Cu $2p_{3/2}$ region only shows a new contribution located about 932.6 eV, which is typical of reduced Cu species (Figure 10A). On the other hand, it is striking that 15Cu-Sep catalyst displays the main contribution at higher binding energy value. This shift is attributed to a change in the electronic charge density of the Cu^0 particles, which is related to the lower reducibility (Figure 7) and higher crystal size of the metallic particles in 15Cu-Sep catalyst (Figures 8 and 9). The analysis of the atomic concentration on the surface of the catalyst (Table 2) confirms the lower amount of Cu exposed on the surface for 15Cu-Sep, suggesting a poorer dispersion of available Cu species.

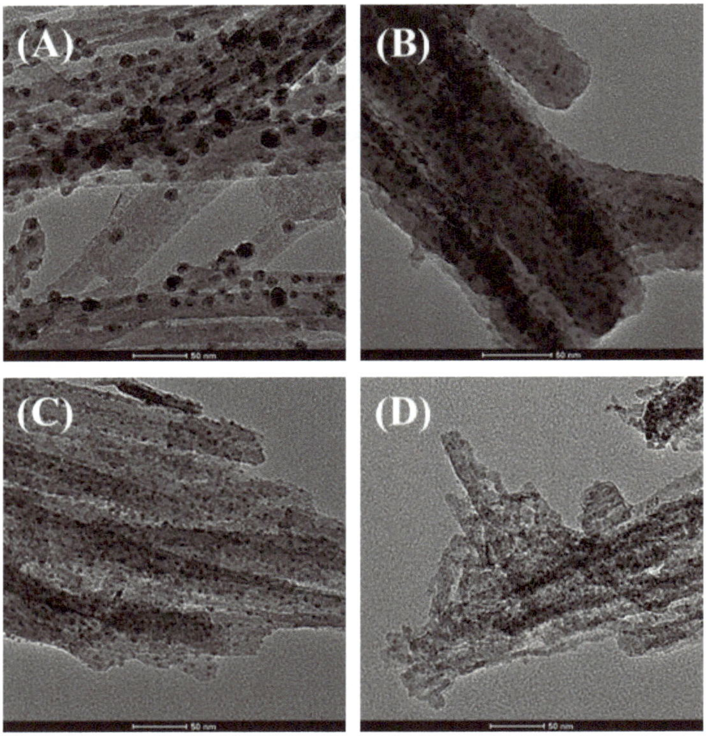

Figure 9. TEM micrographs of the Cu-based catalysts supported on raw sepiolite and sepiolite modified with microwave-assisted acid treatment at different times. 15Cu-Sep (**A**), 15Cu-Sep-2 (**B**), 15Cu-Sep-8 (**C**) and 15Cu-Sep-16 (**D**). Scale 50 nm.

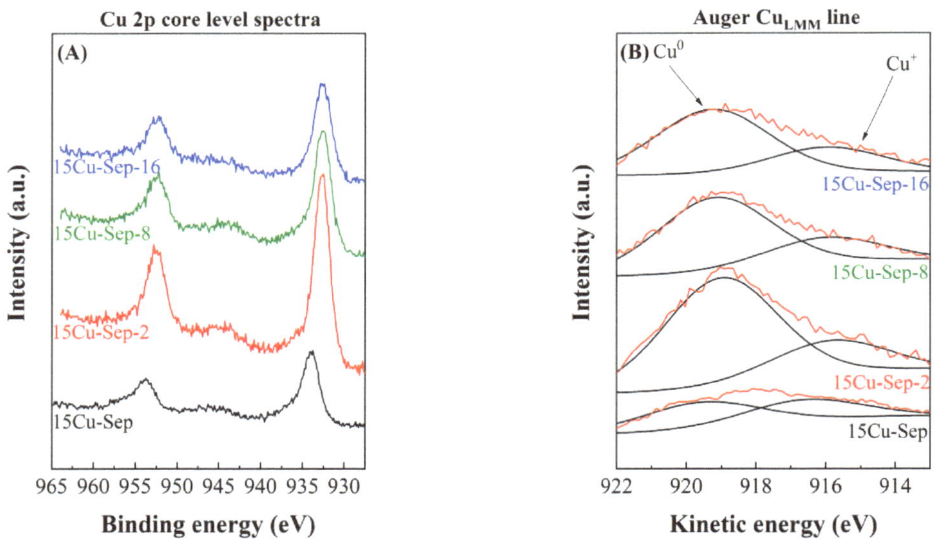

Figure 10. Cu 2p core level spectra (**A**) and Cu_{LMM} Auger line (**B**) of the Cu-based catalysts supported on raw sepiolite and sepiolite modified with microwave-assisted acid treatment at different times.

Table 2. Atomic concentrations on the surface obtained from the XPS spectra of the Cu-based catalysts supported on raw sepiolite and sepiolite modified with microwave-assisted acid treatment at different times.

Sample	Atomic Concentrations (%)					
	C 1s	O 1s	Mg 2p	Al 2s	Si 2p	Cu 2p
15Cu-Sep	22.54	45.66	9.13	4.33	17.40	0.29
15Cu-Sep-2	17.44	48.13	6.53	4.32	20.71	2.91
15Cu-Sep-8	13.21	53.36	4.22	4.35	22.02	2.85
15Cu-Sep-16	8.94	56.10	1.21	4.44	28.72	1.91

From the Cu 2p core level spectra, it is not possible to discern between the reduced Cu species (Cu^0 and Cu^+). However, the analysis of the Cu_{LMM} Auger line allows to differentiate between both reduced species (Figure 10B). The analysis of this region displays a main contribution located about 919 eV, which is assigned to the presence of Cu^0 species while the contribution with lower intensity, located around 916 eV, is attributed to the existence of Cu^+ species [34–36] although the H_2-TPR signals confirmed the total reduction in Cu species (Figure 7). Nevertheless, the presence of Cu^+ species can be considered beneficial for the FUR reaction since it could provide acid sites, which are involved in the hydrogenolysis of FOL to MF [63,64].

4. Catalytic Results

Once the Cu-based catalysts were characterized, the samples were assayed in gas-phase in the FUR hydrogenation to obtain valuable products. For this study cyclopenthyl methyl ether (CPME) was selected as solvent since this compound is considered environmentally friendly, being used in some biomass valorization reactions as the dehydration of xylans to obtain FUR [65]. The thermochemical stability of CPME was studied considering similar catalytic conditions to those performed in the gas-phase of FUR hydrogenation. The obtained results hardly show FUR conversion, discarding its reactivity in the catalytic conditions used in the present study.

In the first study, Cu-based catalysts (15 wt.% of Cu) supported in sepiolite, which was subjected to microwave-assisted acid treatment at different times, was studied in the FUR hydrogenation at 190 °C for 5 h (Figure 11A). Despite all catalysts displaying similar Cu content, the catalytic behavior is different between them, suggesting that the modification in the chemical composition of the support plays an important role in the catalytic behavior. Thus, the catalyst 15Cu-Sep catalyst achieves the lowest conversion values (51% after 1 h of reaction, decreasing this conversion until 34% after 5 h of reaction at 190 °C). The use of sepiolite modified under microwave-assisted acid treatment as support causes a clear improvement of the FUR conversion, reaching a full conversion after 1 h of reaction when the sepiolite is only treated for 2 min (15Cu-Sep-2 catalyst). In the same way, the deactivation is more limited along the time since 15Cu-Sep-2 catalyst still maintains a conversion of 93% after 5 h of reaction at 190 °C. However, the use of a support where sepiolite has been subjected to a longer acid treatment causes a progressively worsens in the catalytic behavior. Thus, 15Cu-Sep-16 catalyst shows a decrease in the FUR conversion from 88% after 1 h or reaction to 61% after 5 h of reaction at 190 °C. As was described in previous studies, the progressive decrease in the FUR conversion along the time on stream is attributed to the high reactivity of the FUR molecules since this molecule is prone to suffer polymerization reactions, mainly in gas-phase, causing a strong deactivation of the available active sites due to the deposition of carbonaceous deposits on the surface of the catalyst [34,35].

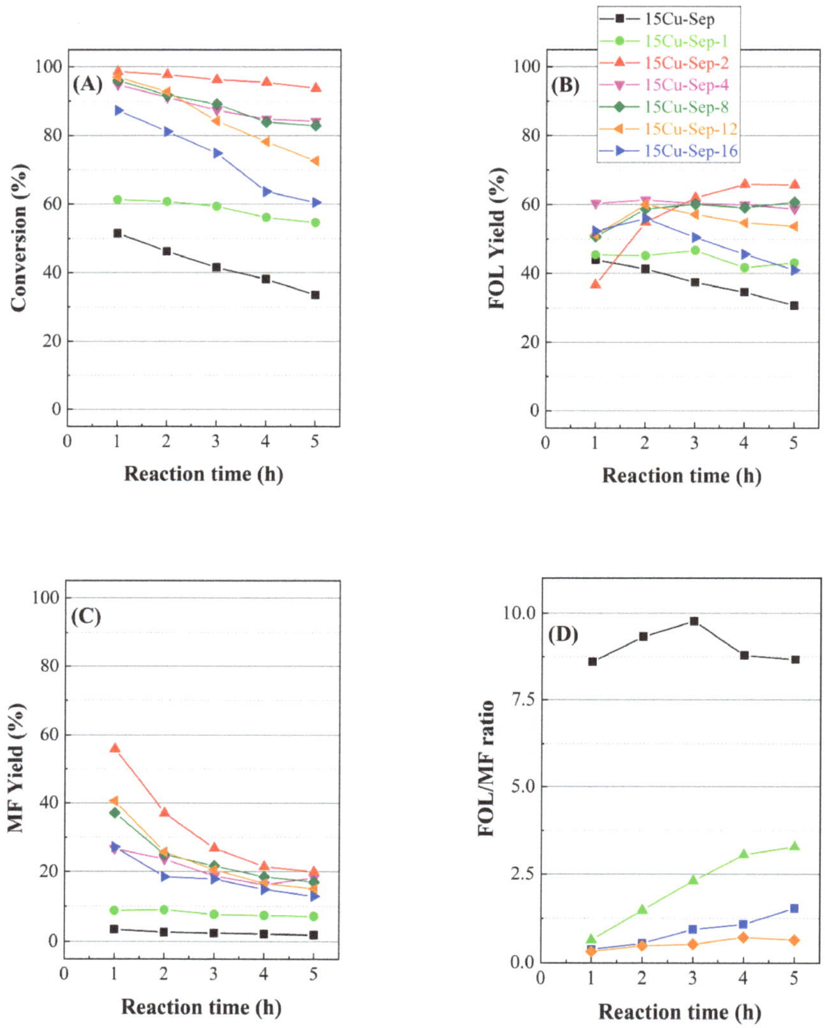

Figure 11. FUR conversion (**A**), FOL yield (**B**), MF yield (**C**) and FOL/MF molar ratio (**D**) the FUR hydrogenation in gas phase using Cu-based catalysts supported on sepiolite modified with an acid treatment. Experimental conditions: mass of catalyst = 0.15 g, reaction temperature = 190 °C, reaction time = 1–5 h, pressure = 0.1 MPa, H_2 flow = 10 mL/min, and feed flow = 2.3 mmol FUR/h.

The poorest catalytic behavior of the 15Cu-Sep sample attributed to this catalyst displays bigger metallic particles in comparison with other catalysts where sepiolite was subjected to microwave-assisted acid treatment. This implies a lower amount of Cu sites exposed on its surface (Figure 10 and Table 2). The acid treatment causes a progressive increase in the specific surface area of the support, which could promote a higher dispersion of the active phase due to the loss of the octahedral sheet by leaching of the Mg species. The characterization of these catalysts reported a smaller particle size of Cu^0 (Figure 9), which implies a higher dispersion of the available hydrogenating sites, promoting a higher conversion value. The presence of these smaller Cu^0 crystals is observed in all catalysts where the sepiolite is treated with acid treatment at different times. Despite the catalyst displaying an improvement in the textural properties, the use of longer acid treatment promotes a deterioration of the catalytic behavior probably due to modifications in the interactions between metallic sites and support. In this sense, previous studies have reported that the presence of Mg species provides basic sites, which has a positive effect in the catalytic behavior [66,67]. Considering those catalysts where the crystal size of Cu^0 is similar and well dispersed, the catalytic results seem to indicate that the formation of catalysts with lower levels of Mg species worsens the catalytic behavior, so the presence of these Mg species seems to be a beneficial effect on the FUR conversion, as was observed in previous studies [68].

Regarding the selectivity pattern, furfuryl alcohol (FOL) and 2-methylfuran (MF) were the obtained products (Figure 11B,C). It is striking how the most active catalysts achieve higher proportions of MF in shorter times. However, the yield towards this product notably decreases along the time on stream. This decrease is accompanied by a progressive increase in the FOL content, which suggests the presence of consecutive reactions in these reaction conditions. Several studies have proposed the mechanism of this reaction. Thus, the first studies, carried out by Sitthisa et al., pointed out that the interaction between the Cu sites and FUR molecule to form FOL by a hydrogenation reaction takes place through the oxygen atom of the carbonyl group with a η^1-(O)-aldehyde configuration [27,69], which was confirmed by DFT studies later [28]. In a second step, FOL interacts with the active phase through the oxygen atom of the hydroxyl group to form MF via hydrogenolysis reaction [28]. This step requires acid sites although the acidity must be modulated since the presence of high number of acid sites with high strength promotes the formation of the humins, which causes a fast deactivation of the catalyst. In this sense, previous authors have pointed out that presence of low acidity can be enough to promote this hydrogenolysis reaction [30]. In the same way, the presence of small particles of Cu^0 particles as well as the presence of Cu^+-species can act as Lewis acid sites to promote the hydrogenolysis reaction [30,34,36]. The profile of MF and FOL yield suggests that the sites involved in the hydrogenolysis reaction (FOL $\rightarrow$ MF) are more prone to suffer deactivation in comparison to the sites involved in the hydrogenation reaction (FUR $\rightarrow$ FOL). Considering these premises, the catalytic results show how the catalyst with lower dispersion and bigger crystal size, i.e., 15Cu-Sep sample, hardly displays yield towards MF. These data agree with those reported in the literature since the presence of bigger Cu^0-crystals promotes the hydrogenation reaction (FUR $\rightarrow$ FOL) [36]. However, the presence of a higher dispersion of Cu^0 and the existence of Cu^+-species on the surface promotes the hydrogenolysis reaction (FOL $\rightarrow$ MF) [34,36]. Among the samples, the most active is 15Cu-Sep-2 catalyst, which achieves a MF yield of 56% after 1 h of reaction at 190 °C. The biggest MF yield can be ascribed to the higher amount of Cu species exposed on its surface as well as the presence of Mg species, which promotes an appropriate interaction active-phase support where the presence of collateral reaction and the formation of undesired by-products are minimized [34,68]. Nevertheless, the yield decreases notably until reaching a value of 20% after 5 h of reaction at 190 °C. This decrease is accompanied by a clear increase in the FOL yield, obtaining a maximum value of 65% after 5 h of reaction, as was indicated in Figure 11D. In summary, the crystal size of the Cu species (Cu^0 and Cu^+) has a determining role in the selectivity pattern since the catalyst with bigger crystal size tends to promote the hydrogenation reaction (FUR $\rightarrow$ FOL). However, the presence of smaller crystals are prone to the hydrogenation and a subsequent hydrogenolysis reaction to form MF (FUR $\rightarrow$ FOL $\rightarrow$ MF), although these sites are more sensitive to suffer deactivation processes in such a way the reaction is retained in the hydrogenation reaction and does not evolve towards MF, forming FOL as main product (Figure 11D).

As 15Cu-Sep-2 catalyst reached the highest activity and stability along the time on stream, this sample was selected to evaluate the influence of the reaction temperature in the catalytic behavior (Figure 12A). These data show how FUR conversion decreases along the time when the reaction takes place at milder temperature (170 °C) from 78% after 1 h of reaction and to 63% after 5 h of reaction. The use of a higher temperature leads to almost full conversion along the time on stream. Despite this similar activity, the yields towards FOL and MF differ between them (Figure 12B,C). The analysis of the yields at shorter reaction times reveals that FOL is preferably formed at lower temperature, obtaining a maximum yield of 63% after 1 h when the reaction takes place at 170 °C. This trend is hardly modified along the time on stream since the reaction at 170 °C is very stable obtaining FOL as main product along the reaction (Figure 12D). However, the catalytic stability for the tests carried out at higher temperature is lower due to the yield towards MF decreases throughout the reaction while FOL levels increase (Figure 12D). From these data, it can be inferred that the hydrogenolysis reaction is promoted at higher temperatures although these sites are prone to be deactivated under these conditions along the reaction, remaining the reaction in hydrogenation of FUR to FOL.

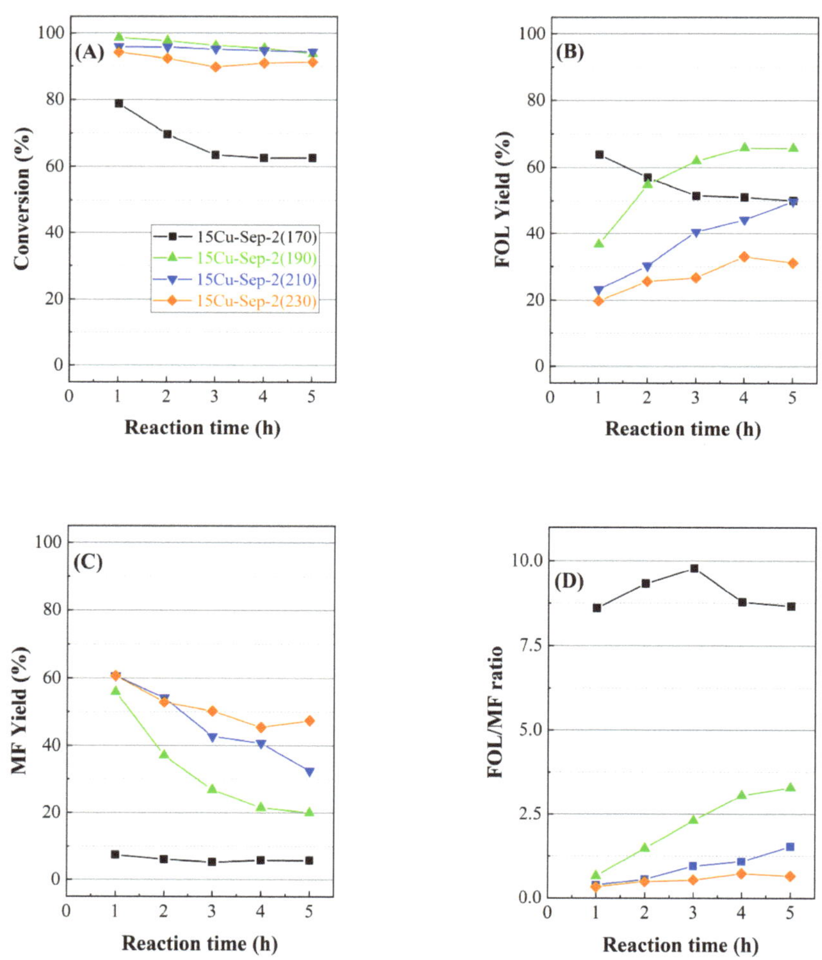

Figure 12. FUR conversion (**A**), FOL yield (**B**), MF yield (**C**) and FOL/MF molar ratio (**D**) in the FUR hydrogenation in gas phase using Cu-based catalysts supported on sepiolite modified with an acid treatment. Experimental conditions: mass of catalyst = 0.15 g, reaction temperature = 170–230 °C, reaction time = 1–5 h, pressure = 0.1 MPa, H_2 flow = 10 mL/min, and feed flow = 2.3 mmol FUR/h.

In a final catalytic study, the influence of Cu content was evaluated considering the best catalytic support (Sep-2) (Figure 13). Catalytic data reveal that the reaction is sensitive to the Cu content (Figure 13A). Thus, an increase in the Cu content causes a significant improvement in the catalytic behavior, since the 5Cu-Sep-2 and 10Cu-Sep-2 catalysts reach a conversion of 40% and 72%, respectively, after 1 h of reaction while the FUR conversion is practically complete along 5 h of reaction when the catalyst displays a Cu content higher than 15 wt.%. Regarding the obtained products (Figure 13B,C), a low Cu content promotes the hydrogenation of FUR to FOL although the conversion is poorer than that observed for other catalysts with higher Cu loading. However, those catalysts with a higher Cu content promote hydrogenation and a subsequent hydrogenolysis step to form MF. In fact, the data reported in Figure 13D shows how the 5Cu-Sep-2 catalyst, where the hydrogenation reaction towards FOL is promoted, maintains a more stable FOL/MF ratio along the time on stream. However, the catalysts with higher Cu content, which are more selective towards MF in shorter reaction times, modify the FOL/MF ratio along the time on stream. This confirms the faster deactivation of the hydrogenolysis sites.

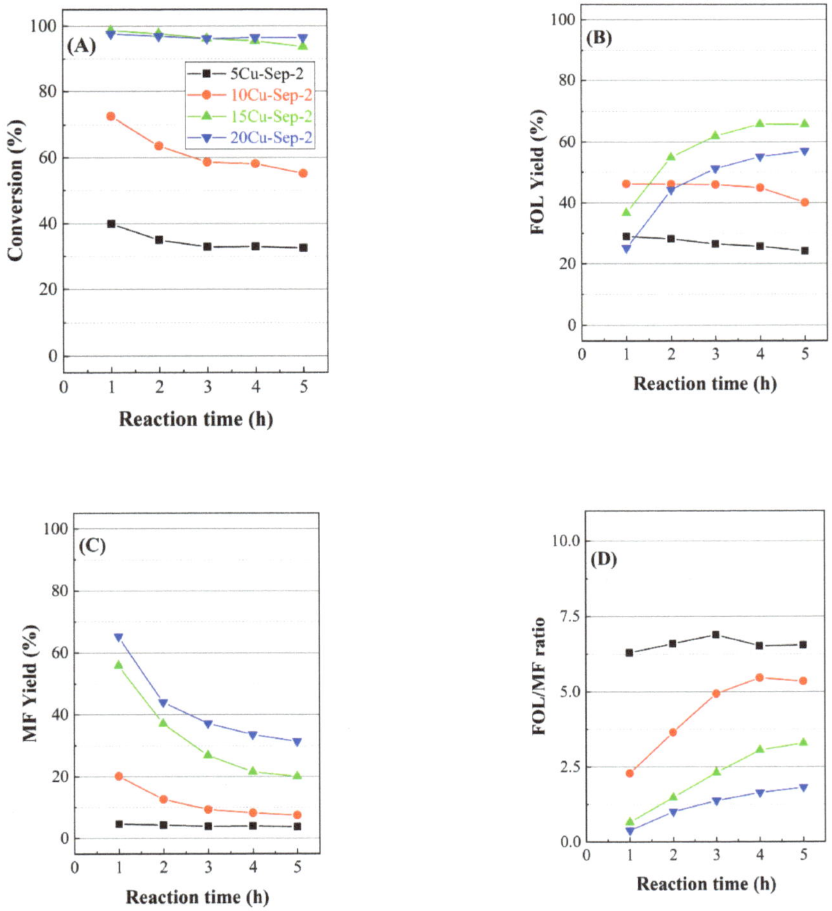

Figure 13. FUR conversion (**A**), FOL yield (**B**), MF yield (**C**) and FOL/MF molar ratio (**D**) in the FUR hydrogenation in gas phase using Cu-based catalysts supported on sepiolite modified with an acid treatment. Experimental conditions: mass of catalyst = 0.15 g, reaction temperature = 190 °C, reaction time = 1–5 h, pressure = 0.1 MPa, H_2 flow = 10 mL/min, and feed flow = 2.3 mmol FUR/h.

Finally, the catalysts with different Cu content supported on Sep-2 were characterized to understand the differences in the catalytic behavior. Firstly, the reducibility of the Cu-based precursors was analyzed by H_2-TPR (Figure 14A). These data reveal that the H_2-consumption increases with the Cu content, which is expected due to the reduction in a bigger amount of Cu^{2+}-species to Cu^0. However, the most representative data of these profiles is related to the similar temperature at which the maximum H_2-consumption takes place. This suggests that the interaction precursor-support hardly suffers modification when the content of the Cu incorporated is modified. The study of the samples with different Cu content by XRD (Figure 14B) reveals that all catalysts, except that sample with the highest Cu content, do not show any reflections lines ascribed to Cu species. In the case of 20Cu-Sep-2 sample, the main diffraction peak ascribed to Cu^0 is observed at 2θ (°) of 43.2 (PDF N°: 00-002-1225), although this signal is relatively wide in such a way that the metallic particles must be relatively small even for high Cu content (20 wt.%). This fact was confirmed by the analysis of their TEM micrographs (Figure 14C and Supplementary Material, Figure S11) since the particles are homogeneous and spherical with diameter lower than 5 nm in all catalysts, confirming the high dispersion of the active phase in all cases. The analysis of the samples by XPS (Figure 14D,E) reveals that the Cu exposed on the surface increases directly with the Cu content. In the same way, the analysis of the Cu 2p core level spectra show the existence of reduced Cu species in all catalysts exclusively (Figure 14D). The detailed analysis of the Cu_{LMM} Auger line displays the presence of two contributions (Figure 14E). The main contribution, located about 919 eV, is assigned to Cu^0 while the lower contribution about 916 eV is attributed to Cu^+ species [34,35]. Despite presenting the same contributions in the Cu regions, the ratio of products obtained is variable for the catalysts with different Cu loading. The results obtained suggest that the hydrogenolysis sites involved in the formation of MF seems to be deactivated more quickly. Thus, the lower proportion of hydrogenolysis sites in 5Cu-Sep-2 catalyst implies a faster deactivation of these sites in such a way this catalyst displays low proportion of MF while the FOL yield is more stable along the time on stream. On the opposite, the catalyst with higher amount of Cu exposed on the surface (Cu^0 and Cu^+), i.e., higher amount of hydrogenation and hydrogenolysis sites promotes the formation of MF during longer time although these sites are also prone to be deactivated along the time on stream, as indicates Figure 13D.

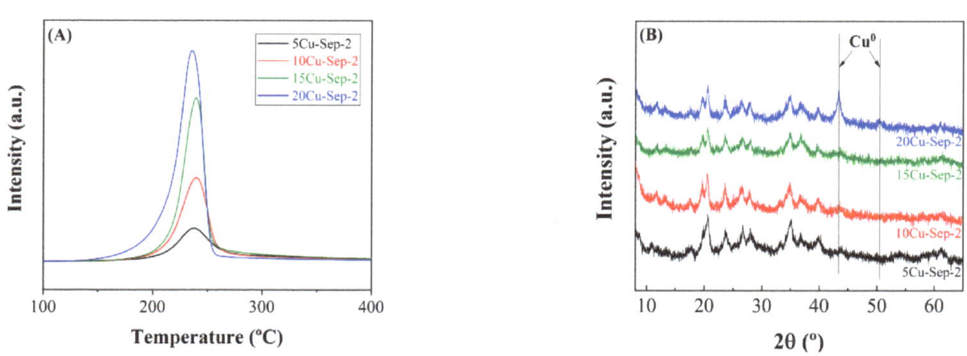

Figure 14. *Cont.*

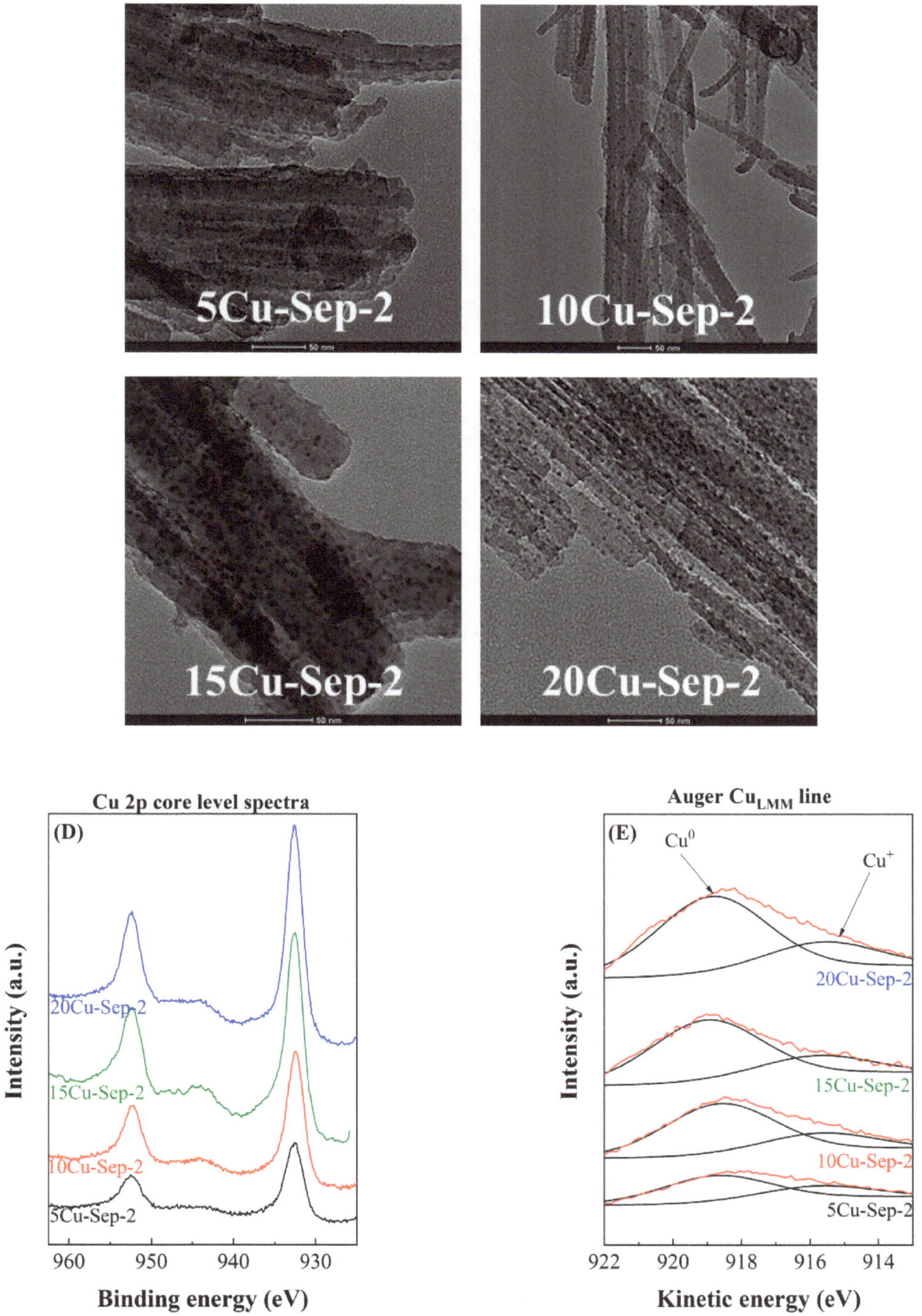

Figure 14. H_2-TPR profiles (**A**), X-ray diffractograms (**B**), TEM micrographs (Scale: 50 nm) (**C**), Cu 2p core level spectra (**D**) and Cu_{LMM} Auger line (**E**) of 5Cu-Sep-2, 10Cu-Sep-2, 15Cu-Sep-2 and 20Cu-Sep-2 catalysts.

5. Conclusions

In this work, a fibrous clay mineral, sepiolite, has been modified through a microwave-assisted acid treatment. This treatment has caused a progressive leaching of the Mg^{2+} species located in the octahedral layer, leading to an increase in the surface area and pore volume. These samples with different textural properties have been used as supports for the precipitation-deposition of Cu species on its surface. The characterization of the samples revealed that CuO particles were dispersed on the surface of the raw sepiolite, while $Cu_2(OH)_3^+$ species have been deposited over the sepiolite subjected to microwave-assisted acid treatment in these sites where Mg^{2+}-species were previously leached by the microwave-assisted acid treatment. These Cu species are highly dispersed on the surface of these modified supports even after calcination and reduction treatment, obtaining Cu^0 particles with a size less than 5 nm. In contrast, the Cu^0 particles supported on raw sepiolite display a much larger particle size between 15 and 25 nm.

Once the materials were characterized, the samples were assayed in the gas phase hydrogenation of furfural. Generally, the difference in Cu^0 crystal size was reflected in its catalytic behavior. These results showed that the catalyst with a larger Cu^0 crystal size (15Cu-Sep) is selective to the hydrogenation reaction, obtaining FOL as main product. However, the catalysts where the sepiolite was subjected to different microwave-assisted acid treatments and smaller Cu^0 particle size are more active and promote two consecutive reactions. Firstly, FOL is obtained through a hydrogenation reaction of FUR. In a second step, MF is obtained from a hydrogenolysis reaction of FOL through mild acid sites. Between the two reactions, the catalytic results indicated that the sites involved in the hydrogenolysis reaction are more prone to be deactivated, while the sites involved in the hydrogenation reaction are more resistant to deactivation although these sites are less active.

The results obtained reveal that the presence of small Cu crystals has a determining role in the catalytic behavior. However, the use of short acid treatment times, where a proportion of Mg^{2+}-species remain in the fibrous structure, also has a positive effect in the catalytic behavior, since the presence of Mg in the support also has a positive effect on the FUR hydrogenation reaction in gas phase.

Supplementary Materials: The following supporting information can be downloaded at: https://www.mdpi.com/article/10.3390/ceramics8010021/s1, Table S1. Chemical composition of the raw sepiolite and sepiolite subjected at different times of microwave-assisted acid treatment; Table S2. Atomic concentrations on the surface obtained from the XPS spectra for the raw sepiolite and sepiolite subjected at different times of microwave-assisted acid treatment with Cu-species deposited on its surface; Figure S1. N_2 adsorption-desorption isotherms of the raw sepiolite and sepiolite subjected at different times of microwave-assisted acid treatment; Figure S2. Pore size distribution of the raw sepiolite and sepiolite subjected at different times of microwave-assisted acid treatment; Figure S3. Zeta potential of the raw sepiolite and sepiolite subjected at different times of microwave-assisted acid treatment; Figure S4. X-Ray diffractogram of the solid obtained after the treatment of $Cu(NO_3)_2$ $3H_2O$ in urea solution at 95 °C; Figure S5. N 1s core level spectra of the raw sepiolite and sepiolite subjected at different times of microwave-assisted acid treatment with Cu-species deposited on its surface (estimated by XPS); Figure S6. Images of Prec-15CuO-Sep (A), Prec-15CuO-Sep-2 (B), Prec-15CuO-Sep-8 (C) and Prec-15CuO-Sep-16 (D); Figure S7. Images of 15CuO-Sep (A), 15CuO-Sep-2 (B), 15CuO-Sep-8 (C) and 15CuO-Sep-16 (D); Figure S8. Method of band gap energy (eV) determination from theTauc plot of the raw sepiolite and sepiolite subjected at different times of microwave-assisted acid treatment with Cu-species deposited on its surface; Figure S9. EDX mapping of the Cu-based catalysts supported on raw sepiolite and sepiolite modified with microwave-assisted acid treatment at different times. 15Cu-Sep (A), 15Cu-Sep-2 (B), 15Cu-Sep-8 (C) and 15Cu-Sep-16 (D). Scale 100 nm; Figure S10. Distribution of the Cu crystals for 15Cu-Sep, 15Cu-Sep-2, 15Cu-Sep-8 and 15Cu-Sep-16;

Figure S11. EDX mapping of the Cu-based catalysts supported on raw sepiolite and sepiolite modified with microwave-assisted acid treatment at different times. 5Cu-Sep (A), 10Cu-Sep-2 (B), 15Cu-Sep-2 (C) and 20Cu-Sep-2 (D). Scale 100 nm.

Author Contributions: Conceptualization, J.A.C. and P.J.M.-T.; methodology, J.A.C., R.M.-T. and P.J.M.-T.; software, A.G.-T., R.M.-S. and J.A.C.; validation, J.A.C., R.M.-T. and C.G.-S.; formal analysis, A.G.-T., R.M.-S. and J.A.C.; investigation, A.G.-T., R.M.-S. and J.A.C.; resources, R.M.-T. and P.J.M.-T.; data curation, J.A.C., R.M.-T., J.J.Q.-S. and C.G.-S.; writing—original draft preparation, J.A.C.; writing—review and editing, J.A.C., C.G.-S. and P.J.M.-T.; visualization, J.A.C.; supervision, J.J.Q.-S., R.M.-T., P.J.M.-T.; project administration, R.M.-T. and P.J.M.-T.; funding acquisition, R.M.-T. and P.J.M.-T. All authors have read and agreed to the published version of the manuscript.

Funding: This research was funded by the Spanish Ministry of Science and Innovation (PID2021-122736OB-C42) and FEDER (European Union) funds.

Institutional Review Board Statement: Not applicable.

Informed Consent Statement: Not applicable.

Data Availability Statement: The data presented in this study are available on request from the corresponding author.

Acknowledgments: The authors thank Servicio Central de Apoyo a la Investigación (SCAI—Universidad de Málaga) for its availability in the characterization. The authors also thank Tolsa to supply the starting sepiolite.

Conflicts of Interest: The authors declare no conflicts of interest.

References

1. Bian, Z.; Kawi, S. Preparation, characterization and catalytic application of phyllosilicate: A review. *Catal. Today* **2020**, *339*, 3–23. [CrossRef]
2. Ruíz, A.I.; Ruíz-García, C.; Ruíz-Hitzky, E. From old to new inorganic materials for advanced applications: The paradigmatic example of the sepiolite clay mineral. *Appl. Clay Sci.* **2023**, *235*, 106874. [CrossRef]
3. Galán, E.; Pozo, M. Palygorskite and sepiolite deposits in continental environments. description, genetic patterns and sedimentary settings. *Develop. Clay Sci.* **2011**, *3*, 125–173. [CrossRef]
4. Suárez, M.; García-Romero, E. Advances in the crystal chemistry of sepiolite and palygorskite. *Dev. Clay Sci.* **2011**, *3*, 33–65. [CrossRef]
5. Jones, B.F.; Galán, E. Sepiolite and palygorskite. *Rev. Mineral.* **1988**, *19*, 631–674. [CrossRef]
6. Cecilia, J.A.; Vilarrasa-García, E.; Cavalcante, C.L., Jr.; Azevedo, D.C.S.; Franco, F.; Rodríguez-Castellón, E. Evaluation of two fibrous clay minerals (sepiolite and palygorskite) for CO_2 capture. *J. Environ. Chem. Eng.* **2018**, *6*, 4573–4587. [CrossRef]
7. Álvarez, A. Sepiolite: Properties and uses. *Dev. Sedimentol.* **1984**, *37*, 253–287. [CrossRef]
8. Song, N.; Hursthouse, A.; McLellan, I.; Wang, Z. Treatment of environmental contamination using sepiolite: Current approaches and future potential. *Environ. Geochem. Health* **2021**, *43*, 2679–2697. [CrossRef]
9. Hamid, Y.; Tang, L.; Hussain, B.; Usman, M.; Liu, L.; Ulhassan, Z.; He, Z.; Yang, X. Sepiolite clay: A review of its applications to immobilize toxic metals in contaminated soils and its implications in soil–plant system. *Environ. Technol. Innov.* **2021**, *23*, 101598. [CrossRef]
10. Biddeci, G.; Spinelli, G.; Colomba, P.; Di Blasi, F. Halloysite nanotubes and sepiolite for health applications. *Int. J. Mol. Sci.* **2023**, *24*, 4801. [CrossRef]
11. Tang, Y.; Yang, D.; Ting, V.P.; Hamerton, I.; Van Duijneveldt, J.S.; Rochat, S. Modified sepiolite nanoclays in advanced composites for engineering applications. *ACS Appl. Nano Mater.* **2024**, *7*, 19221–19232. [CrossRef] [PubMed]
12. Franco, F.; Pozo, M.; Cecilia, J.A.; Benítez-Guerrero, M.; Pozo, E.; Martín Rubí, J.A. Microwave assisted acid treatment of sepiolite: The role of composition and "crystallinity". *Appl. Clay Sci.* **2014**, *102*, 15–27. [CrossRef]
13. Jiménez-López, A.; López-González, J.D.; Ramírez-Sáenz, A.; Rodríguez-Reinoso, F.; Valenzuela-Calahorro, C.; Zurita-Herrera, L. Evolution of surface area in a sepiolite as a function of acid and heat treatments. *Clay Miner.* **1978**, *13*, 375–385. [CrossRef]
14. Balci, S. Effect of heating and acid pre-treatment on pore size distribution of sepiolite. *Clay Miner.* **1999**, *34*, 647–655. [CrossRef]
15. Yebra-Rodríguez, A.; Martín-Ramos, J.D.; del Rey, F.; Viseras, C.; López-Galindo, A. Effect of acid treatment on the structure of sepiolite. *Clay Miner.* **2003**, *38*, 353–360. [CrossRef]

16. Surendra, B.S.; Veerabhadraswamy, M.; Kumar, H.G.A.; Kendagannaswamy, B.K.; Nagaswarupa, H.P.; Prashanth, S.C. Microwave assisted physico-chemical modification of Bentonite clay: Characterization and photocatalytic activity. *Mater. Today* **2017**, *4*, 11727–11736. [CrossRef]
17. Wu, C.R.; Hong, Z.Q.; Zhan, B.J.; Tang, W.; Cui, S.C.; Kou, S.C. Effect of acid treatment on the reactivity of natural sepiolite used as a supplementary cementitious material. *Constr. Build. Mater.* **2022**, *316*, 125860. [CrossRef]
18. Jaswal, A.; Singh, P.P.; Mondal, T. Furfural—A versatile, biomass-derived platform chemical for the production of renewable chemicals. *Green Chem.* **2022**, *24*, 510–551. [CrossRef]
19. Yong, K.J.; Wu, T.Y.; Lee, C.B.T.L.; Lee, Z.J.; Liu, Q.; Jahim, J.M.; Zhou, Q.; Zhang, L. Furfural production from biomass residues: Current technologies, challenges and future prospects. *Biomass Bioenergy* **2022**, *161*, 106458. [CrossRef]
20. Jatoi, A.S.; Abbasi, S.A.; Hashmi, Z.; Shah, A.K.; Alam, M.S.; Bhatti, Z.A.; Maitlo, G.; Hussain, S.; Khandro, G.A.; Usto, M.A.; et al. Recent trends and future perspectives of lignocellulose biomass for biofuel production: A comprehensive review. *Biomass Conv. Bioref.* **2023**, *13*, 6457–6469. [CrossRef]
21. Adhami, W.; Richel, A.; Len, C. A review of recent advances in the production of furfural in batch system. *Mol. Catal.* **2023**, *545*, 113178. [CrossRef]
22. Wang, Y.; Zhao, D.; Rodríguez-Padrón, D.; Len, C. Recent advances in catalytic hydrogenation of furfural. *Catalysts* **2019**, *9*, 796. [CrossRef]
23. Adkins, H.; Connor, R. The catalytic hydrogenation of organic compounds over copper chromite. *J. Am. Chem. Soc.* **1931**, *53*, 1091–1095. [CrossRef]
24. Rao, R.; Dandekar, A.; Baker, R.T.K.; Vannice, M.A. Properties of copper chromite catalysts in hydrogenation reactions. *J. Catal.* **1997**, *171*, 406–419. [CrossRef]
25. Liu, D.; Zemlyanov, D.; Wu, T.; Lobo-Lapidus, R.J.; Dumesic, J.A.; Miller, J.T.; Marshall, C.L. Deactivation mechanistic studies of copper chromite catalyst for selective hydrogenation of 2-furfuraldehyde. *J. Catal.* **2013**, *299*, 336–345. [CrossRef]
26. Zhang, H.; Lei, Y.; Kropf, A.J.; Zhang, G.; Elam, J.W.; Miller, J.T.; Sollberger, F.; Ribeiro, F.; Akatay, M.C.; Stach, E.A.; et al. Enhancing the stability of copper chromite catalysts for the selective hydrogenation of furfural using ALD overcoating. *J. Catal.* **2014**, *317*, 284–292. [CrossRef]
27. Sitthisa, S.; Sooknoi, T.; Ma, Y.; Balbuena, P.B.; Resasco, D.E. Kinetics and mechanism of hydrogenation of furfural on Cu/SiO_2 catalysts. *J. Catal.* **2011**, *277*, 1–13. [CrossRef]
28. Shi, Y.; Zhu, Y.; Li, Y.W.; Jiao, H. Exploring furfural catalytic conversion on Cu(111) from computation. *ACS Catal.* **2015**, *5*, 4020–4032. [CrossRef]
29. Yan, K.; Wu, G.; Lafleur, T.; Jarvis, C. Production, properties and catalytic hydrogenation of furfural to fuel additives and value-added chemicals. *Renew. Sustain. Energy Rev.* **2014**, *38*, 663–676. [CrossRef]
30. Dong, F.; Zhu, Y.; Zheng, H.; Zhu, Y.; Li, X.; Li, Y. Cr-free Cu-catalysts for the selective hydrogenation of biomass-derived furfural to 2-methylfuran: The synergistic effect of metal and acid sites. *J. Mol. Catal. A Chem.* **2015**, *398*, 140–148. [CrossRef]
31. Jiménez-Gómez, C.P.; Cecilia, J.A.; Durán-Martín, D.; Moreno-Tost, R.; Santamaría-González, J.; Mérida-Robles, J.; Mariscal, R.; Maireles-Torres, P. Gas-phase hydrogenation of furfural to furfuryl alcohol over Cu/ZnO catalysts. *J. Catal.* **2016**, *336*, 107–115. [CrossRef]
32. Nagaraja, B.M.; Kumar, V.S.; Shasikala, V.; Padmasri, A.H.; Sreedhar, S.; Raju, B.D.; Rao, K.S.R. A highly efficient Cu/MgO catalyst for vapour phase hydrogenation of furfural to furfuryl alcohol. *Catal. Commun.* **2003**, *4*, 287–293. [CrossRef]
33. Yang, X.; Chen, H.; Meng, Q.; Zheng, H.; Zhu, Y.; Li, Y.W. Insights into influence of nanoparticle size and metal–support interactions of Cu/ZnO catalysts on activity for furfural hydrogenation. *Catal. Sci. Technol.* **2017**, *7*, 5625–5634. [CrossRef]
34. Jiménez-Gómez, C.P.; Cecilia, J.A.; Moreno-Tost, R.; Maireles-Torres, P. Selective production of 2-methylfuran by gas-phase hydrogenation of furfural on copper incorporated by complexation in mesoporous silica catalysts. *ChemSusChem* **2017**, *10*, 1448–1459. [CrossRef]
35. Jiménez-Gómez, C.P.; Cecilia, J.A.; Alba-Rubio, A.C.; Cassidy, A.; Moreno-Tost, R.; García-Sancho, C.; Maireles-Torres, P. Tailoring the selectivity of Cu-based catalysts in the furfural hydrogenation reaction: Influence of the morphology of the silica support. *Fuel* **2022**, *319*, 123827. [CrossRef]
36. García-Sancho, C.; Mérida-Robles, J.M.; Cecilia-Buenestado, J.A.; Moreno-Tost, R.; Maireles-Torres, P.J. The role of copper in the hydrogenation of furfural and levulinic acid. *Int. J. Mol. Sci.* **2023**, *24*, 2443. [CrossRef]
37. Franco, F.; Pozo, M.; Cecilia, J.A.; Benítez-Guerrero, M.; Lorente, M. Effectiveness of microwave assisted acid treatment on dioctahedral and trioctahedral smectites. The influence of octahedral composition. *Appl. Clay Sci.* **2016**, *120*, 70–80. [CrossRef]
38. Korichi, S.; Elias, A.; Mefti, A. Characterization of smectite after acid activation with microwave irradiation. *Appl. Clay Sci.* **2009**, *42*, 432–438. [CrossRef]
39. Pardo, L.; Cecilia, J.A.; López-Moreno, C.; Hernández, V.; Pozo, M.; Bentabol, M.J.; Franco, F. Influence of the structure and experimental surfaces modifications of 2:1 clay minerals on the adsorption properties of methylene blue. *Minerals* **2018**, *8*, 359. [CrossRef]

40. Myriam, M.; Suárez, M.; Martín-Pozas, J.M. Structural and textural modifications of palygorskite and sepiolite under acid treatment. *Clays Clay Miner.* **1998**, *46*, 225–231. [CrossRef]
41. Frost, R.L.; Locos, O.B.; Ruan, H.; Kloprogge, J.T. Near-infrared and mid-infrared spectroscopic study of sepiolites and palygorskites. *Vib. Spectrosc.* **2001**, *27*, 1–13. [CrossRef]
42. Vilarrasa-García, E.; Cecilia, J.A.; Moura, P.A.S.; Azevedo, D.C.S.; Rodríguez-Castellón, E. Assessing CO_2 adsorption on amino-functionalized mesocellular foams synthesized at different aging temperatures. *Front. Chem.* **2020**, *8*, 591766. [CrossRef]
43. Thommes, M.; Kaneko, K.; Neimark, A.V.; Olivier, J.P.; Rodríguez-Reinoso, F.; Rouquerol, J.; Sing, K.S.W. Physisorption of gases, with special reference to the evaluation of surface area and pore size distribution (IUPAC Technical Report). *Pure Appl. Chem.* **2015**, *87*, 1051–1069. [CrossRef]
44. Brunauer, S.; Emmett, P.H.; Teller, E. Adsorption of gases in multimolecular layers. *J. Am. Chem. Soc.* **1938**, *60*, 309–319. [CrossRef]
45. Komadel, P. Acid activated clays: Materials in continuous demand. *Appl. Clay Sci.* **2016**, *131*, 84–99. [CrossRef]
46. Guerrero-Torres, A.; Jiménez-Gómez, C.P.; Cecilia, J.A.; García-Sancho, C.; Franco, F.; Quirante-Sánchez, J.J.; Maireles-Torres, P. Ni supported on sepiolite catalysts for the hydrogenation of furfural to value-added chemicals: Influence of the synthesis method on the catalytic performance. *Top. Catal.* **2019**, *62*, 535–550. [CrossRef]
47. Sarmet, J.; Taviot-Gheho, C.; Thirouard, R.; Leroux, F.; Douard, C.; Gaalich, I.; Brousse, T.; Toussaint, G.; Stevens, P. Electrochemical behavior of morphology-controlled copper (II) hydroxide nitrate nanostructures. *Cryst. Growth Des.* **2023**, *23*, 2634–2643. [CrossRef]
48. Luo, Y.H.; Huang, J.; Jin, J.; Peng, X.; Schmitt, W.; Ichinose, I. Formation of positively charged copper hydroxide nanostrands and their structural characterization. *Chem. Mater.* **2006**, *18*, 1795–1802. [CrossRef]
49. Van der Grift, C.J.G.; Elberse, P.A.; Mulder, A.; Geus, J.W. Preparation of silica-supported copper catalysts by means of deposition-precipitation. *Appl. Catal.* **1990**, *59*, 275–289. [CrossRef]
50. Lee, S.H.; Her, Y.S.; Matijevic, E. Preparation and growth mechanism of uniform colloidal copper oxide by the controlled double-jet precipitation. *J. Colloid Interface Sci.* **1997**, *186*, 193–202. [CrossRef]
51. Toupance, T.; Kermarec, M.; Lambert, J.F.; Louis, C. Conditions of formation of copper phyllosilicates in silica-supported copper catalysts prepared by selective adsorption. *J. Phys. Chem. B* **2002**, *106*, 2277–2286. [CrossRef]
52. Toupance, T.; Kermarec, M.; Louis, C. Metal particle size in silica-supported copper catalysts. Influence of the conditions of preparation and of thermal pretreatments. *J. Phys. Chem. B* **2000**, *104*, 965–972. [CrossRef]
53. Trouillet, L.; Toupance, T.; Villain, F.; Louis, C. In situ characterization of the coordination sphere of CuII complexes supported on silica during the preparation of Cu/SiO_2 catalysts by cation exchange. *Phys. Chem. Chem. Phys.* **2000**, *2*, 2005–2014. [CrossRef]
54. Zhang, G.; Long, J.; Wang, X.; Zhang, Z.; Dai, W.; Liu, P.; Li, Z.; Wu, L.; Fu, X. Catalytic role of Cu sites of Cu/MCM-41 in phenol hydroxylation. *Langmuir* **2010**, *26*, 1362–1371. [CrossRef]
55. Gomulka, A.; Iwaszko, O.; Kowalczyk, A.; Piwowarska, Z.; Rutkowska, M.; Chmielarz, L. Catalytic functionalization of MCM-41 and MCM-48 with copper by modified template ion-exchange method for low-temperature NH_3-SCR process: The role of complexing agents for copper dispersion. *Res. Chem. Intermed.* **2024**, *50*, 5399–5426. [CrossRef]
56. Giordano, F.; Vennestrøm, P.N.R.; Lundegaard, L.F.; Stappen, F.N.; Mossin, S.; Beato, P.; Bordiga, S.; Lamberti, C. Characterization of Cu-exchanged SSZ-13: A comparative FTIR, UV-Vis, and EPR study with Cu-ZSM-5 and Cu-β with similar Si/Al and Cu/Al ratios. *Dalton Trans.* **2013**, *42*, 12741–12761. [CrossRef]
57. Bezerra, S.M.; Levchenko, V.; Piccinelli, V.; Bettinelli, M.; Longo, R.L.; Malta, O.L.; Reisfeld, R. Amplification of light emission of lanthanide complexes by copper and copper oxide nanoparticles. *J. Lumin.* **2022**, *251*, 119217. [CrossRef]
58. Yin, A.; Guo, X.; Dai, W.L.; Li, H.; Fan, K. Highly active and selective copper-containing HMS catalyst in the hydrogenation of dimethyl oxalate to ethylene glycol. *Appl. Catal. A Gen.* **2008**, *31*, 91–99. [CrossRef]
59. Di, W.; Cheng, J.; Tian, S.; Li, J.; Chen, J.; Sun, Q. Synthesis and characterization of supported copper phyllosilicate catalysts for acetic ester hydrogenation to ethanol. *Appl. Catal. A Gen.* **2016**, *510*, 244–259. [CrossRef]
60. Ye, R.P.; Lin, L.; Yang, J.X.; Sun, M.L.; Li, F.; Li, B.; Yao, Y.G. A new low-cost and effective method for enhancing the catalytic performance of Cu–SiO_2 catalysts for the synthesis of ethylene glycol via the vapor-phase hydrogenation of dimethyl oxalate by coating the catalysts with dextrin. *J. Catal.* **2017**, *350*, 122–132. [CrossRef]
61. Arango-Díaz, A.; Cecilia, J.A.; Moretti, E.; Talon, A.; Núñez, P.; Marrero-Jerez, J.; Jiménez-Jiménez, J.; Jiménez-López, A.; Rodríguez-Castellón, E. Comparative study of CuO supported on CeO_2, $Ce_{0.8}Zr_{0.2}O_2$ and $Ce_{0.8}Al_{0.2}O_2$ based catalysts in the CO-PROX reaction. *Int. J. Hydrogen Energy* **2014**, *39*, 4102–4108. [CrossRef]
62. Ayastuy, J.L.; Gurbani, A.; González-Marcos, M.P.; Gutiérrez-Ortiz, M.A. Selective CO oxidation in H_2 streams on $CuO/Ce_xZr_{1-x}O_2$ catalysts: Correlation between activity and low temperature reducibility. *Int. J. Hydrogen Energy* **2012**, *37*, 1993–2006. [CrossRef]
63. Xia, Z.; Li, Y.; Wu, J.; Huang, Y.C.; Zhao, W.; Lu, Y.; Pan, Y.; Yue, X.; Wang, Y.; Dong, C.L.; et al. Promoting the electrochemical hydrogenation of furfural by synergistic Cu^0-Cu^+ active sites. *Sci. China Chem.* **2022**, *65*, 2588–2595. [CrossRef]

64. Zhang, J.; Ding, G.; Jin, Y.; Wei, L.; Li, X.; Wang, D.; Zhu, Y.; Li, Y. Stabilizing the interfacial Cu^0-Cu^+ dual sites toward furfural hydrodeoxygenation to 2-methylfuran via fabricating nest-like copper phyllosilicate precursor. *Fuel* **2023**, *337*, 127212. [CrossRef]
65. Campos Molina, M.J.; Mariscal, R.; Ojeda, M.; López Granados, M. Cyclopentyl methyl ether: A green co-solvent for the selective dehydration of lignocellulosic pentoses to furfural. *Bioresour. Technol.* **2012**, *126*, 321–327. [CrossRef]
66. Nagaraja, B.M.; Padmasri, A.H.; David Raju, B.; Rama Rao, K.S. Vapor phase selective hydrogenation of furfural to furfuryl alcohol over Cu–MgO coprecipitated catalysts. *J. Mol. Catal. A Chem.* **2007**, *265*, 90–97. [CrossRef]
67. Ghashghaee, M.; Sadjadi, S.; Shirvani, S.; Farzaneh, V. A novel consecutive approach for the preparation of Cu-MgO catalysts with high activity for hydrogenation of furfural to furfuryl alcohol. *Catal. Lett.* **2017**, *147*, 318–327. [CrossRef]
68. Guerrero-Torres, A.; Jiménez-Gómez, C.P.; Cecilia, J.A.; García-Sancho, C.; Quirante-Sánchez, J.J.; Mérida-Robles, J.M.; Maireles-Torres, P. Influence of the incorporation of basic or amphoteric oxides on the performance of Cu-based catalysts supported on sepiolite in furfural hydrogenation. *Catalysts* **2019**, *9*, 315. [CrossRef]
69. Sitthisa, S.; Resasco, D.E. Hydrodeoxygenation of furfural over supported metal catalysts: A comparative study of Cu, Pd and Ni. *Catal. Lett.* **2011**, *141*, 784–791. [CrossRef]

ceramics

MDPI

Article

Ultra-Broadband Plasmon Resonance in Gold Nanoparticles Precipitated in $ZnO-Al_2O_3-SiO_2$ Glass

Georgiy Shakhgildyan [1,*], Leon Avakyan [2], Grigory Atroshchenko [1], Maxim Vetchinnikov [1], Alexandra Zolikova [1], Elena Ignat'eva [1], Mariam Ziyatdinova [1], Elena Subcheva [3], Lusegen Bugaev [2] and Vladimir Sigaev [1]

1 Department of Glass and Glass-Ceramics, Mendeleev University of Chemical Technology, 125047 Moscow, Russia
2 Physical Faculty, Southern State University, 344090 Rostov-on-Don, Russia
3 Biomaterials Resource Center, Sirius University of Science and Technology, 354340 Sirius, Russia
* Correspondence: shakhgildian.g.i@muctr.ru

Abstract: Optical materials with a tunable localized surface plasmon resonance (LSPR) are of great interest for applications in photonics and optoelectronics. In the present study, we explored the potential of generating an LSPR band with an ultra-broad range of over 1000 nm in gold nanoparticles (NPs), precipitated through a thermal treatment in $ZnO-Al_2O_3-SiO_2$ glass. Using optical absorption spectroscopy, we demonstrated that the LSPR band's position and shape can be finely controlled by varying the thermal treatment route. Comprehensive methods including Raman spectroscopy, X-ray diffraction, and high-resolution transmission electron microscopy were used to study the glass structure, while computational approaches were used for the theoretical description of the absorption spectra. The obtained results allowed us to suggest a scenario responsible for an abnormal LSPR band broadening that includes a possible interparticle plasmonic coupling effect taking place during the liquid–liquid phase separation of the heat-treated glass. The formation of gold NPs with an ultra-broad LSPR band in glasses holds promise for sensitizing rare earth ion luminescence for new photonics devices.

Keywords: gold nanoparticles; glass; surface plasmon resonance; LSPR; plasmonics; phase separation; microstructure

1. Introduction

Gold nanoparticles (NPs) have emerged as a central figure in the advancement of nanotechnology and materials science, owing to their unique optical properties, particularly the phenomenon of localized surface plasmon resonance (LSPR) [1]. This phenomenon, which results from the collective oscillation of conduction band electrons in gold NPs under light excitation, has fueled extensive research due to its potential applications across a broad spectrum of fields including photonics, optoelectronics, biomedicine, and sensing technologies [2,3]. The optical properties of gold NPs, such as their absorption and scattering spectra, are governed by LSPR, which can be finely tuned across the visible and near-infrared spectrum by adjusting the NPs' physical parameters. Research has shown that the size of gold NPs plays a critical role in determining their LSPR characteristics, with smaller particles predominantly absorbing light and larger particles scattering it. For instance, it was demonstrated that the LSPR wavelength redshifts as the size of the gold nanosphere increases, highlighting the tunability of NPs for specific applications based on size adjustments alone [4]. Furthermore, the shape of gold NPs significantly impacts their optical properties, with non-spherical nanoparticles such as rods, triangles, and stars exhibiting multiple plasmon resonance peaks corresponding to their various geometrical features. It was shown that the shape and geometry of gold nanorods influence their optical absorption and local field enhancement, crucial for applications in surface-enhanced Raman spectroscopy [5]. The surrounding dielectric environment also modifies the LSPR of gold NPs, affecting their optical sensitivity and enabling the design of environment-responsive systems. Djorović et al. introduced a method to enhance the characterization of gold NPs through the measurement of the extinction to absorption ratio, offering a new way to assess the volume and dielectric function of NPs with increased accuracy [6].

This tunability of the LSPR, coupled with the strong electromagnetic field enhancement near the nanoparticle surface, makes gold NPs ideal candidates for enhancing the performance of photonic devices, especially in the field of new optical media [7]. The enhancement of rare earth ions (REIs) luminescence through the LSPR of gold NPs presents a sophisticated interplay of multifactorial dependencies critical for the advancement of photonic and optoelectronic devices. This enhancement process is primarily governed by the energy overlap between the LSPR of the gold NPs and the absorption energies of the REIs, which are suitable for their excitation [8]. Hence, the precise tuning of the LSPR band for specific REIs is crucial for the development of advanced optical materials.

One of the most suitable media for the development of optical materials utilizing the advantages of gold NPs and REI interactions is glass. The incorporation of gold NPs into glass matrices provides a promising platform for exploiting the LSPR phenomena, enabling novel applications in optics, sensing, and photonics [9]. Hence, beyond their aesthetic appeal, glasses embedded with gold NPs have garnered significant interest for applications in photonics and nonlinear optics [10,11]. The integration of gold NPs into glasses combines the desirable mechanical and chemical properties of the glass matrix with the unique plasmonic properties of the NPs. This synergy is being explored for the development of efficient laser materials for enhancing the third-order nonlinear optical susceptibility and for surface-enhanced Raman spectroscopy. As in other media, the efficacy of these applications of gold NPs formed in glass is critically dependent on the characteristics of the LSPR band, which in turn are influenced by the abovementioned factors. Recent experiments have demonstrated the formation of gold NPs in glass with a red-shifted LSPR band in different glass systems [12–14]. Laser irradiation of glass also marks a significant leap forward in the domain of the formation of NPs with tunable LSPR bands [15–17]. These studies collectively underline the significant potential of gold NPs in glass matrices for enhancing and manipulating LSPR for various technological applications. But the ability to tune the plasmonic properties of NPs in a wide range remains a challenge, the solution to which can open new opportunities for the development of advanced optical devices and sensors, exploiting the unique interactions between light and matter at the nanoscale.

Hence, in this study, we explored for the first time the potential of gold NPs precipitation in ZnO-Al_2O_3-SiO_2 glass and showed the possibility of generating an ultra-broad LSPR band. The obtained results allowed us to suggest a scenario responsible for such abnormal LSPR band broadening, which included a possible interparticle plasmonic coupling effect taking place during the liquid–liquid phase separation of the glass under the heat treatment. The formation of gold NPs with an ultra-broad LSPR band in glasses holds promise for sensitizing REIs for new photonics devices.

2. Materials and Methods

2.1. Glass Synthesis

In this work, we synthesized glass with the following nominal composition (mol.%): 1.0 SnO_2; 25.0 Al_2O_3; 25.0 ZnO; and 49.0 SiO_2, with 0.01% of $AuCl_3$, which was introduced in addition to the 100%. The glass batch calculated to produce 500 g of glass was prepared using high-purity raw materials (SiO_2, $Al(OH)_3$, ZnO, SnO_2, $AuCl_3$) thoroughly mixed together. The glass was melted at 1590 °C for 6 h in a 600 mL silica crucible in the laboratory electric furnace with SiC heating elements. The glass was casted in the preheated steel mold and subsequently annealed at 600 °C for 4 h in the muffle furnace. The resulted glass casting exhibited high quality, a significant part of which was free of striae and bubbles. The most homogeneous parts of obtained glass were cut into plates and polished. To study the effect of gold NPs formation, synthesized glass samples were heat-treated in the muffle furnace at various temperatures in the 750–850 °C range for 5 and 20 h. All of the heat-treated samples will be referred to further as "temperature-time" (e.g., 750-20).

2.2. Glass Characterization

The visual appearance of the samples was captured by a digital camera in transmitted and reflected light. For the determination of the glass transition temperature (Tg) and the crystallization temperature (T_C), differential scanning calorimetry (DSC) was used. A bulk glass sample of about 20 mg was heated in a Pt crucible using the simultaneous thermal analyzer STA 449 F3 Jupiter (NETZSCH, Selb, Germany) with a dynamic flow atmosphere of Ar. The temperature range was from room temperature to 1200 °C with a heating rate of 10 °C/min. X-ray diffraction (XRD) patterns of the powdered samples were recorded by means of a diffractometer D2 Phaser (Bruker, Billerica, MA, USA) employing nickel-filtered CuKα radiation. Raman spectra were collected from the polished glass samples using an NTEGRA Spectra spectrometer, (NT-MDT Co., Moscow, Russia) with an argon (Ar) laser beam (488 nm excitation wavelength). The optical absorption and reflectance spectra of the samples were recorded in the 200–1650 nm spectral range by the spectrophotometer UV-3600 (Shimadzu, Kyoto, Japan). The bulk glass refractive index was determined at 589.3 nm by an DR-M4 Abbe refractometer (ATAGO, Tokyo, Japan).

The microstructure of the samples was studied by a high-resolution transmission electron microscopy (HRTEM) with the transmission electron microscope JEM-2100Plus (JEOL, Peabody, MA, USA) in the 200 kV mode. Bulk glass samples were grounded in an agate mortar to fine powders and dispersed in ethanol. The obtained solution was dropped on a microscope grid, which was dried for 20 min. Images were obtained in the bright field mode on different parts of the grid loaded with the material, providing a fairly complete and objective visualization of the samples. The HRTEM images were analyzed with ImageJ software version 1.53n (direct measurement of particle sizes and interparticle distances (number of measured units exceed 50); and measurement of the lattice spacings of crystal particles on the Fourier-filtered region of the image) [18,19].

3. Results

3.1. Optical Properties of Glasses upon Heat Treatment

The DSC curve of the synthesized glass shows two exothermic peaks corresponding to the crystallization of different phases; the Tg temperature of the glass is 750 °C (Figure S1). The glass composition chosen in this work is close to one of the compositions (namely ZAS50.25) from a recent study that investigated the glass formation and crystallization region in the ZnO-Al_2O_3-SiO_2 system [20]. A comparison of the Tg temperatures and number of exothermic peaks shows a general similarity, while some differences in the values of crystallization temperatures (T_C) can be related to the different conditions of glass synthesis in this work and in [20].

The raw glass was transparent without visible signs of crystallization or coloration, and its absorption spectrum also did not show any bands in the visible and near-infrared spectral region (Figure 1). The refractive index n_D of the glass was 1.569. Heat treatment of glass at temperatures above Tg led to a noticeable coloration of the samples, which is expected for glasses containing gold and a small addition of SnO_2 [21]. However, in contrast to the red (ruby) coloring typical for silicate glasses with gold, heat treatment of glasses for 5 h in the range of 800–850 °C led to the appearance of a turquoise color which, with an increase in the treatment temperature, changed to blue (Figure 1, inset). It should be noted that the above colors were observed when the samples were studied in transmitted light; at the same time, when the angle of illumination was changed and the samples were observed in reflected light, the color of the samples changed to brown. The intensity of this effect increases with increasing treatment temperature: for sample 800-5, the effect is weakly pronounced (not distinguishable on the photo), and for sample 850-5 with maximum treatment temperature, the effect is clearly visible. The observed effect can be attributed to a dichroism phenomenon (color change in transmitted and reflected light). In previous works, the dichroism effect was obtained in glasses of phosphate systems containing both gold and copper NPs [22–24]; in tellurite glasses with gold NPs [14]; and in glasses of the K_2O-B_2O_3-Sb_2O_3-ZnO system, also containing gold NPs [25]. The nature of dichroism in glasses with NPs is complex and has not yet been unequivocally established.

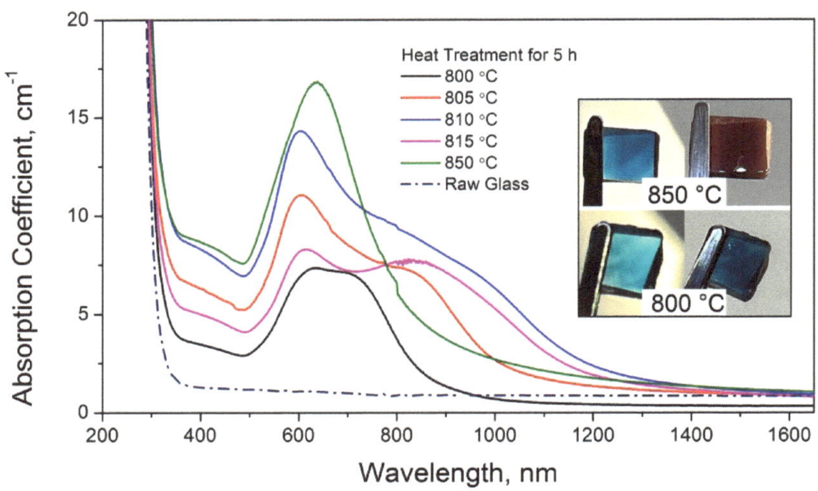

Figure 1. Absorption spectra of the raw glass and glasses heat-treated in the 800–850 °C range for 5 h. Insets show visual appearance of glasses treated at 800 and 850 °C in transmitted and reflected light (left and right panel, respectively). Each glass sample size is ≈1 cm × 1 cm.

The absorption spectra of glasses heat-treated for 5 h (Figure 1) show the presence of bands, the shape and intensity of which strongly depend on the treatment temperature. Because the absorption spectra of synthesized glasses of the same composition but without gold and heat-treated under the same regimes show no absorption bands (the spectra are not shown in this work), we attribute the bands in Figure 1 to the LSPR of gold NPs. An analysis of the experimental spectra shows that at the minimum treatment temperature (800 °C), a broad band with an asymmetric plateau extending from ≈630 to 750 nm is formed. Increasing the treatment temperature leads to an increase in the intensity of the band and strengthening of asymmetry: a maximum at 603 nm and a broad shoulder with a maximum at ≈810 nm are formed.

A further increase in temperature (up to 810 °C) intensifies the short-wave band, the maximum of which shifts to ≈610 nm; the long-wave shoulder broadens, and its intensity slightly increases. The increase in treatment temperature to 815 °C leads to the formation of a band with two maxima that are almost equal in intensity (at 613 and 823 nm). A further increase in temperature up to 850 °C leads to the formation of a symmetric band with one maximum at 638 nm. Such a sequence of changes in the LSPR bands of gold NPs in glasses seems atypical and has not been previously described in the literature. To study this phenomenon in more detail, we carried out a series of heat treatments of glass samples at a longer exposure time (20 h).

Figure 2 shows the absorption spectra of glasses after the heat treatment in the range of 750–820 °C for 20 h, as well as the photographs of some samples. Analysis of the appearance of the samples obtained after the temperature treatment shows the following dynamics: Treatment at Tg temperature (750 °C) leads to a weak coloration of the glass sample in a color close to violet, and the effect of dichroism is not manifested. Increased treatment temperature leads to coloring of the glass in a turquoise-blue color, and with rising temperature, the intensity of coloring increases. The dichroism effect in these samples appears and intensifies with increasing treatment temperature (Inset in Figure 2).

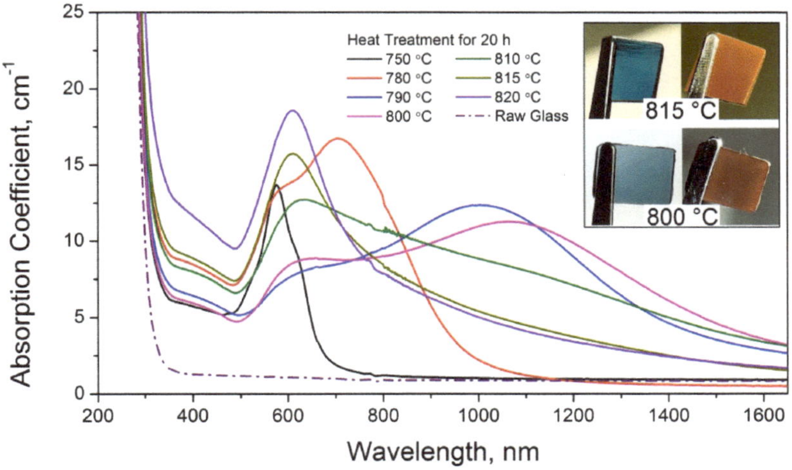

Figure 2. Absorption spectra of the raw glass and glasses heat-treated in the 750–820 °C range for 20 h. Insets show the visual appearance of glasses treated at 800 and 815 °C in transmitted and reflected light (left and right panel, respectively). Each glass sample size is ≈1 cm × 1 cm.

Increasing the treatment temperature of the glasses initiates significant changes in the absorption spectra (Figure 2). The LSPR band of sample 750-20 is slightly asymmetric, with a clear maximum at 576 nm and a small long-wavelength shoulder up to 650 nm. For sample 780-20, changes in the band shape are noticeable: the long-wavelength shoulder turns into a broad band with a maximum at 705 nm, and a short-wavelength shoulder at 595 nm also appears. A further increase in the glass treatment temperature up to 800 °C leads to an extraordinary broadening of the LSPR band with the formation of two maxima at 658 and ≈1150 nm. At the same time, the long-wavelength shoulder of the band extends beyond 1600 nm. For sample 810-20, there is a sharp change in the shape of the band: there is one maximum at 630 nm, and a broad long-wavelength shoulder is formed, also extending into the IR region. At the maximum treatment temperatures (815 and 820 °C), the band shape becomes similar to sample 850-5 (Figure 1), with the difference being that the band maximum is fixed at 610 nm. To the best of our knowledge, the results obtained for sample 800-20 for the broadening of the LSPR band (more than 1000 nm) are record-breaking.

The pronounced changes in the shape and position of the LSPR band during the heat treatment of glasses in a fairly narrow temperature range indicate significant changes that occur in the structure of glasses. To understand the mechanisms responsible for the changes in the optical properties of glasses, we studied their microstructure.

3.2. Structural Study of Glasses

Figure 3 shows the results of the XRD and Raman spectroscopy analyses of glasses heat-treated for 5 and 20 h. The XRD patterns (Figure 3a,b) show that regardless of the heat treatment conditions, the glasses remain X-ray amorphous, which indicates the absence of crystalline inclusions that can be detected by this method. The absence of the peaks related to Au NPs can be explained by the low concentration of $AuCl_3$ in the glass composition. Moreover, in [20], in the Au-free glass of similar composition, HT-XRD analysis at 850 °C revealed the presence of a ZnO-Al_2O_3-SiO_2 solid solution (ZAS-s.s.) phase (ICDD card #00-032-1455), which was subsequently replaced by a gahnite phase (ICDD card #01-074-1138) at 950 °C. Apparently, in our case, the absence of ZAS-s.s. reflections in the XRD patterns of the heat-treated glass samples is due to the higher temperature of the first T_{C1} exopeak compared with the above work (992 °C in our case and 907 °C in [20]). However, it can be assumed that at high treatment temperatures in the glasses studied by us, the formation of the ZAS-s.s. phase can begin, the sizes of which are so small that they cannot be detected by XRD methods.

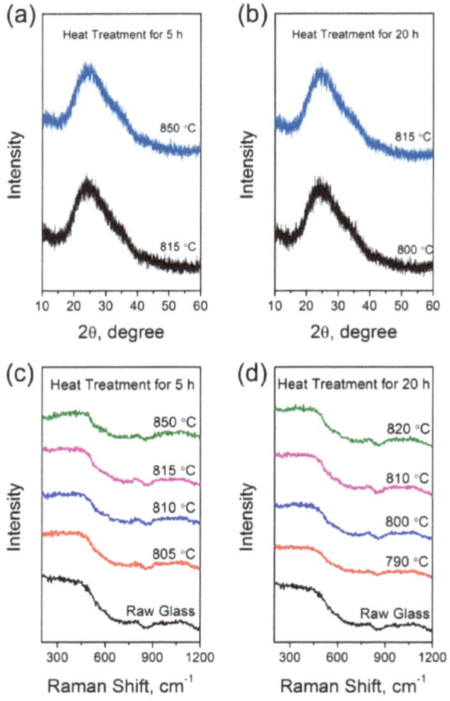

Figure 3. XRD patterns of glasses heat-treated for 5 h (**a**) and 20 h (**b**); Raman spectra of raw glass and glasses heat-treated for 5 h (**c**) and 20 h (**d**). Heat treatment temperatures are indicated.

The Raman spectra (Figure 3c,d) also demonstrate that there are no changes resulting from the heat treatment of the glasses, indicating the absence of bulk crystallization in the studied samples.

To study the microstructure of the glasses in detail, we analyzed images obtained by high-resolution TEM (HRTEM), which allowed us to distinguish the visualization of phase and chemical contrast that is not possible by other methods [26]. Figure 4 shows TEM images for samples 815-5 and 850-5. For sample 815-5, one can clearly observe the presence of zones on the order of 10–15 nm in size that stand out significantly in chemical contrast but on closer inspection have no discernible crystal planes. Such regions were previously described in glasses of the ZnO-Al_2O_3-SiO_2 system and attributed to regions of amorphous phase separation, wherein the glass matrix is enriched in SiO_2 and the amorphous phase inhomogeneities are enriched in ZnO [27]. The analysis of HRTEM images of sample 810-5 demonstrates that NPs of the order of 5 nm in size are formed within and between the amorphized regions; a detailed analysis of the interplanar distances in the NPs shows values of 0.23 and 0.2 nm, which correspond to the (111) and (200) planes in the Au crystal lattice. It should be noted that a similar pattern was obtained for sample 800-5. At the same time, for the sample 850-5, amorphous regions are not detected; moreover, even on the large scales, crystalline particles immediately become visible, which on closer examination were then identified by us to be gold NPs (Figure 4b).

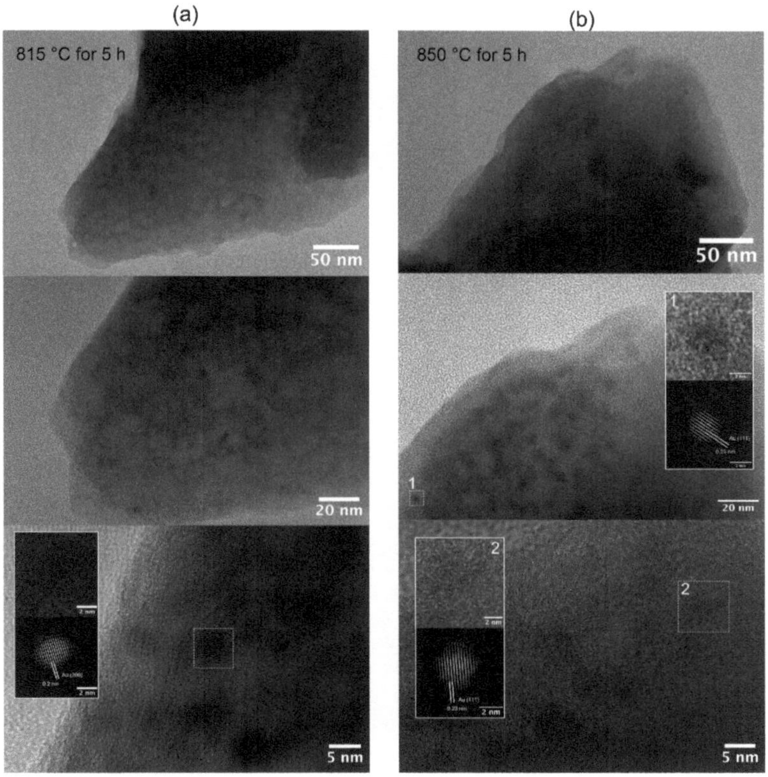

Figure 4. TEM and HRTEM images of glass samples heat-treated for 5 h at 815 °C (**a**) and 850 °C (**b**). Insets show enlarged images and Fourier-filtered images of the marked zones with corresponding numbers.

Figure 5a–c show the size distributions of NPs obtained by analyzing images of glass samples heat-treated for 5 h. It can be seen that when the treatment temperature is increased from 800 to 815 °C, the size of NPs does not change much and is approximately equal to 5 nm. At the same time, growth of the treatment temperature up to 850 °C leads to an increase in the size of NPs up to 6 nm on average. The analysis of TEM images (Figure 4) also demonstrates a more clustering tendency of nanoparticles rather than a uniform distribution in the glass matrix. In this regard, we analyzed the distance between the particles for each of the samples (Figure 5d–f). It can be seen that for the samples 800-5 and 815-5, the particle spacing is on average within 6 nm, while for the sample 850-5, the spacing increases and is on average within 8–10 nm. Such changes in the particle size and spacing for sample 850-5 compared with samples 800-5 and 815-5 are in some agreement with changes in the shape of the LSPR band in the absorption spectra of these samples. To study these relationships in more detail, we also studied the microstructure of samples heat-treated for 20 h.

Figure 6 shows TEM images for sample 800-20. As in the previous case, the images show clearly visible areas of amorphous phase separation, the sizes of which can be estimated within 15 nm on average. Analysis of the HRTEM images demonstrates the presence of gold NPs in the microstructure, which is confirmed by the size of interplanar distances in crystalline microdomains corresponding to the (111) and (200) planes in the Au crystal lattice. The distribution of NPs is also characterized by their clustering, but the analysis of the images does not allow us to draw an unambiguous conclusion as to whether this occurs between the phase separation regions or within them.

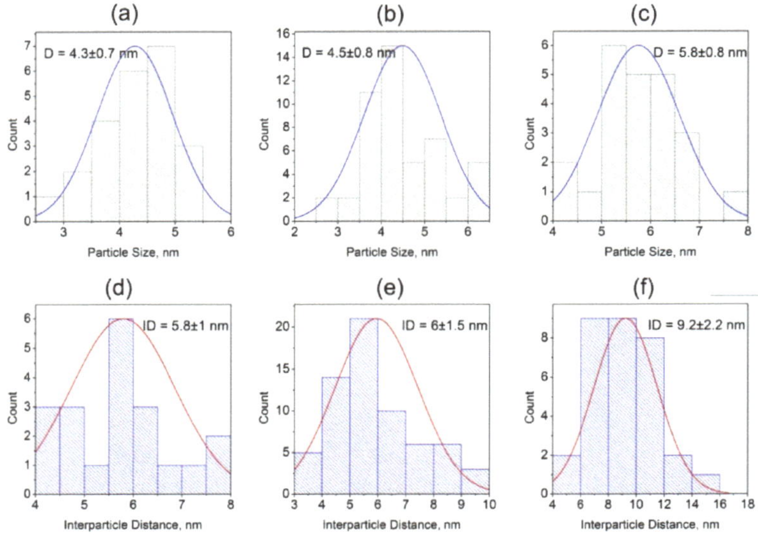

Figure 5. Particle size distribution (**a–c**) and interparticle distance (**d–f**) obtained from the TEM images of the glass samples treated for 5 h at 800, 815, and 850 °C (from left to right).

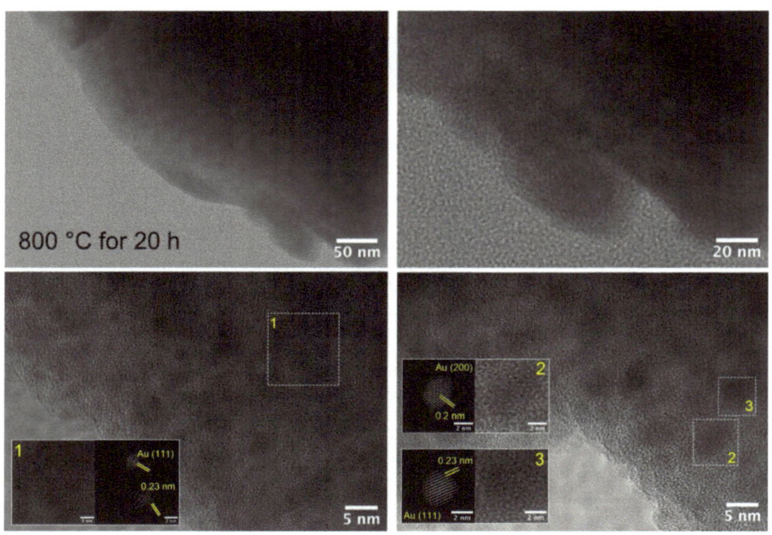

Figure 6. TEM and HRTEM images of glass samples heat-treated for 20 h at 800 °C. Insets show enlarged images and Fourier-filtered images of the marked zones with corresponding numbers.

A similar situation is observed for the sample 810-20 (Figure 7). TEM images show that the glass microstructure is characterized by an amorphous phase separation, with microdomain sizes on the order of 15 nm. Image analysis demonstrates that most of the NPs are located inside the liquation zones, which are presumably enriched with ZnO.

For the glass sample 815-20, TEM images show several striking differences in the microstructure (Figure 8). First, the images lack the pronounced zones of phase inhomogeneity that are evident in the images of samples 800-20 and 810-20. Secondly, the analysis of interplanar distances in crystal particles shows that along with gold NPs, crystals of another nature are formed in the glass; the values of the interplanar distances in them are in the range of 0.27–0.3 nm. The exact determination of the nature of these crystals is possible with the use of methods of higher resolution microscopy and local chemical analysis on the scale of units of nanometers, which is beyond the capabilities of this study. At the same time, it can be assumed that the observed crystalline inclusions may belong to the ZAS-s.s. phase, which according to [20] is formed in glasses of this composition at close treatment temperatures.

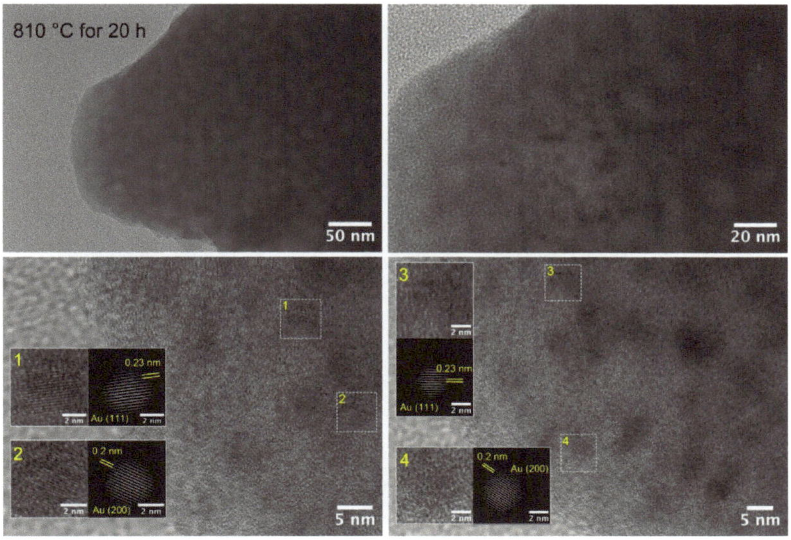

Figure 7. TEM and HRTEM images of glass samples heat-treated for 20 h at 810 °C. Insets show enlarged images and Fourier-filtered images of the marked zones with corresponding numbers.

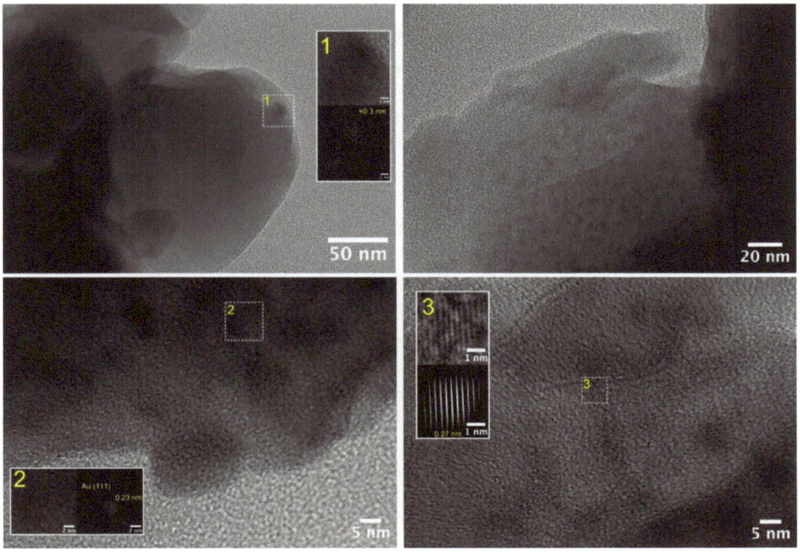

Figure 8. TEM and HRTEM images of glass samples heat-treated for 20 h at 815 °C. Insets show enlarged images and Fourier-filtered images of the marked zones with corresponding numbers.

Figure 9a–c shows the NP size distributions for glasses heat-treated for 20 h. It can be seen that with increasing treatment temperature, the average particle size increases from ≈4 nm to ≈7 nm. At the same time, the average distance between particles also increases from ≈5 nm at 800 °C to ≈11 nm when treated at 815 °C (Figure 9d–f).

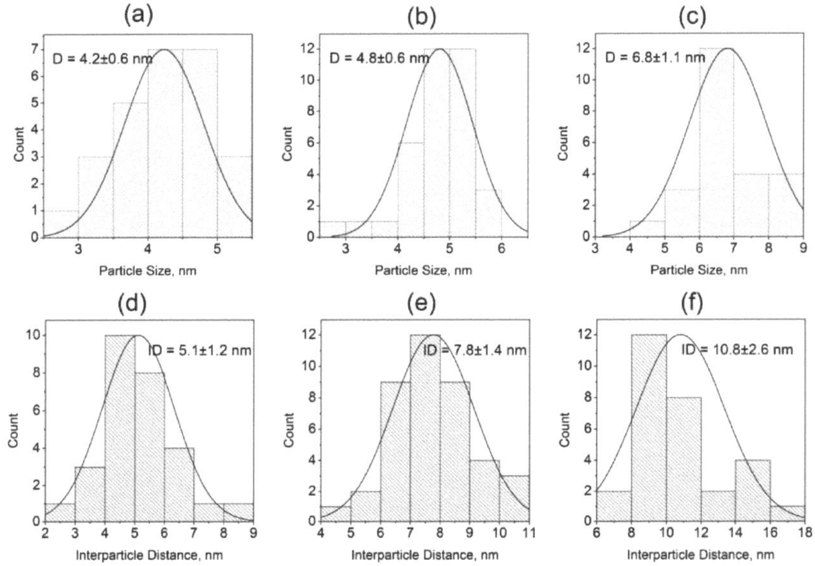

Figure 9. Particle size distribution (**a**–**c**) and interparticle distance (**d**–**f**) obtained from the TEM images of the glass samples treated for 20 h at 800, 810, and 815 °C (from left to right).

4. Discussion

The analysis of the obtained results allows us to conclude that in this work, during the heat treatment of ZAS glass containing Au additives, it was possible to form gold NPs, which have an ultra-wide LSPR band (over 1000 nm); meanwhile, changing the parameters of heat treatment allowed us to vary the shape and position of the band. In our opinion, these results are of interest both from the fundamental scientific point of view and in terms of finding certain practical applications. However, it is also important to determine the mechanism responsible for such a change in optical properties.

The factors that influence the change in the shape and position of the LSPR band of metal NPs (Au, Ag, Cu) in glasses include: (i) change in the size of nanoparticles; (ii) change in the shape of nanoparticles; (iii) change in the refractive index of the medium around the nanoparticle; and (iv) the presence of interactions between closely spaced nanoparticles. Each of these factors is further discussed in detail below.

(i) It is known that an increase in the size of NPs in glass leads to a change in the color of the glass caused by an increase in the intensity and change in the shape of the LSPR band, as well as a shift of the band maximum to the red region of the spectrum [11,28]. Thus, in many works using different glass-forming systems, it has been shown that by increasing the temperature of glass treatment, there is a process of increasing the size of NPs; the associated first increase is in the intensity, and then the red shift of the LSPR band occurs [28–30]. In [14], it was shown that increasing the size of gold NPs formed in tellurite glass leads to both broadening of the LSPR band and the manifestation of a dichroism effect similar in description to that obtained in our work. At the same time, in our work, despite the fact that the TEM data (Figures 5 and 9) show an increase in the size of gold NPs with an increasing treatment temperature, the dynamics of the change in the position and shape of the LSPR bands do not correspond to the previously described results. Moreover, the results of the calculations of the LSPR band of gold NPs in ZAS glass medium does not describe the experimental spectra in any way. Figure 10 shows that in spite of the nanoparticle size, the calculated spectra are far away from the experimental results. Thus, only the change in the size of gold NPs in the process of the heat treatment does not allow us to describe the mechanism of the ultra-broadening of the LSPR band.

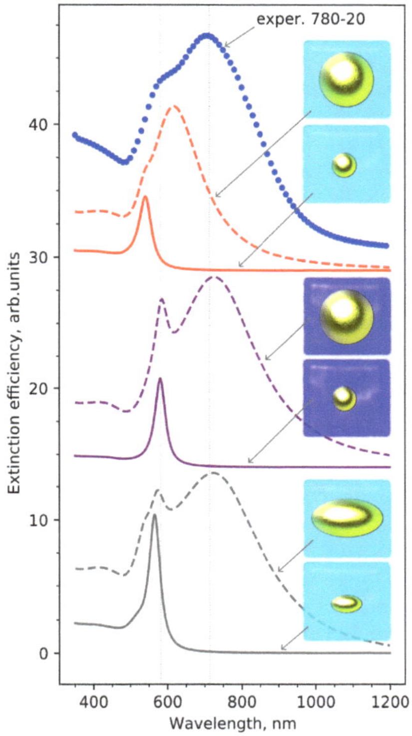

Figure 10. The comparison of the experimental absorption spectrum of the sample 780-20 (blue dots) with the calculated spectra for different models of single gold NPs: top panel—50 nm spherical particle (red dashes) and 10 nm spherical particle (solid red), both in the in matrix with $n_D = 1.6$; middle panel—50 nm spherical particle (purple dashed) and 10 nm spherical particle (solid purple), both in the in matrix with $n_D = 2.0$; bottom panel—60 nm oblate spheroid particle and 10 nm oblate spheroid particle, both in the in matrix with $n_D = 1.6$. The smaller 10 nm particles cannot explain the experimental data, while big particles (50–60 nm) are able to describe the experiment but are not presented in the sample.

(ii) Changing the shape of NPs to a shape other than spherical has a significant effect on the LSPR band and the position of the LSPR maximum: elongated particles are characterized by the splitting of the LSPR band into two components (with two maxima: one each in the short-wave and long-wave regions of the spectrum), which become dependent on the polarization of the incident light [31,32]. The formation of elongated NPs in glasses has been demonstrated using a variety of systems and methods, including mechanical deformation of glasses where spherical NPs are formed [33–35] and as a result of intensive laser irradiation of glasses with NPs [36,37]. In a series of papers by Som T. and Karmakar B., Sb_2O_3-based glasses with elliptical gold NPs with aspect ratios ranging from 1.2 to 2.1 were synthesized. The glasses exhibited a dichroism effect similar to that described in our work, where the LSPR band for compositions with high Au content in the cited works was broadened, the band maximum shifted to the red region, and no splitting of the band into two component's was observed. The authors attributed the mechanism of deformation of gold NPs to the high viscosity of the glass melt without providing additional explanations; the authors also did not provide the calculated spectra of the LSPR bands for the experimentally determined sizes of NPs, so it is not possible to validate the proposed model [25,38]. An analysis of the TEM images in our work demonstrates that the shape of most NPs is close to spherical and the presence of elongated or elliptical particles could not be detected. Fitting of the experimental spectrum of the sample 780-20 with elliptical particles (Figure 10 bottom panel) allows us to describe the spectrum, but the size of NPs corresponding to the calculated spectrum should be larger than 58 nm with an aspect ratio of 1.63, which is not confirmed by the TEM data or by the data of the XRD and Raman spectroscopy; the presence of such large particles should be detected in X-ray diffraction patterns and Raman spectra, which was not observed experimentally (Figure 3).

(iii) The refractive index of the medium around NPs has a great influence on the position of the LSPR band. As the refractive index increases, the maximum of the band shifts to the red region of the spectrum; this effect is the basis for sensors that can detect trace amounts of chemical substances [39]. Using this effect, noble-metal NPs were synthesized in amorphous films based on SiO_2 containing different contents of oxides with high refractive indexes (TiO_2, ZrO_2). At the same time, the position of the LSPR band maximum in these NPs was tuned in the range of up to 600 nm by varying the content of the oxides with high refractive indexes, thus changing the average value of the refractive index of the medium around NPs [40,41]. Moreover, based on this effect, in our previous works [42,43], we have shown the possibility for tuning the position of the maximum of the LSPR band of gold NPs in glasses of the ZnO-MgO-Al_2O_3-SiO_2 system containing TiO_2 and ZrO_2 additives. During heat treatment in glasses along with the growth of NPs, the processes of phase separation first took place with the formation of amorphized zones enriched in TiO_2 and ZrO_2 (components with high refractive index); with the further increase in the temperature of the treatment, $ZnAl_2O_4$ crystals were formed in glasses, while TiO_2 and ZrO_2 components were uniformly distributed in the matrix. Amorphized zones enriched in TiO_2 and ZrO_2 increased the local refractive index of the medium around NPs, which influenced the position of the LSPR band, the maximum of which was shifted by more than 100 nm to the red region of the spectrum.

The analysis of TEM images in the current work also demonstrates the presence of phase separation zones (Figures 6 and 7), but due to the fact that the glass composition does not contain TiO_2 and ZrO_2, the chemical composition of the liquation zones seems to be enriched only in ZnO [27]. The refractive index n_D for ZnO is in the range of 1.6–1.78 according to the research data [44,45], which is not very different from the refractive index value for glass (n_D = 1.569); however, at the same time, it may have an effect on the LSPR band shift. The variation of the refraction index in the simulated optical spectra lead to a very high values, such as above 2.0 (Figure 10 middle panel). Such high refraction index values are not expected in the glass under the study as the densest component, ZnO, has a refraction index that is not higher than 2.0, but this phenomenon may have a partial effect on the overall LSPR band's position and shape.

(iv) The interaction between closely spaced plasmonic NPs can also be considered as a factor that has a great influence on the optical properties of the whole system [46]. It was experimentally demonstrated that a systematic change in the distance between gold NPs on the SiO_2 surface from 15 to 0.5 nm leads to a nonlinear broadening of the LSPR band. At a minimum distance, the dipole–dipole interaction between gold NPs leads to a maximum broadening of the band [47], and the authors showed that such a system is well described within the Maxwell–Garnett theory. In another work, gold NPs formed by a femtosecond laser beam on the Al_2O_3 substrate showed a broadened LSPR band at the highest irradiation energies, which was also explained by the reduction of the distance between NPs and their intense dipole–dipole interaction [48]. Jiménez J.A. et al., in a series of works with phosphate glasses containing NPs of silver, gold, and copper, also described the effect of the nonlinear broadening of the LSPR band in the framework of the dipole and plasmonic interaction of closely spaced NPs [22,23,49]. For these glasses, the authors also described a dichroism effect similar to the effect obtained in the glasses in this work. Laser irradiation of the initial glass with NPs acted as a driver for the formation of zones with increased content of closely located silver NPs, which led to the formation of "super-nucleation" domains separated by amorphous zones of 15–20 nm in size, not containing NPs [49].

A comparative analysis of the change in the position of the LSPR band of the studied glass samples heat-treated for 20 h, as well as their calculated sizes and distances between the particles depending on the treatment temperature, is shown in Figure 11. It can be seen that the shift of the maximum of the LSPR band from ≈1100 nm to ≈600 nm with increasing treatment temperature is accompanied by an increase in both NPs sizes and distances between them. This suggests that the effect of the ultra-broadening of the LSPR band in glasses observed in this work can be related to the plasmonic interaction between closely spaced NPs.

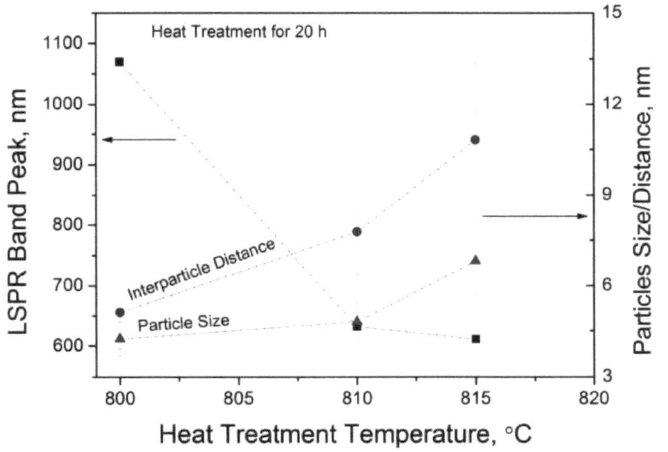

Figure 11. Evolution with heat treatment temperature at 20 h for the following parameters: LSPR band peak position (squares), particle size (triangles), and interparticle distance (circles).

To verify this hypothesis, we calculated the optical absorption spectra for agglomerates of NPs using a multiple spheres T-matrix approach, as implemented in MSTM code [50,51]. In the calculations, the NPs were represented as spheres with dielectric function as calculated by the Riuox formula [52] in a medium with a refraction index of 1.6. To simulate the closely packed agglomerate, the centers of the spheres were located at the nodes of a face-centered lattice with spacing so that the gap between NPs was 1 nm. The size of NPs of 10 nm and gap between them of 1 nm were selected as the smallest possible value for which the classical electrodynamic calculations are valid. The consideration of smaller sizes requires an application of size-corrected dielectric functions, and smaller gaps lead to the rise of electron tunneling.

Figure 12a shows the results of the calculations of the absorption spectra of an isolated gold nanoparticle of 10 nm and agglomerates of NPs of the same size containing 19, 43, and 87 NPs. It can be seen that as the number of particles in the agglomerate increases from 1 to 87, there is a shift in the main absorption maximum. However, the magnitude of this shift is on the order of 50 nm, which does not allow us to describe the observed shifts in resonance from 600 to 750 nm or more. However, in addition to the shift in the position of the main peak, a tendency to decrease its intensity and increase the intensity of the satellite long-wavelength peak as the number of NPs in the agglomerate increases is noticeable. In case of the agglomerate of 87 NPs, the intensities of these peaks are equalized. The trend is such that as the number of NPs in the agglomerate increases, one would expect to reproduce the experimental curve. However, the quantitative description requires computation resources that are too extensive. Nevertheless, based on the obtained data, we assume that the agglomeration of NPs may be the main contributor to the visible splitting of the LSPR bands and their significant broadening.

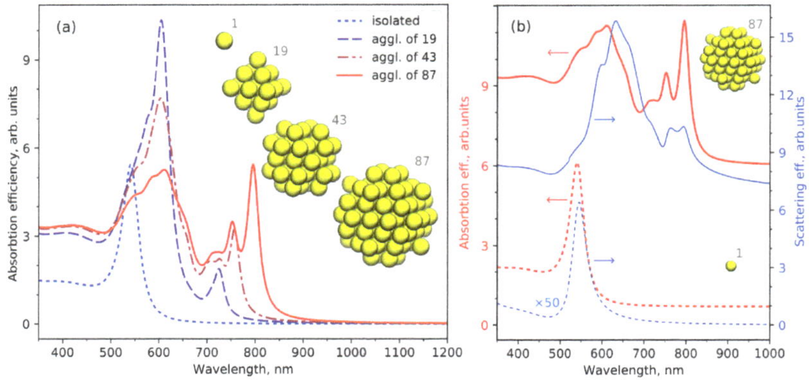

Figure 12. (**a**) The absorption efficiencies calculated for isolated gold nanoparticle and agglomerates containing 19, 43, or 87 nanoparticles. (**b**) The absorption (left y axis) and scattering (right y axis) efficiencies calculated for an agglomerate of 87 nanoparticles (solid curves) and isolated nanoparticles (short dashes).

Figure 12b compares the absorption and scattering spectra calculated for the isolated nanoparticle and the agglomerate of 87 NPs. The wavelength position of the absorption and scattering peaks are almost identical so that there is no difference in the transmitted and reflected color of the sample. In contrast, the scattering and absorption spectra for agglomerates are very different, which alter the color of material depending on the sample illumination. The experimental reflectance spectra of the studied glasses (Figure S2) are in line with the theoretical one showing intense reflection in the 650–750 nm range. This explains the observed dichroism of the samples by the presence of closely packed agglomerates of NPs.

Thus, the scenario of the changes in the optical properties of glasses observed in this work can be explained as follows. Phase separation processes take place in the glass during the heat treatment in a narrow temperature range. As a result, the formation of the localized amorphous regions on the order of 15 nm in size, apparently enriched in ZnO, occurs. In these regions, the formation of NPs can be energetically more favorable, so it leads to the formation of a kind of "plasmonic domains" with a high concentration of closely located particles, which is expressed in their plasmonic interaction and broadening of the LSPR spectrum. The presence of high concentrations of ZnO inside the domains can also lead to an increase in the local refractive index around the NPs and an additional redshift of the band. As the treatment temperature increases, the phase separation process is gradually replaced by the process of nucleation and growth of the crystalline phase, which is disintegrating the "plasmonic domains" and the distance between the nanoparticles increasing, which leads to a decrease in the plasmonic interaction effect and is expressed in the narrowing and blue shift of the LSPR band. At the same time, the increasing dichroism effect in the glass samples obtained by the heat treatment at higher temperature seems to also be related to the interactions between nanoparticles as well as formed nanocrystals of the ZAS-s.s. phase.

5. Conclusions

In conclusion, in this work, we have shown for the first time that it is possible to form gold NPs with an ultra-broad plasmon resonance (more than 1000 nm) in ZnO-Al_2O_3-SiO_2

glass. Precise thermal treatments of the glass allow for the tuning of the position and range of the LSPR band. The optical absorption spectroscopy results demonstrate that the LSPR band's position and shape can be finely controlled by varying the temperature during the treatment. The combination of Raman spectroscopy and XRD analysis reveals that the glasses remain completely amorphous despite prolonged treatment at temperatures above Tg. However, HRTEM results indicate that gold NPs with a size of approximately 5 nm are formed in the glass samples, along with amorphous areas of phase separation with sizes of around 15 nm. The mechanism responsible for the ultra-broadening of the LSPR spectrum and its redshift is attributed to the interparticle plasmonic coupling effect that is due to the formation of agglomerates containing a tenth of NPs with small distance between them, caused by the phase separation zones, which hinder the thermodiffusion of NPs. The presence of nanoparticle agglomerates is responsible for the observed dichroism. At higher treatment temperatures, the crystal nucleus (presumably ZAS-s.s.) forms in the sites of the amorphous phase separation zones, increasing the distance between NPs and significantly narrowing and shifting the LSPR band towards the blue region. The formation of gold NPs with an ultra-broad LSPR band in glasses holds promise for enhancing the luminescence of rare earth ions and advancing the development of novel photonic devices.

Supplementary Materials: The following supporting information can be downloaded at: https://www.mdpi.com/article/10.3390/ceramics7020037/s1, Figure S1: DSC curve of the synthesized glass; Figure S2: Reflectance spectra of the glass samples heat-treated for 20 h at temperatures in the 795–805 °C range.

Author Contributions: Conceptualization, G.S. and L.A.; methodology, G.A.; software, L.A.; validation, L.B., V.S. and M.V.; investigation, G.A., M.V., E.I., A.Z., M.Z. and E.S.; writing—original draft preparation, G.S.; writing—review and editing, L.A., E.I. and L.B.; visualization, L.A.; supervision, V.S.; funding acquisition, G.S. All authors have read and agreed to the published version of the manuscript.

Funding: This study was funded by the Russian Science Foundation grant No. 22-73-00236.

Data Availability Statement: Data are available upon request.

Acknowledgments: We thank N.V. Golubev for the XRD measurements and O.S. Dymshits for the fruitful discussion.

Conflicts of Interest: The authors declare no conflicts of interest.

References

1. Amendola, V.; Pilot, R.; Frasconi, M.; Maragò, O.; Iatì, M. Surface Plasmon Resonance in Gold Nanoparticles: A Review. *J. Phys. Condens. Matter* **2017**, *29*, 203002. [CrossRef] [PubMed]
2. Ferrari, E. Gold Nanoparticle-Based Plasmonic Biosensors. *Biosensors* **2023**, *13*, 411. [CrossRef] [PubMed]
3. Sarfraz, N.; Khan, I. Plasmonic Gold Nanoparticles (AuNPs): Properties, Synthesis and Their Advanced Energy, Environmental and Biomedical Applications. *Chem. Asian J.* **2021**, *16*, 720–742. [CrossRef] [PubMed]
4. Shafiqa, A.R.; Abdul Aziz, A.; Mehrdel, B. Nanoparticle Optical Properties: Size Dependence of a Single Gold Spherical Nanoparticle. *J. Phys. Conf. Ser.* **2018**, *1083*, 012040. [CrossRef]
5. Núñez-Leyva, J.M.; Kolosovas-Machuca, E.S.; Sánchez, J.; Guevara, E.; Cuadrado, A.; Alda, J.; González, F.J. Computational and Experimental Analysis of Gold Nanorods in Terms of Their Morphology: Spectral Absorption and Local Field Enhancement. *Nanomaterials* **2021**, *11*, 1696. [CrossRef] [PubMed]
6. Djorović, A.; Oldenburg, S.J.; Grand, J.; Le Ru, E.C. Extinction-to-Absorption Ratio for Sensitive Determination of the Size and Dielectric Function of Gold Nanoparticles. *ACS Nano* **2020**, *14*, 17597–17605. [CrossRef] [PubMed]
7. Yu, H.; Peng, Y.; Yang, Y.; Li, Z.-Y. Plasmon-Enhanced Light–Matter Interactions and Applications. *NPJ Comput. Mater.* **2019**, *5*, 45. [CrossRef]
8. Srabionyan, V.V.; Vetchinnikov, M.P.; Rubanik, D.S.; Durymanov, V.A.; Viklenko, I.A.; Avakyan, L.A.; Zinina, E.M.; Shakhgildyan, G.Y.; Sigaev, V.N.; Bugaev, L.A. Local Electric Field Enhancement in the Vicinity of Aggregates of Ag, Au, Rb Containing Nanoparticles in Oxide Glasses. *J. Non Cryst. Solids* **2024**, *631*, 122927. [CrossRef]
9. Shakhgildyan, G.; Lipatiev, A.; Lotarev, S.; Fedotov, S.; Sigaev, V. Glass: Home of the Periodic Table. *Front. Chem.* **2020**, *8*, 384. [CrossRef] [PubMed]
10. Shakhgil'dyan, G.Y.; Ziyatdinova, M.Z.; Kovgar, V.V.; Lotarev, S.V.; Sigaev, V.N.; Prusova, I.V. Effect of Gold Nanoparticles on the Spectral Luminescence Properties of Eu^{3+}-Doped Phosphate Glass. *Glass Ceram.* **2019**, *76*, 121–125. [CrossRef]

11. Savinkov, V.I.; Shakhgil'Dyan, G.Y.; Paleari, A.; Sigaev, V.N. Synthesis of Optically Uniform Glasses Containing Gold Nanoparticles: Spectral and Nonlinear Optical Properties. *Glass Ceram.* **2013**, *70*, 143–148. [CrossRef]
12. Som, T.; Karmakar, B. Plasmon Tuning of Nano-Au in Dichroic Devitrified Antimony Glass Nanocomposites by Refractive Index Control. *Chem. Phys. Lett.* **2009**, *479*, 100–104. [CrossRef]
13. Chen, F.; Dai, S.; Xu, T.; Shen, X.; Lin, C.; Nie, Q.; Liu, C.; Heo, J. Surface-Plasmon Enhanced Ultrafast Third-Order Optical Nonlinearities in Ellipsoidal Gold Nanoparticles Embedded Bismuthate Glasses. *Chem. Phys. Lett.* **2011**, *514*, 79–82. [CrossRef]
14. Wei, Y.; Zhao, J.; Fuhrmann, S.; Sajzew, R.; Wondraczek, L.; Ebendorff-Heidepriem, H. Controlled Formation of Gold Nanoparticles with Tunable Plasmonic Properties in Tellurite Glass. *Light Sci. Appl.* **2023**, *12*, 293. [CrossRef] [PubMed]
15. Lipat'ev, A.S.; Shakhgil'dyan, G.Y.; Lipat'eva, T.O.; Lotarev, S.V.; Fedotov, S.S.; Vetchinnikov, M.P.; Ignat'eva, E.S.; Golubev, N.V.; Sigaev, V.N.; Kazanskii, P.G. Formation of Luminescent and Birefringent Microregions in Phosphate Glass Containing Silver. *Glass Ceram.* **2016**, *73*, 277–282. [CrossRef]
16. Srabionyan, V.V.; Heinz, M.; Kaptelinin, S.Y.; Avakyan, L.A.; Sukharina, G.B.; Skidanenko, A.V.; Pryadchenko, V.; Abdulvakhidov, K.G.; Mikheykin, A.; Durymanov, V.; et al. Effect of Thermal Post-Treatment on Surface Plasmon Resonance Characteristics of Gold Nanoparticles Formed in Glass by UV Laser Irradiation. *J. Alloys Compd.* **2019**, *803*, 354–363. [CrossRef]
17. Heinz, M.; Srabionyan, V.V.; Avakyan, L.A.; Bugaev, A.L.; Skidanenko, A.V.; Pryadchenko, V.V.; Ihlemann, J.; Meinertz, J.; Patzig, C.; Dubiel, M.; et al. Formation and Implantation of Gold Nanoparticles by ArF-Excimer Laser Irradiation of Gold-Coated Float Glass. *J. Alloys Compd.* **2018**, *736*, 152–162. [CrossRef]
18. Schneider, C.A.; Rasband, W.S.; Eliceiri, K.W. NIH Image to ImageJ: 25 Years of Image Analysis. *Nat. Methods* **2012**, *9*, 671–675. [CrossRef] [PubMed]
19. Martín-Yerga, D.; Yu, X.; Terekhina, I.; Henriksson, G.; Cornell, A. In Situ Catalyst Reactivation for Enhancing Alcohol Electro-Oxidation and Coupled Hydrogen Generation. *Chem. Commun.* **2020**, *56*, 4011–4014. [CrossRef]
20. Cormier, L.; Delbes, L.; Baptiste, B.; Montouillout, V. Vitrification, Crystallization Behavior and Structure of Zinc Aluminosilicate Glasses. *J. Non Cryst. Solids* **2021**, *555*, 120609. [CrossRef]
21. Shakhgildyan, G.Y.; Ziyatdinova, M.Z.; Vetchinnikov, M.P.; Lotarev, S.V.; Savinkov, V.I.; Presnyakova, N.N.; Lopatina, E.V.; Vilkovisky, G.A.; Sigaev, V.N. Thermally-Induced Precipitation of Gold Nanoparticles in Phosphate Glass: Effect on the Optical Properties of Er3+ Ions. *J. Non Cryst. Solids* **2020**, *550*, 120408. [CrossRef]
22. Jiménez, J.A. Photoluminescence of Sm^{3+} Ions in Au-Doped Plasmonic and Dichroic Phosphate Glass. *Eur. Phys. J. D* **2023**, *77*, 124. [CrossRef]
23. Jiménez, J.A. Dichroism in Plasmonic Cu Nanocomposite Glass: Selective Enhancement of the Orange-Red Emission from Sm^{3+}. *Opt. Mater. X* **2019**, *1*, 100002. [CrossRef]
24. Singh, S.P.; Nath, M.; Karmakar, B. Quantum and Dielectric Confinements of Sub-10 Nm Gold in Dichroic Phosphate Glass Nanocomposites. *Mater. Chem. Phys.* **2014**, *146*, 198–203. [CrossRef]
25. Som, T.; Karmakar, B. Surface Plasmon Resonance and Enhanced Fluorescence Application of Single-Step Synthesized Elliptical Nano Gold-Embedded Antimony Glass Dichroic Nanocomposites. *Plasmonics* **2010**, *5*, 149–159. [CrossRef]
26. Montazerian, M.; Mancini, M.; Mauro, J.C. Advanced Tools for Unveiling Nucleation in Nanostructured Glass-Ceramics. *Crit. Rev. Solid State Mater. Sci.* **2023**, *48*, 411–439. [CrossRef]
27. Jiazhi, L.; Ying, S.; Guanging, H. An Investigation of Relationship between Phase Separation and Crystallization of $ZnO-Al_2O_3-SiO_2$ Glasses. *J. Phys. Colloq.* **1982**, *43*, C9-231–C9-234. [CrossRef]
28. Pellerin, N.; Blondeau, J.-P.; Noui, S.; Allix, M.; Ory, S.; Veron, O.; De Sousa Meneses, D.; Massiot, D. Control of Selective Silicate Glass Coloration by Gold Metallic Nanoparticles: Structural Investigation, Growth Mechanisms, and Plasmon Resonance Modelization. *Gold Bull.* **2013**, *46*, 243–255. [CrossRef]
29. Sigaev, V.N.; Savinkov, V.I.; Lotarev, S.V.; Shakhgildyan, G.Y.; Lorenzi, R.; Paleari, A. Spatially Selective Au Nanoparticle Growth in Laser-Quality Glass Controlled by UV-Induced Phosphate-Chain Cross-Linkage. *Nanotechnology* **2013**, *24*, 225302. [CrossRef] [PubMed]
30. Simo, A.; Polte, J.; Pfänder, N.; Vainio, U.; Emmerling, F.; Rademann, K. Formation Mechanism of Silver Nanoparticles Stabilized in Glassy Matrices. *J. Am. Chem. Soc.* **2012**, *134*, 18824–18833. [CrossRef] [PubMed]
31. Grand, J.; Adam, P.-M.; Grimault, A.-S.; Vial, A.; Lamy de la Chapelle, M.; Bijeon, J.-L.; Kostcheev, S.; Royer, P. Optical Extinction Spectroscopy of Oblate, Prolate and Ellipsoid Shaped Gold Nanoparticles: Experiments and Theory. *Plasmonics* **2006**, *1*, 135–140. [CrossRef]
32. Pérez-Juste, J.; Pastoriza-Santos, I.; Liz-Marzán, L.M.; Mulvaney, P. Gold Nanorods: Synthesis, Characterization and Applications. *Coord. Chem. Rev.* **2005**, *249*, 1870–1901. [CrossRef]
33. Berger, A. Prolate Silver Particles in Glass Surfaces. *J. Non Cryst. Solids.* **1993**, *163*, 185–194. [CrossRef]
34. Mennig, M.; Berg, K.-J. Determination of Size Shape and Concentration of Spheroidal Silver Colloids Embedded in Glass by VIS-Spectroscopy. *Mater. Sci. Eng. B* **1991**, *9*, 421–424. [CrossRef]
35. Suszynska, M.; Krajczyk, L.; Capelletti, R.; Baraldi, A.; Berg, K.J. Microstructure and Silver Nanoparticles in Ion-Exchanged and Deformed Soda-Lime Silicate Glasses. *J. Non Cryst. Solids* **2003**, *315*, 114–123. [CrossRef]
36. Stalmashonak, A.; Matyssek, C.; Kiriyenko, O.; Hergert, W.; Graener, H.; Seifert, G. Preparing Large-Aspect-Ratio Prolate Metal Nanoparticles in Glass by Simultaneous Femtosecond Multicolor Irradiation. *Opt. Lett.* **2010**, *35*, 1671. [CrossRef] [PubMed]

37. Shakhgildyan, G.Y.; Lipatiev, A.S.; Fedotov, S.S.; Vetchinnikov, M.P.; Lotarev, S.V.; Sigaev, V.N. Microstructure and Optical Properties of Tracks with Precipitated Silver Nanoparticles and Clusters Inscribed by the Laser Irradiation in Phosphate Glass. *Ceram. Int.* **2021**, *47*, 14320–14329. [CrossRef]
38. Som, T.; Karmakar, B. Enhanced Frequency Upconversion of Sm^{3+} Ions by Elliptical Au Nanoparticles in Dichroic Sm3+: Au-Antimony Glass Nanocomposites. *Spectrochim. Acta Part A Mol. Biomol. Spectrosc.* **2010**, *75*, 640–646. [CrossRef]
39. Mayer, K.M.; Hafner, J.H. Localized Surface Plasmon Resonance Sensors. *Chem. Rev.* **2011**, *111*, 3828–3857. [CrossRef]
40. De, S.; Medda, S.K.; De, G. Refractive Index Controlled Plasmon Tuning of Au Nanoparticles in SiO_2 -ZrO_2 Film Matrices. *J. Nanosci. Nanotechnol.* **2008**, *8*, 3868–3876. [CrossRef] [PubMed]
41. De, G.; Medda, S.K.; De, S.; Pal, S. Metal Nanoparticle Doped Coloured Coatings on Glasses and Plastics through Tuning of Surface Plasmon Band Position. *Bull. Mater. Sci.* **2008**, *31*, 479–485. [CrossRef]
42. Shakhgildyan, G.; Durymanov, V.; Ziyatdinova, M.; Atroshchenko, G.; Golubev, N.; Trifonov, A.; Chereuta, O.; Avakyan, L.; Bugaev, L.; Sigaev, V. Effect of Gold Nanoparticles on the Crystallization and Optical Properties of Glass in ZnO-MgO-Al_2O_3-SiO_2 System. *Crystals* **2022**, *12*, 287. [CrossRef]
43. Shakhgildyan, G.; Avakyan, L.; Ziyatdinova, M.; Atroshchenko, G.; Presnyakova, N.; Vetchinnikov, M.; Lipatiev, A.; Bugaev, L.; Sigaev, V. Tuning the Plasmon Resonance of Gold Nanoparticles in Phase-Separated Glass via the Local Refractive Index Change. *J. Non Cryst. Solids* **2021**, *566*, 120893. [CrossRef]
44. Stelling, C.; Singh, C.R.; Karg, M.; König, T.A.F.; Thelakkat, M.; Retsch, M. Plasmonic Nanomeshes: Their Ambivalent Role as Transparent Electrodes in Organic Solar Cells. *Sci. Rep.* **2017**, *7*, 42530. [CrossRef] [PubMed]
45. Aguilar, O.; de Castro, S.; Godoy, M.P.F.; Rebello Sousa Dias, M. Optoelectronic Characterization of Zn_{1-x} Cd_x O Thin Films as an Alternative to Photonic Crystals in Organic Solar Cells. *Opt. Mater. Express* **2019**, *9*, 3638. [CrossRef]
46. Ghosh, S.K.; Pal, T. Interparticle Coupling Effect on the Surface Plasmon Resonance of Gold Nanoparticles: From Theory to Applications. *Chem. Rev.* **2007**, *107*, 4797–4862. [CrossRef] [PubMed]
47. Ung, T.; Liz-Marzán, L.M.; Mulvaney, P. Optical Properties of Thin Films of Au@SiO $_2$ Particles. *J. Phys. Chem. B* **2001**, *105*, 3441–3452. [CrossRef]
48. Nedyalkov, N.N.; Nakajima, Y.; Takami, A.; Koleva, M.; Karashanova, D.; Terakawa, M. Laser Induced Morphological and Optical Properties Changes in Au Doped Aluminum Oxide and Silicon Oxide Thin Films. *Opt. Laser Technol.* **2016**, *79*, 179–187. [CrossRef]
49. Sendova, M.; Jiménez, J.A. Plasmonic Coupling in Silver Nanocomposite Glasses. *J. Phys. Chem. C* **2012**, *116*, 17764–17772. [CrossRef]
50. Mackowski, D.W.; Mishchenko, M.I. A Multiple Sphere T-Matrix Fortran Code for Use on Parallel Computer Clusters. *J. Quant. Spectrosc. Radiat. Transf.* **2011**, *112*, 2182–2192. [CrossRef]
51. Avakyan, L.A.; Heinz, M.; Skidanenko, A.V.; Yablunovski, K.A.; Ihlemann, J.; Meinertz, J.; Patzig, C.; Dubiel, M.; Bugaev, L.A. Insight on Agglomerates of Gold Nanoparticles in Glass Based on Surface Plasmon Resonance Spectrum: Study by Multi-Spheres T-Matrix Method. *J. Phys. Condens. Matter* **2018**, *30*, 045901. [CrossRef] [PubMed]
52. Rioux, D.; Vallières, S.; Besner, S.; Muñoz, P.; Mazur, E.; Meunier, M. An Analytic Model for the Dielectric Function of Au, Ag, and Their Alloys. *Adv. Opt. Mater.* **2014**, *2*, 176–182. [CrossRef]

MDPI AG
Grosspeteranlage 5
4052 Basel
Switzerland
Tel.: +41 61 683 77 34

Ceramics Editorial Office
E-mail: ceramics@mdpi.com
www.mdpi.com/journal/ceramics

www.ingramcontent.com/pod-product-compliance
Lightning Source LLC
Chambersburg PA
CBHW041353210926
51850CB00013B/3064
* 9 7 8 3 7 2 5 8 4 5 4 8 4 *